基于Java的物联网基础应用开发

虞 芬 王燕贞 徐 杰 吴焕祥 主编

殷 侠 王法强 吴冬燕 荣喜丰 李兰兰 副主编

彭坤容 魏美琴 蔡 敏 黄非娜 邹宗冰 参编

清華大学出版社

北京

内 容 简 介

本书编者总结了十多年的物联网专业教学和指导学生参加竞赛的经验，精心选择物联网方面典型的项目展开分析，根据项目需求设计工作任务，采用任务式结构编写，通过引导读者完成不同的物联网程序任务，对 Java 物联网开发各方面的知识进行讲解。全书共分为 13 个项目，建议学时为 100 学时，不包括项目 12 和项目 13。项目 12 和项目 13 建议采用综合实训课形式。

本书既可以作为高等院校本科和高职物联网工程及相关专业学生的教材使用，又可以作为物联网相关从业者和爱好者的参考用书。

图书在版编目(CIP)数据

基于 Java 的物联网基础应用开发/虞芬等主编. 一北京：清华大学出版社，2021.6（2024.8重印）
ISBN 978-7-302-58342-4

Ⅰ. ①基…　Ⅱ. ①虞…　Ⅲ. ①JAVA 语言—程序设计　Ⅳ. ①TP312.8

中国版本图书馆 CIP 数据核字(2021)第 102337 号

责任编辑：梁媛媛
封面设计：李　坤
责任校对：吴春华
责任印制：杨　艳
出版发行：清华大学出版社
网　　址：https://www.tup.com.cn，https://www.wqxuetang.com
地　　址：北京清华大学学研大厦 A 座　　**邮　　编：**100084
社 总 机：010-83470000　　**邮　　购：**010-62786544
投稿与读者服务：010-62776969, c-service@tup.tsinghua.edu.cn
质量反馈：010-62772015, zhiliang@tup.tsinghua.edu.cn
课件下载：https://www.tup.com.cn , 010-62791865
印 装 者：三河市铭诚印务有限公司
经　　销：全国新华书店
开　　本：185mm×260mm　　**印　张：**23.5　　**字　数：**571 千字
版　　次：2021 年 7 月第 1 版　　**印　次：**2024 年 8 月第 5 次印刷
定　　价：69.00 元

产品编号：093191-01

前　言

物联网被看作是继计算机、互联网与移动通信之后的又一次信息产业浪潮，将成为未来带动中国经济发展的主要生力军。2009年，在美国总统奥巴马与工商业领袖举办的圆桌会议上，IBM首席执行官首次提出了“智慧地球”(Smart Earth)的构想。同年，欧盟发布了物联网研究战略路线图(Internet of Things Strategic Research Roadmap)。在我国，物联网同样得到了高度重视，在2010年“两会”期间物联网已经被写入政府工作报告，确立为国家五大战略新兴产业之一。在2016年国务院印发的《“十三五”国家信息化规划》中特别提出要加快信息化和生态文明建设深度融合，利用新一代信息技术，促进产业链接循环化；推进物联网感知设施规划布局，发展物联网开发应用；实施物联网重大应用示范工程，推进物联网应用区域试点，建立城市级物联网接入管理与数据汇聚平台，深化物联网在城市基础设施、生产经营等环节中的应用。

本书编者总结了十多年的物联网专业教学和指导学生参加竞赛的经验，精心选择物联网方面典型的项目展开分析，根据项目需求设计工作任务，采用任务式结构编写，通过引导读者完成不同的物联网程序任务，对Java物联网开发各方面的知识进行讲解。

本书的特点如下。

1. 紧密结合物联网开发

本书以智慧园区项目为中心，将Java程序设计所需要掌握的知识拆分到不同的物联网开发情景中，让读者能够及时地将所学的知识运用到实际开发中，提升学习兴趣，培养动手能力。

2. 先封装调用、后详细拆解

前期开发中遇到的有关Java高级编程的知识，采用“先封装调用，后详细拆解”的方式，让读者“先使用，后理解”，由易到难、深入浅出地学习Java开发知识。

3. 综合运用、实战检验

最后通过智慧园区的串口篇和云平台篇的综合开发，让读者将各部分知识综合使用、融会贯通，充分掌握Java物联网程序设计基础知识。

本书具体内容介绍如下。

- **项目1**　智慧园区系统项目概述——本项目介绍智慧园区项目的需求、功能模块、运行结果以及技术选型和设备选型等内容(建议学时：2学时)。
- **项目2**　初识Java与物联网——本项目介绍Java与物联网的关系，并在搭建好Java开发环境后，让读者完成“随心所欲亮灯灭灯”的第一个物联网程序(建议学时：6学时)。
- **项目3**　传感数据解析和控制指令生成——本项目介绍如何利用Java的基本语法对ZigBee传感数据进行计算、采集分析、显示，以及ZigBee控制指令的生成(建议学时：16学时)。
- **项目4**　从串口获取传感器数据——本项目介绍串口管理工具类的封装，以及如何通过工具类从串口获取真实的传感器数据，让读者理解 Java 面向对象的编程思想(建议学时：8 学时)。

- **项目 5** 采集传感数据的 API 的构建——本项目介绍如何利用 Java 的继承、接口、多态等机制构建采集 ZigBee 数据的 API，以及串口开发自定义异常的 API(建议学时：6 学时)。
- **项目 6** 认识系统常用类——本项目介绍 ZigBee 控制器命令生成工具的封装及用户注册信息的验证，让读者掌握系统常用类的使用(建议学时：8 学时)。
- **项目 7** 智慧园区系统界面开发和事件处理——本项目介绍如何利用 JavaFX 完成智慧园区系统界面，实现界面之间的跳转，并处理控件的事件监听，让读者初步了解图形化界面的制作过程和原理(建议学时：12 学时)。
- **项目 8** 使用集合——本项目介绍如何使用 List 存储传感器数据日志，使用 Set 实现用户注册功能，使用 Map 存储采集器数据，并让读者掌握 Java 集合的使用(建议学时：12 学时)。
- **项目 9** 使用 IO 流——本项目介绍如何使用 File 类读写用户信息文件，以及如何利用 IO 流保存用户信息和读写系统配置文件，并让读者掌握 IO 编程的基本知识(建议学时：14 学时)。
- **项目 10** 实时更新数据——本项目介绍如何实现可用串口列表的实时更新，园区门禁的实时监测，及实时火警警示，并讲解了 Java 多线程的使用(建议学时：12 学时)。
- **项目 11** 网络与定位技术的使用——本项目介绍如何使用北斗定位模块获取地理位置信息，并将园区位置信息上报到云平台(建议学时：4 学时)。
- **项目 12** 智慧园区系统综合实现(串口篇)——本项目介绍智慧园区串口部分全部功能的综合实现，包含门禁安防模块、室内环境监控模块等功能(建议以实训形式教学)。
- **项目 13** 智慧园区环境实时监测(云平台篇)——本项目介绍如何利用网络编程与云平台连接，综合实现智慧园区室外环境监测模块的功能(建议以实训形式教学)。

本书适合物联网工程以及相关专业的学生使用。

本书的编写得到了北京新大陆时代教育科技有限公司相关人员的大力帮助和支持，在此表示感谢。

由于编者水平有限，书中疏漏之处在所难免，敬请各位读者不吝赐教，以求共同进步，感激不尽。

编　者

(扫一扫，了解本书配套资源目录)

(扫一扫，试看配套的精美课件)

目　录

项目 1　智慧园区系统项目概述

扫码观看视频讲解

【需求描述】

某园区计划进行智能改造，要求实现下列功能。

(1) 在园区入口设置红外对射传感器，红外对射传感器监测到人员由外进入园区时，触发电动推杆打开大门，15 秒后电动推杆自动关闭大门。当园区内人员需要出门时，按下微动开关，触发电动推杆打开大门，15 秒后自动关闭大门。

(2) 室内环境监控：采集光敏数据、温湿度、噪声、二氧化碳的数据，显示在客户端界面上。能根据温度高低自动判断是否需要开/关室内空调，能根据光敏数据自动判断是否需要开/关室内灯。

(3) 园区环境监测：采集到的空气质量、垃圾桶、PM2.5、水质等数据经 LoRa 网关上传到物联网云平台，采集到的甲烷和一氧化碳等传感数据经 NB-IoT 上传到物联网云平台，Java 客户端程序实时从云平台取数据进行展示。

【需求分析】

根据需求描述，可以将用户的需求大致分为三个模块：门禁安防模块、室内环境监控模块、园区环境监测模块。其中，门禁安防模块与室内环境监控模块在智慧园区系统综合实现(串口篇)中完成，园区环境监测模块在智慧园区环境监测(云平台篇)中完成。

1. 门禁安防模块

门禁安防模块采用有线传感网络实现，如图 1-1 所示。

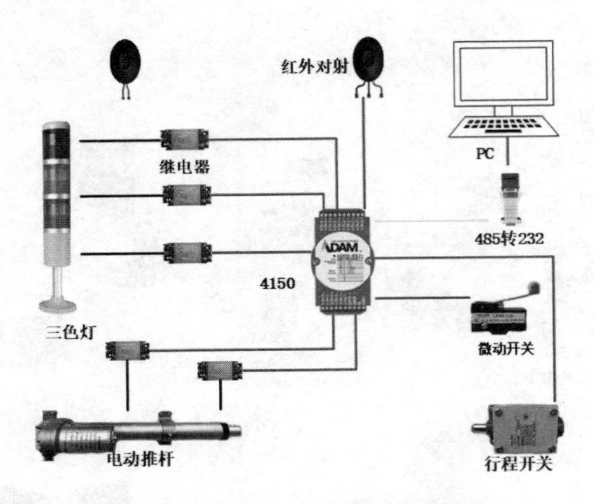

图 1-1 门禁、安防模块拓扑图

(1) 门禁部分硬件组成。

门禁硬件由直流电动推杆(模拟大门)、微动开关、行程开关组成。

(2) 门禁部分软件功能。

门禁部分软件功能包括以下两种。

① 由外入内开门。人体传感器检测到人员由园区外进入园区内，系统控制直流推杆开门。开门停止后，计时 15 秒，软件控制电机关门。

② 由内出外开门。园区人员出园区大门，按下微动开关，园区大门打开，开门后，计时 15 秒，自动关门。

(3) 安防部分硬件组成。

安防部分硬件由人体传感器、报警灯组成。

(4) 安防部分软件功能。

安防部分软件功能如下。

① 设防/撤防。设防时，会触发报警；撤防时，不会触发报警。

② 报警联动。当人体传感器监测到入侵时，报警灯能闪烁显示报警。

2. 室内环境监控模块

(1) 室内环境监控模块硬件组成。

室内环境监控模块采用 ZigBee 传感网实现，如图 1-2 所示。设置有温湿度传感器、噪声传感器以及风扇、照明灯等。

(2) 室内环境监控模块软件功能。

室内环境监控模块软件功能如下。

① 传感数据采集。采集温湿度、噪声数据，显示在界面上。

② 设置联动。当设置联动时，执行下列联动：当采集到的温度超过用户设定温度时，自动打开风扇；根据采集的室外光敏数据，自动开灯。

3. 园区环境监测模块

(1) 园区环境监测模块硬件组成。

园区环境监测模块采用 LoRa 网络和 NB-IoT 网络实现，如图 1-3 所示。

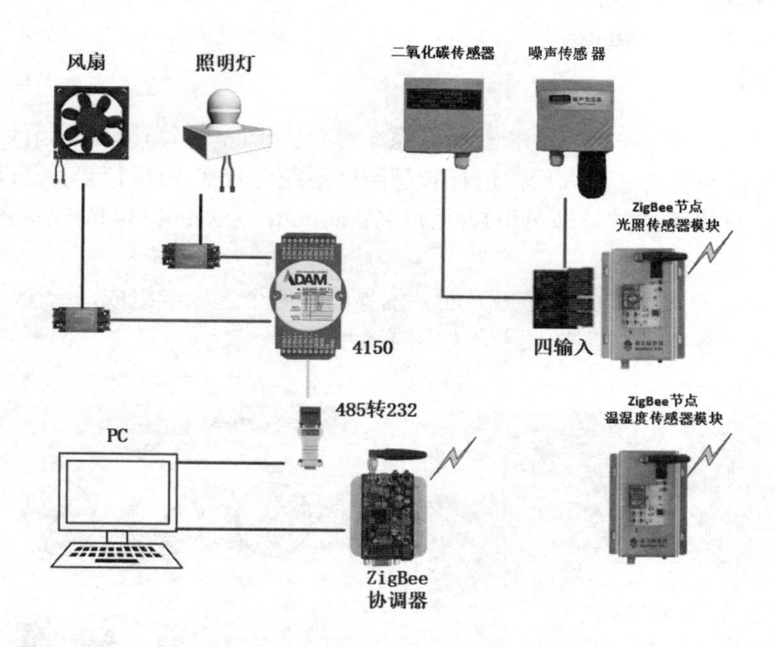

图 1-2 室内环境监控模块拓扑图

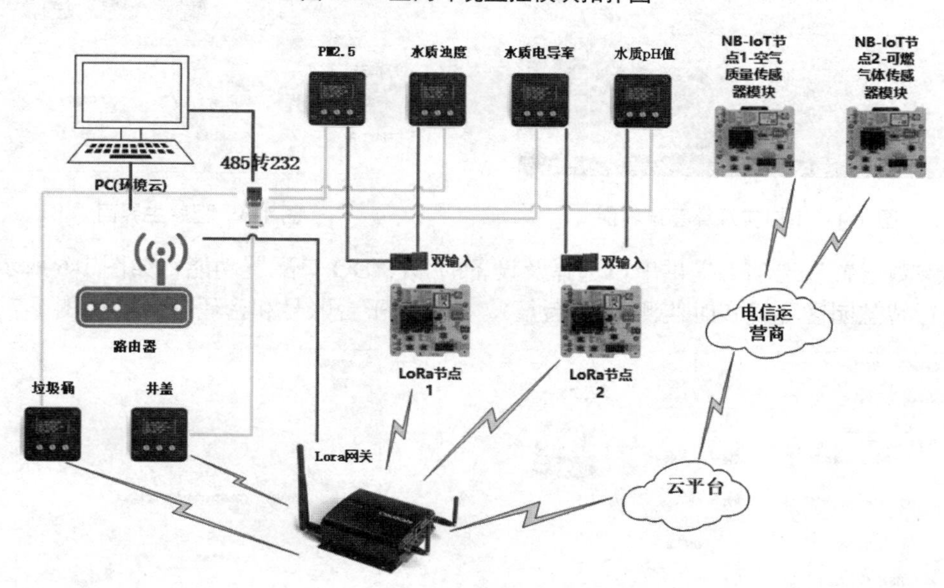

图 1-3 园区环境监测模块拓扑图

(2) 园区环境监测模块软件功能。

园区环境监测模块软件功能如下。

① 云平台账号登录。

② 监测传感器数据：能监测超声波、甲烷、井盖、一氧化碳、空气质量、垃圾桶、PM2.5、可燃气体、水质传感器的数据。

③ 设置云平台设备参数。

【运行效果】

本智慧园区项目有两种实现：智慧园区系统综合实现(串口篇)和智慧园区环境实时监测(云平台篇)。设备安装调试完成后，运行智慧园区系统综合实现(串口篇)源程序。程序的首页为登录界面，读者可通过管理员用户(用户名：admin；密码：123456)登录系统，如图 1-4 所示。

登录成功后，程序跳转到程序主界面。主界面由两个选项卡组成，一个为“设备数据”选项卡，一个为“参数设置”选项卡，如图 1-5 所示。

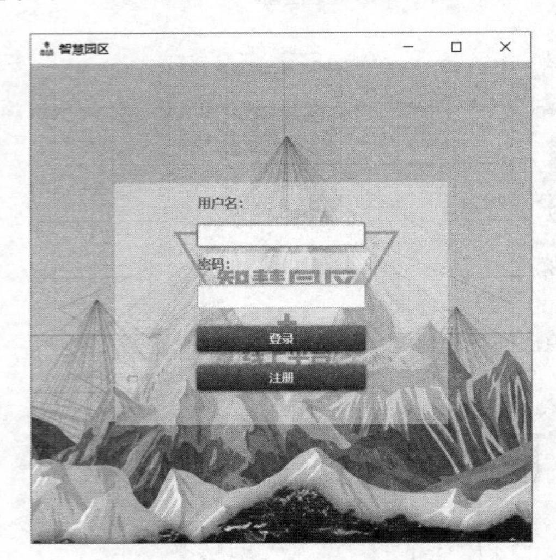

图 1-4　串口实现篇登录界面

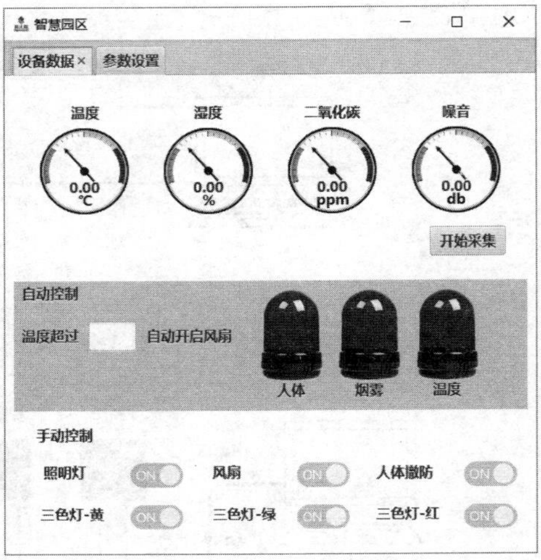

图 1-5　程序主界面

“参数设置”选项卡提供串口选择及设备的 DI、DO 口配置功能，如图 1-6 所示。

运行智慧园区环境实时监测(云平台篇)，使用云平台账号和密码登录，如图 1-7 所示。

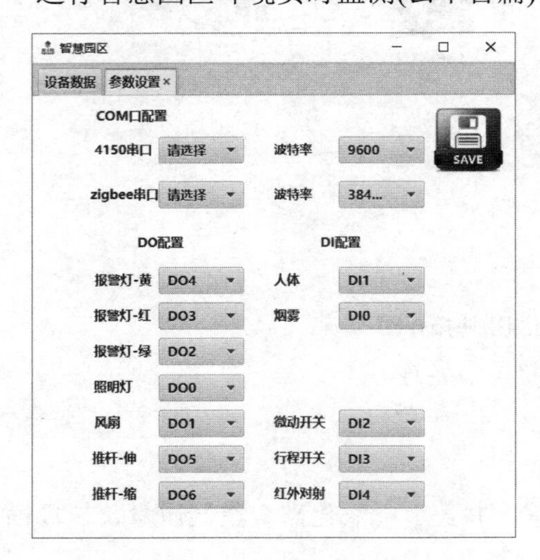

图 1-6　“参数设置”界面

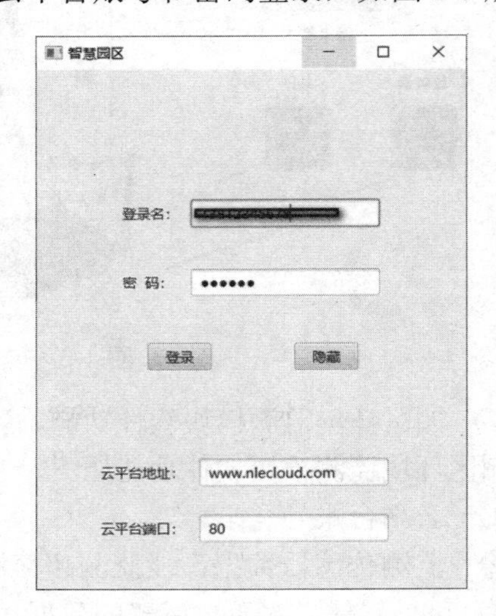

图 1-7　云平台实现篇登录界面

云平台实现篇和串口实现篇类似，主界面由两个选项卡组成，分别是“实时监测”选项卡和“参数设置”选项卡，如图 1-8、图 1-9 所示。

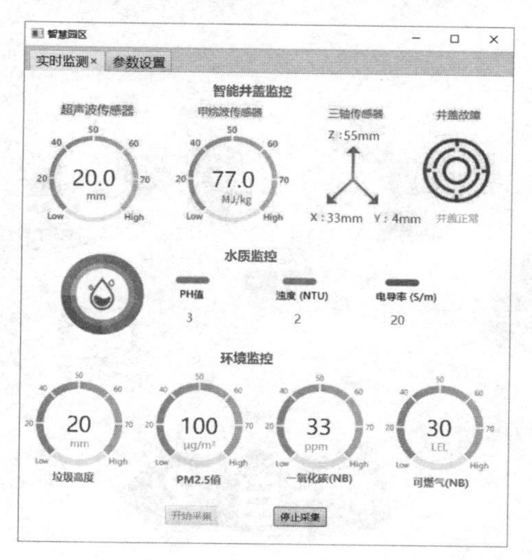

图 1-8 “实时监测”选项卡

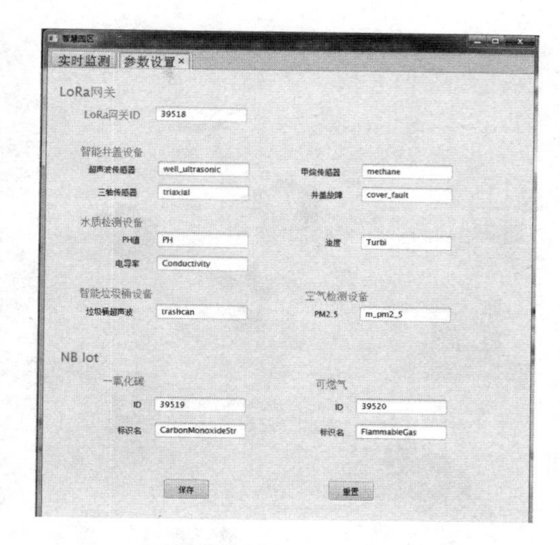

图 1-9 “参数设置”选项卡

【技术选型】

(1) 开发机操作系统：Windows 10。
(2) JDK：1.8.0_201。
(3) Eclipse：Eclipse Photon Release (4.8.0)。
(4) Eclipse 的 JavaFX 开发插件：e(fx)clipse 2.4.0。
(5) JavaFX 界面可视化设计工具：JavaFX Scene Builder 2.0。
(6) Java 串口开发工具包：RXTX 2.2。

【设备选型】

本教材涉及的设备选用新大陆物联网工程实训系统 2.0 套件，在任务描述中有对应的设备清单说明，读者可以按清单说明进行设备的选择和按工程实训中设备的连接要求进行接线。

【知识前提】

学习本教材知识的前提是有物联网工程实训基础，在本书中不对使用的传感器设备、ModBus 协议、ZigBee 协议、ZigBee 烧写和配置、设备连接做过多的阐述。本书任务代码涉及的串口与 DI、DO 口以编写时的数据为准，读者在完成任务时，需根据实际开发环境自行修改。

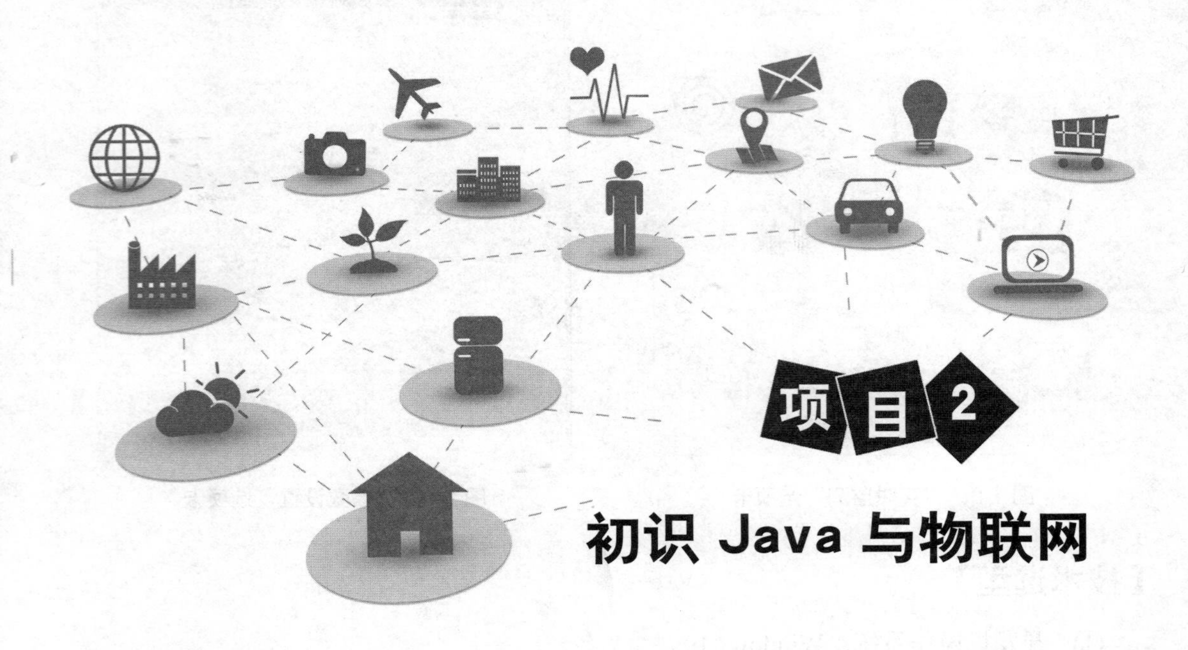

项目 2 初识 Java 与物联网

【项目描述】

Java 是现阶段中国互联网公司中，应用范围最广的研发语言。Java 后端服务，最常见的研发语言是 Java 和 C/C++；移动端 Android，最常见的研发语言是 Java 和 Kotlin；大数据，最常见的研发语言是 Java 和 Python；Java 是成熟的产品，Amazon、Google、eBay、淘宝、京东、阿里和其他的大型电子商务品牌都在使用 Java 做后台处理；Java 的安全性、开放性、稳定性和跨平台性使 Java 可以在各领域持续称霸。因此，如果掌握了 Java 技术体系，不管是在成熟的大公司，快速发展的公司，还是在创业阶段的公司，都能有立足之地。

本项目通过 5 个任务带领大家从 Java 开发环境搭建到写出第一个 Java 物联网应用程序，熟悉集成开发工具 Eclipse 的使用，为后面项目内容的学习做好准备。具体任务列表如图 2-1 所示。

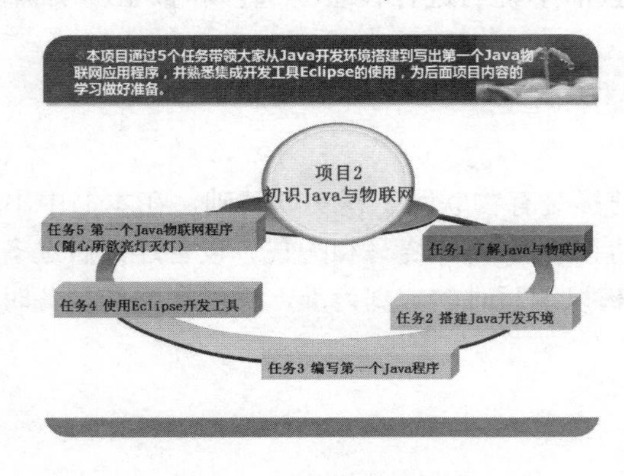

图 2-1 项目 2 任务列表

【学习目标】

知识目标：知道 Java 的历史背景；知道 Java 的语言特点；会搭建 Java 的开发环境；会编写第一个 Java 程序；会使用 Eclipse 开发工具。

技能目标：能安装 JDK1.8 并配置好 path、classpath 环境变量；能开发出第一个 Java 程序并编译和运行；能使用 Eclipse 工具做出第一个 Java 物联网应用程序。

任务 1　了解 Java 与物联网

【任务描述】

本任务主要带领大家在开发 Java 程序之前，先了解 Java 的发展历史、Java 的特点以及 Java 与物联网的关系。任务清单如表 2-1 所示。

表 2-1　任务清单

任务课时	1 课时	任务组员数量	建议 1 人
任务组采用设备	无		

【知识解析】

1. Java 语言介绍

在介绍 Java 语言之前，先认识一下什么是计算机语言。计算机语言(Computer Language)是指用于人与计算机之间通信的语言。它由一些指令组成，开发者可以通过编写指令来指挥计算机进行各种工作。例如，如果在办公室发生了火警，烟雾传感器负责把办公室的烟雾浓度传递给安防系统，当浓度达到设定的危险值后，安防系统启动防御措施。这个判断烟雾是否达到危险值以及获取烟雾传感器数据与发送防御措施的指令就是开发者按某种语言的语法规范编写的计算机语言。

计算机语言的种类非常多，总体来说可以分成机器语言、汇编语言和高级语言三大类。机器语言是由二进制的 0 和 1 组成的编码，但不方便人类的识别与记忆。汇编语言采用英文缩写的标识符，容易识别与记忆。高级语言采用更接近于人类的自然语言进行编程，简化了编程过程。所以，高级语言是绝大多数编程者的选择。因为计算机只能识别机器语言，所以汇编语言和高级语言都要转换成机器语言才能被识别。

Java 是一种高级程序设计语言，它由 Sun 公司(已被 Oracle 公司收购)于 1995 年 5 月推出。目前最新的为 2020 年 4 月份推出的 Java 16 版本。Java 语言简单易用、安全可靠，自问世以来，与之相关的技术和应用发展得非常快，已成为计算机、移动电话、金融行业、物联网、大数据等领域最受欢迎的开发语言之一。

针对不同的市场，Sun 公司将 Java 划分为 3 个技术平台，分别是 JavaSE、JavaEE 和 JavaME。

JavaSE(Java Platform Standard Edition)标准版，是为开发普通桌面应用程序提供的解决方案。同时也是 3 个平台中最核心的部分，是其他两个平台的基础。

JavaEE(Java Platform Enterprise Edition)企业版，是为开发企业级应用程序提供的解决方案。

JavaME(Java Platform Micro Edition)小型版，是为开发电子消费产品和嵌入式设备提供的解决方案。

2. Java 语言的特点

(1) 简单。

Java 语言与 C、C++风格接近。Java 语言不使用 C、C++语言中很难懂的指针，而是使用引用，并提供自动管理内存的机制，使开发者不必为内存管理而担忧。

(2) 面向对象。

Java 是一种面向对象的语言。一切操作都以对象为基本单元，要完成什么功能行为先找有哪个对象可以完成，如果找不到合适的对象，就自己创建新类，把功能写进类内的方法里，再通过对象去调用这个方法完成功能。面向对象技术使得应用程序的开发变得简单易用，节省代码。

(3) 安全。

安全可以分为四个层面，即语言级安全、编译时安全、运行时安全、可执行代码安全。语言级安全指 Java 的数据结构是完整的对象，这些封装过的数据类型具有安全性。编译时安全指编译时要进行 Java 语言和语义的检查，保证每个变量对应一个相应的值，编译后生成 Java 类。运行时安全指运行时 Java 类需要由类加载器载入，并经由字节码校验器校验之后才可以运行。可执行代码安全指 Java 类在网络上使用时，对它的权限进行了设置，保证了被访问用户的安全。

(4) 跨平台。

用 Java 语言编写的程序可以在各种平台上运行。Java 程序的运行要依赖 Java 虚拟机。同一个 Java 程序经过编译生成字节码后可由不同平台的 Java 虚拟机在不同平台下解释运行。所以 Java 是与平台无关的。

(5) 支持多线程。

多线程可以简单地理解为程序中有多个任务可以并发执行，这样可以提高程序的执行效率，Java 语言支持多线程。

3. Java 与物联网

物联网的英文名称是 Internet of Things(IoT)，是新一代信息技术的重要组成部分，也是“信息化”时代的重要发展阶段。

物联网就是物物相连的互联网，其核心和基础仍然是互联网。物联网是互联网的应用拓展，与其说物联网是网络，不如说物联网是业务和应用，以用户体验为核心的创新应用是物联网发展的核心，是物联网发展的灵魂。

物联网的四大核心技术是 RFID 射频识别技术、传感器技术、云计算和网络通信技术。利用这些技术融合，物联网应用的重点领域如图 2-2 所示。

互联网是人与人之间、信息与信息之间的互联通信，是一个庞大的信息连接群体，信息大部分通过网页、App 的形式传递。

物联网则是人与物之间、指令与执行之间的互联通信，是基于互联网将人表达出的信息传给指定的物去执行，物体将自身信息反馈给人。相对来讲，互联网更为广泛，而物联网更具有针对性和业务性。互联网的 Java 程序倾向于信息化的处理、分析、展示等，而物联网的 Java 程序可能更倾向于对信息的操控。

图 2-2　物联网应用重点领域

任务 2　搭建 Java 开发环境

扫码观看视频讲解

【任务描述】

本任务完成 Java 开发环境的搭建，包括下载、安装和配置环境变量。任务清单如表 2-2 所示。

表 2-2　任务清单

任务课时	1 课时	任务组员数量	建议 1 人
任务组采用设备	无		

【知识解析】

1. JDK 简介

JDK 是 Java 的软件开发工具包，是整个 Java 的核心，如图 2-3 所示，包括如下内容。

JDK(Java Development Kit)是 Java 开发工具包。

JRE(Java Runtime Environment)是 Java 运行环境。所有的 Java 程序都要在 JRE 下才能运行，包括 JVM 和 Java 核心类库和支持文件。

JVM(Java Virtual Machine)是 Java 虚拟机。JVM 是 JRE 的一部分，是一个虚构出来的计算机，是通过在实际的计算机上仿真模拟各种计算功能来实现的。JVM 的主要工作是解释自己的指令集(即字节码)并映射到本地 CPU 的指令集或 OS 的系统调用。Java 语言是跨平台运行的，其实就是不同的操作系统，使用不同的 JVM 映射规则，让其与操作系统无关，完成跨平台性。JVM 对上层的 Java 源文件是不关心的，它关注的只是由源文件生成的类文件(class file)。类文件的组成包括 JVM 指令集、符号表以及一些辅助信息。

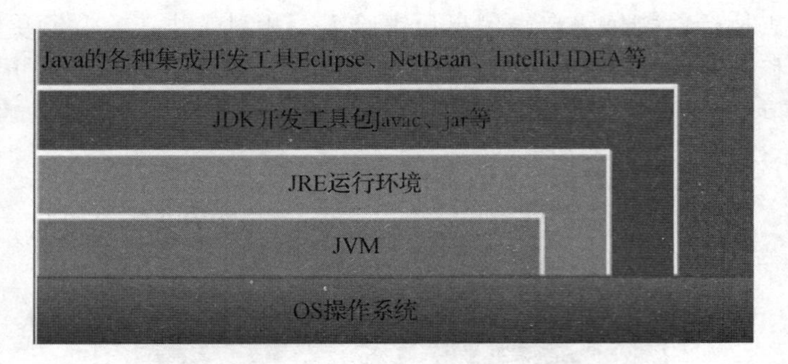

图 2-3　JDK、JRE、JVM 关系图

在实际开发中，利用 JDK(调用 Java API)开发 Java 程序后，通过 JDK 中的编译程序 Javac.exe 把 Java 源程序编译成 Java 字节码，在 JRE 上运行这些 Java 字节码，JVM 解析这些字节码，映射到 CPU 指令集或 OS 的系统调用。

2. 下载并安装 JDK

目前主流的 JDK 是 Sun 公司发布的，还有很多公司和组织都开发了属于自己的 JDK。下面用 Sun 公司的 JDK 给大家进行讲解。

JDK 可以到 Oracle 的官网上进行下载：

https://www.oracle.com/technetwork/java/javase/downloads/index-jsp-138363.html

目前最新的版本是 Java SE12。实际上目前还有很多公司在使用 JDK 8，所以接下来以 64 位的 Windows 系统为例来演示 JDK 8 的安装过程。不同版本的 JDK 安装包可以自行下载，如图 2-4 所示，也可以从随书资料中获取。

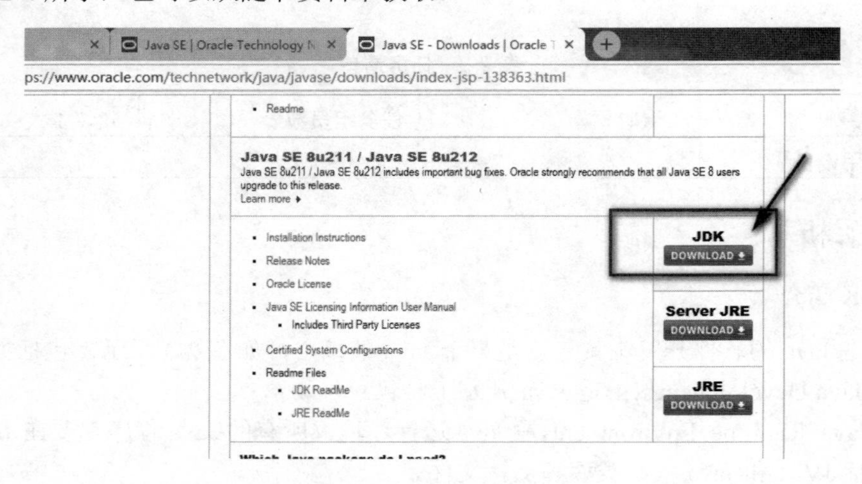

图 2-4　JDK 下载点

第一步：双击下载的 jdk-8u121-windows-x64.exe，进入 JDK 的安装界面，如图 2-5 所示。

第二步：单击“下一步”按钮，如图 2-6 所示，读者可以在此界面中选择要安装的功能以及更改安装路径。本示例使用默认功能和默认路径，然后单击“下一步”按钮，进入安装，如图 2-7 所示。安装成功界面如图 2-8 所示。

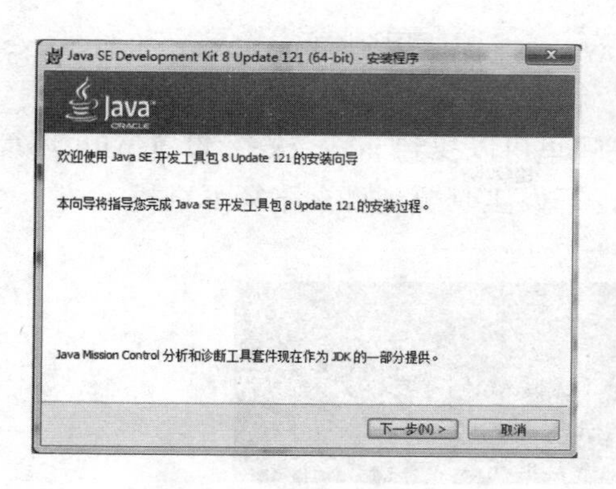

图 2-5　JDK 安装向导

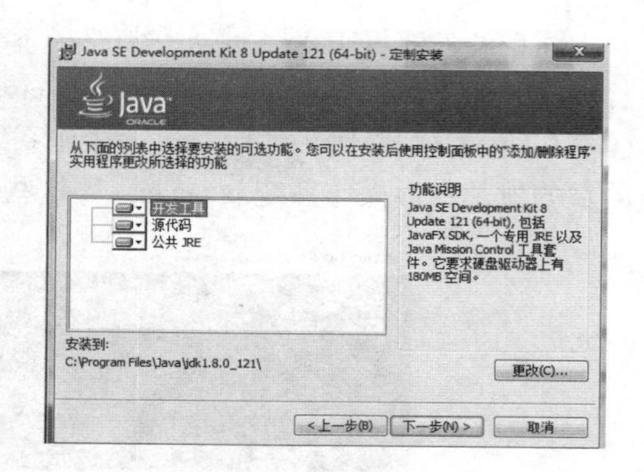

图 2-6　JDK 自定义安装界面

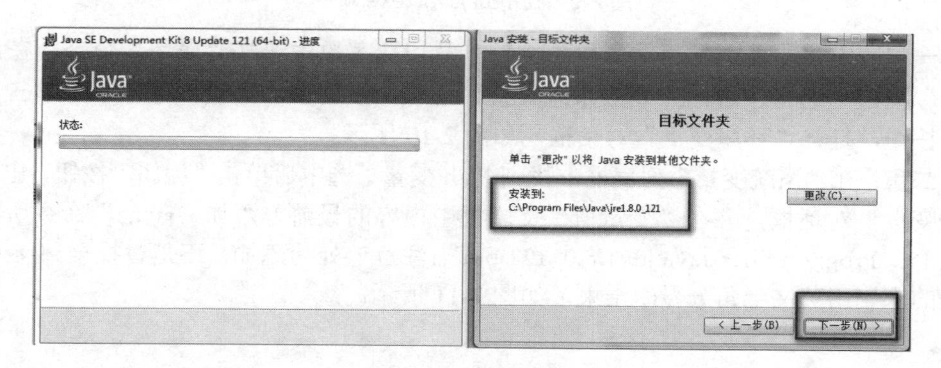

图 2-7　安装 JRE 界面

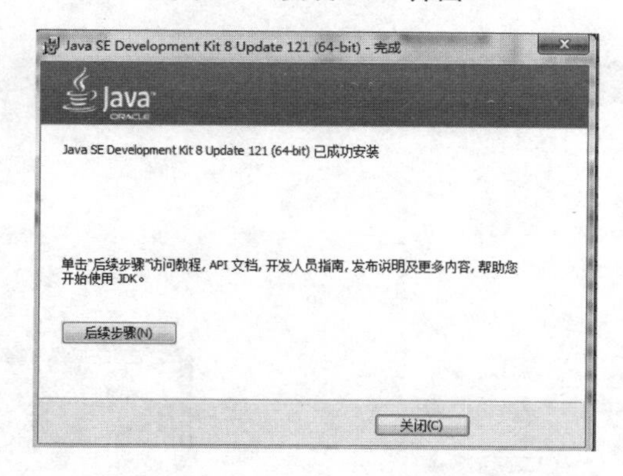

图 2-8　JDK 成功安装界面

3. 配置环境变量

JDK 安装好了以后，需要配置系统环境变量 path 和 classpath，接下来分别对它们进行讲解。

(1) path 环境变量。

path 环境变量是系统环境变量的一种，用于保存一系列的路径，每个路径之间以分号分隔。当在命令行窗口运行一个可执行文件时，操作系统首先会在当前目录下查找该文件是否存在，如果不存在，会继续在 path 环境变量中定义的路径下寻找这个文件，如果仍未找到，系统会报错。

上面已经安装好 JDK，默认安装路径下有 Java 开发需要用到的 javac.exe 和 java.exe 可执行程序。但是，当在命令行窗口输入“javac”命令，并按 Enter 键时，会看到错误提示，如图 2-9 所示。那是因为还没有告诉操作系统在哪里可以找到 javac.exe，设置 path 环境变量的目的就是告诉操作系统 javac.exe 等这些 Java 开发工具要到哪个路径下查找。

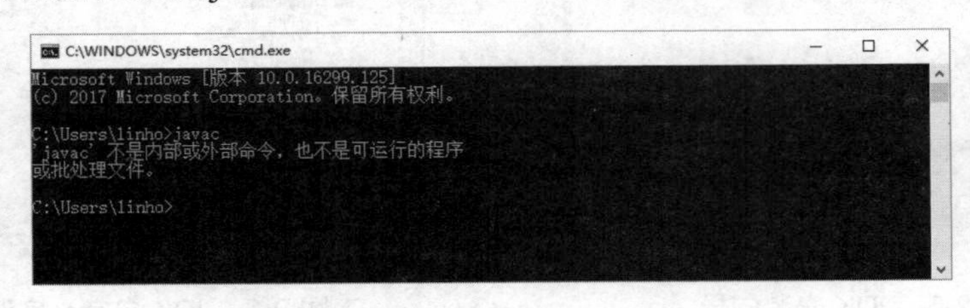

图 2-9　找不到 javac.exe 命令

第一步：右击桌面上的“计算机”图标，从快捷菜单中选择“属性”命令，在出现的“系统”窗口中选择左边的“高级系统设置”选项，接着在“高级”选项卡中单击“环境变量”按钮，打开“环境变量”对话框，如图 2-10 所示。

第二步：在“系统变量”列表框中找到 Path 变量，选中后单击“编辑”按钮，出现“编辑系统变量”对话框，并在“变量值”文本框中内容的最前方添加“javac”命令所在的目录路径“C:\Program Files\Java\jdk1.8.0_121\bin”(注意该处为添加，不是替换全部文本)，并在路径的末尾用英文半角分号(;)结束，如图 2-11 所示。

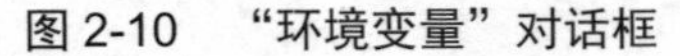

图 2-10　“环境变量”对话框

图 2-11　“编辑系统变量”对话框

添加完成后，依次单击对话框中的“确定”按钮，完成设置。

第三步：验证 Path 系统环境变量。

重新打开一个 CMD，在命令行窗口输入“javac”命令，并按 Enter 键，如果出现图 2-12 所示的信息，说明 Path 变量配置成功。

(2)　classpath 环境变量。

classpath 环境变量是用来保存 Java 虚拟机运行时所需的 class 文件和类包的路径。通常设置为当前路径，所以值为“.”(它代表当前路径)，如图 2-13 所示。

图 2-12 输入 javac.exe 命令结果

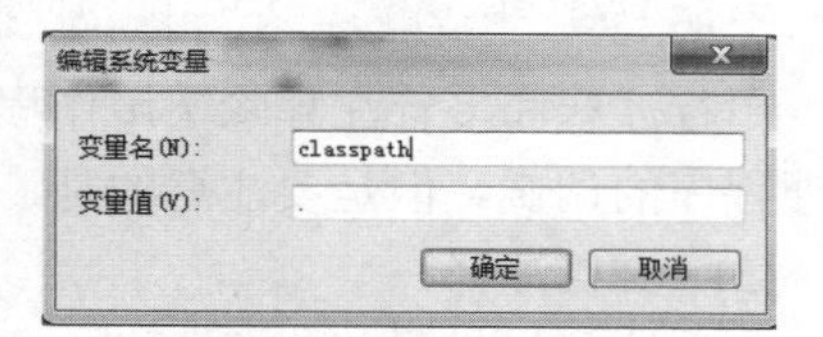

图 2-13 设置 classpath 变量

任务 3 编写第一个 Java 程序

【任务描述】

本任务完成第一个 Java 程序的编写，并学会编译、运行程序。任务清单如表 2-3 所示。

表 2-3 任务清单

任务课时	1 课时	任务组员数量	建议 1 人
任务组采用设备	无		

【知识解析】

1. Java 中的注释

注释是对程序中的某个功能或者某行代码的解释说明，它只在 Java 源文件中有效，在编译程序时，注释不会编译到 class 字节码文件中，会被编译器自动忽略。因此注释是给开发人员看的，有助于开发人员理解程序中的代码功能。

Java 中的注释有三种类型，分别为单行注释、多行注释以及文档注释。

(1) 单行注释。

单行注释使用符号“//”开头，后面为注释内容。一般用于对某行代码进行解释说明，可放在需要注释的代码的后面或者放在它的前一行(不同公司的 Java 代码规范可能对单行注释放置的位置要求不一样)。

```
1.    System.out.println("Hello World!");// 这行代码执行会在控制台输出内容 Hello World!
```

或者

```
1.    // 这行代码执行会在控制台输出内容 Hello World!
2.    System.out.println("Hello World!");
```

(2) 多行注释。

多行注释以符号“/*”开头，以符号“*/”结尾，开头符号和结尾符号之间为多行注释内容。多行注释顾名思义就是，注释的内容可以为多行。

```
1.   /*下面的两行代码执行后会输出两行内容,
2.   * 第一行内容为: Hello World!
3.   * 第二行内容为: 你好 java
4.   */
5.   System.out.println("Hello World!");
6.   System.out.println("你好 java!");
```

单行注释和多行注释除了用作程序的解释说明之外，还可以用来注释暂时不用的代码，方便程序的调试。但是要注意的是，在最后完成项目后，调试的代码需要删除。

(3) 文档注释。

文档注释以符号“/**”开头，以符号“*/”结尾，开头符号和结尾符号之间为文档注释内容。文档注释用于对类或者类方法做解释说明。

```
1.   /**
2.    * ADAM-4150 I/O 控制器
3.    *
4.    */
5.   public class ADAM4150 {
6.
7.       /**
8.        * 分析 ADAM-4150 采集到的数据并存放在 actionStatus 中
9.        * @param data 从串口读取到的数据
10.       * @return 是否有采集到结果
11.       */
12.      private boolean flashStatus(byte[] data) {
13.       return false;
14.      }
15.  }
```

2. Java 中的标识符

在编程语言中，标识符就是指开发人员自己规定的具有特殊含义的词，如包名、类名、方法名、参数名、变量名等。标识符可以由字母、数字、下划线(_)和美元符号($)组成，但标识符不能以数字开头，不能是 Java 关键字。

合法的标识符如下:

```
apple apple123  user_name _username $username  你好
```

不合法的标识符如下:

```
123apple  class   yes/no   user name
```

标识符必须严格遵守上面的规范。除了规范外，为了增强代码的可读性，建议读者在定义标识符时应遵循以下规则。

- 尽量使用有意义的英文单词来定义标识符。例如，用户名 userName、密码 pssword。
- 包名中所有的字母一律小写。例如，com.example.ui。
- 类名和接口名的每个单词的首字母都要大写。例如，Student、ArrayList。
- 常量名的所有字母都大写，单词直接使用下划线连接。例如，MAX_COUNT。
- 变量名和方法名的第 1 个单词首字母小写，从第 2 个单词开始，每个单词首字母大写。例如：userName、getName。

3. Java 中的关键字

在编程语言中，关键字是指事先定义好并赋予了特殊含义的单词。图 2-14 列举出 Java 语言中所有的关键字。

abstract	continue	for	new	switch
assert	default	goto	package	synchronized
boolean	do	if	private	this
break	double	implements	protected	throw
byte	else	import	public	throws
case	enum	instanceof	return	transient
catch	extends	int	short	try
char	final	interface	static	void
class	finally	long	strictfp	volatile
const	float	native	super	while

图 2-14　Java 的关键字

上面列举的关键字中，每个都有特殊的作用。例如，class 关键字用于类的声明，int 关键字用于整型变量的声明。这些关键字在后面的学习中会逐步介绍，这里只做了解即可。

使用 Java 关键字时，需要注意以下四点。

- 所有的关键字都是小写的。
- 标识符不能以关键字命名。
- const 和 goto 是保留关键字，虽然目前在 Java 中没有任何意义，但是保留关键字的地位。
- true、false 和 null 不属于关键字，但是它们是有具体意义的字面量，不能作为标识符。

【任务实施】

1. 编写 Java 程序

在 C 盘根目录下创建文件夹 project2，在 project2 中新建一个 HelloWorld 文件，将文件名的后缀修改为.java(注意修改文件名后缀的时候，请显示文件扩展名)，如图 2-15 所示。

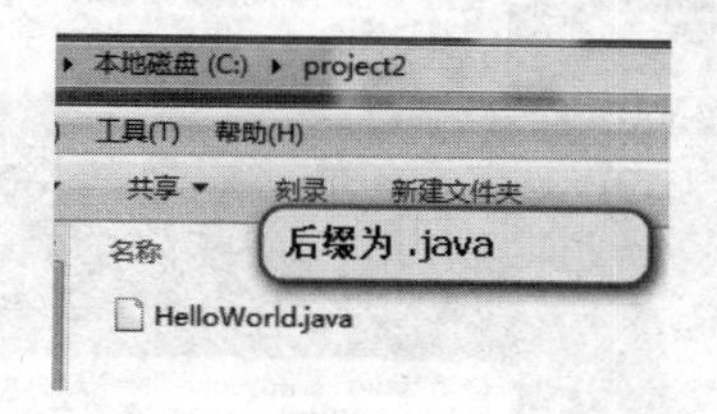

图 2-15　新建 Java 文件

打开 Java 文件，编写文件内容如下：

```
1.   /**
2.    * 这是第一个 Java 程序
3.    */
4.   public class HelloWorld {
5.       /**
6.        * 这是入口方法
```

```
7.        *
8.        * @param args 命令行参数
9.        */
10.      public static void main(String[] args) {
11.          System.out.println("HelloWorld");
12.      }
13.
14.  }
```

对上面程序做一个简要的分析。

- 第 4 行代码为声明一个 HelloWorld 类，public、class 都是关键字，代表这是一个公开的类，这里只需知道定义一个类的基本格式即可，关于类的知识在后面的面向对象相关内容会详细介绍。
- 第 1～3 行是类的注释，第 5～9 行是 main 方法的注释。
- 第 10 行是 Java 程序的入口方法 main，Java 程序从 main 方法开始执行。
- 第 11 行代码将会输出引号里面的内容 Hello World!，程序中出现的所有符号均为英文输入状态下的符号，切勿使用中文字符；、 “ ” 等。
- 标识符：类名 HelloWorld 的首字母大写，方法名 main 的首字母小写。

2. 运行 Java 程序

了解 Java 的运行机制有助于初学者更好地理解 Java 程序的运行过程。Java 程序运行时，必须经过编译和运行两个步骤。首先将文件名后缀为.java 的文件编译生成.class 字节码文件，然后 Java 虚拟机将字节码文件进行解释，并显示结果，如图 2-16 所示。

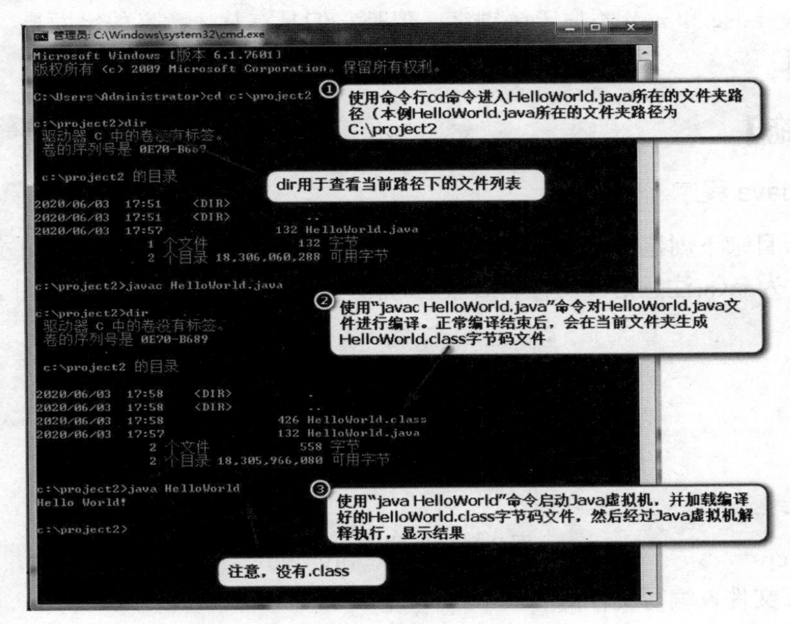

图 2-16　编译和运行 Java 程序

通过上面的分析不难看出，Java 程序是由 Java 虚拟机负责解释运行的，而并非操作系统。这样做的好处可以实现 Java 程序的跨平台。也就是说，在不同的操作系统上，可以运行相同的 Java 程序，只要安装不同的 Java 虚拟机即可，如图 2-17 所示。

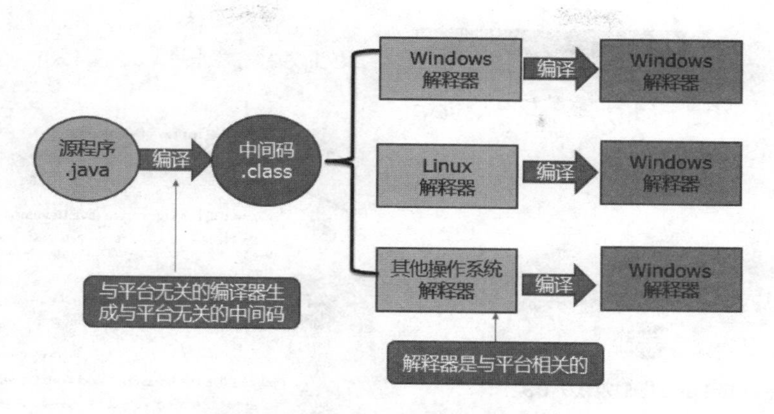

图 2-17 Java 跨平台原理图

从图 2-17 可以看出，不同的操作系统需要使用不同版本的虚拟机，这种方式使得 Java 语言具有“一次编写，处处运行”的特征，这样可以大大降低程序开发和维护的成本。

任务 4 使用 Eclipse 开发工具

扫码观看视频讲解

【任务描述】

在实际的项目开发过程中，开发人员很少使用记事本来编写代码，而是使用编写效率更好、更容易排错的集成开发环境(IDE，Integrated Development Environment)来进行 Java 程序开发。Eclipse 是由 IBM 公司开发的一款 IDE，它是开源的、基于 Java 的可扩展开发平台，是目前最流行的 Java 语言开发工具之一。Eclipse 具有强大的代码编排功能，可以帮助开发人员完成语法修正、代码修正、文字补全、信息提示等编码工作，大大提高了程序开发的效率。

本节任务要求使用 Eclipse 进行 Java 的开发。任务清单如表 2-4 所示。

表 2-4 任务清单

任务课时	1 课时	任务组员数量	建议 1 人
任务组采用设备	无		

【任务实施】

1. 安装 Eclipse 开发工具

登录 Eclipse 官方网站 https://www.eclipse.org/downloads 可以下载 Eclipse，如图 2-18 所示。

本教程所用的 Eclipse 不是最新版的，而是采用 64 位安装版 eclipse-inst-win64.exe，Eclipse 版本为 Photon Release (4.8.0)。

运行安装包 eclipse-inst-win64.exe 可以看到 Eclipse 的安装向导页面，这里选择第一项进行安装，如图 2-19 所示。

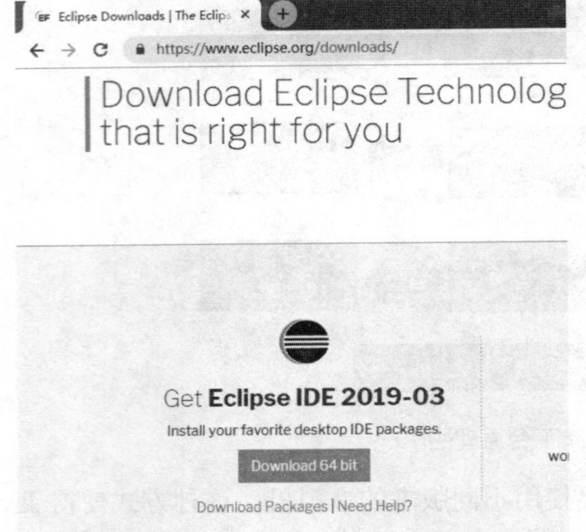

图 2-18　Eclipse 的版本

图 2-19　Eclipse 安装向导

接下来的安装步骤使用默认步骤即可，直到 Eclipse 安装成功。

2. 启动 Eclipse 开发工具

(1) Eclipse 安装成功后，可以通过双击 Windows 桌面上的 Eclipse Java Photon 或在 Eclipse 安装路径中运行 eclipse.exe 打开 Eclipse，启动界面如图 2-20 所示。

(2) Eclipse 启动后会弹出一个对话框，提示选择工作空间(Workspace)，如图 2-21 所示。

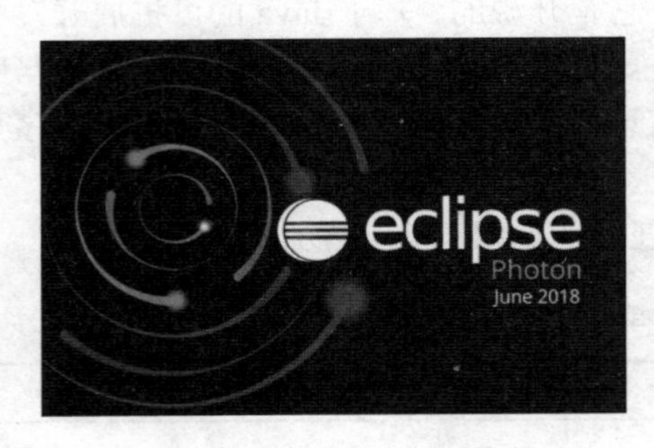

图 2-20　Eclipse 启动窗口

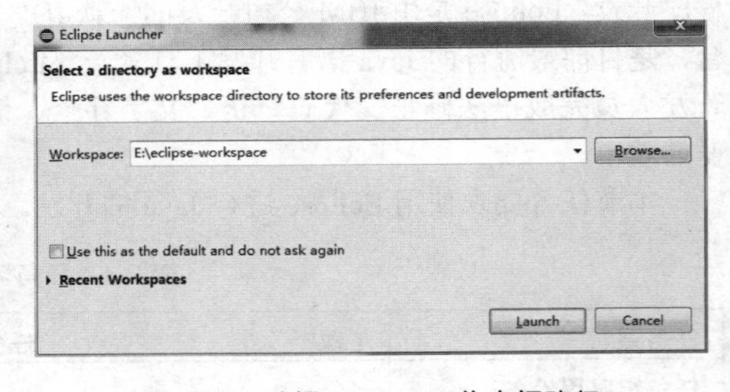

图 2-21　选择 Eclipse 工作空间路径

工作空间用于保存 Eclipse 中创建的项目和相关配置。建议自己选择工作空间目录，工作空间的目录应该设置为非系统盘。设置好后，单击 Launch 按钮即可。

需要注意的是，每次启动 Eclipse 都会显示选择工作空间的对话框，如果不想每次都选择工作空间，可以将图 2-21 中的 Use this as the default and do not ask again 复选框选中，Eclipse 每次启动都会以此工作空间作为默认的工作空间。

(3) 进入 Eclipse 后，会显示 Eclipse 欢迎界面，如图 2-22 所示，如果你不想下次进入 Eclipse 时显示欢迎界面，可以取消选中 Always show Welcome at start up 复选框。

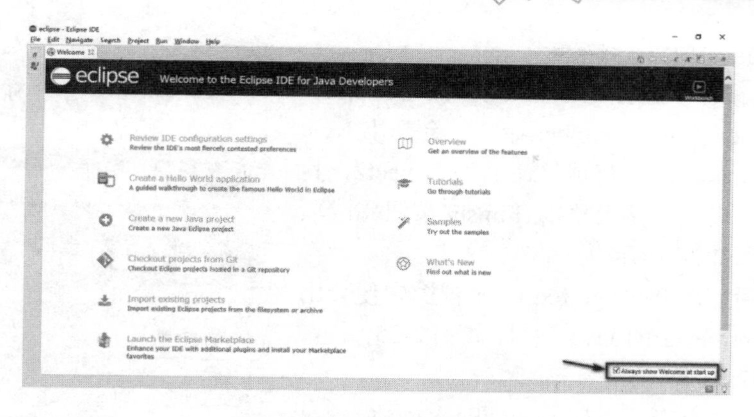

图 2-22 Eclipse 欢迎界面

3. Eclipse 工作台界面

关闭 Eclipse 欢迎界面，就进入 Eclipse 工作台界面。Eclipse 工作台主要由标题栏、菜单栏、工具栏、透视图等部分组成，如图 2-23 所示。

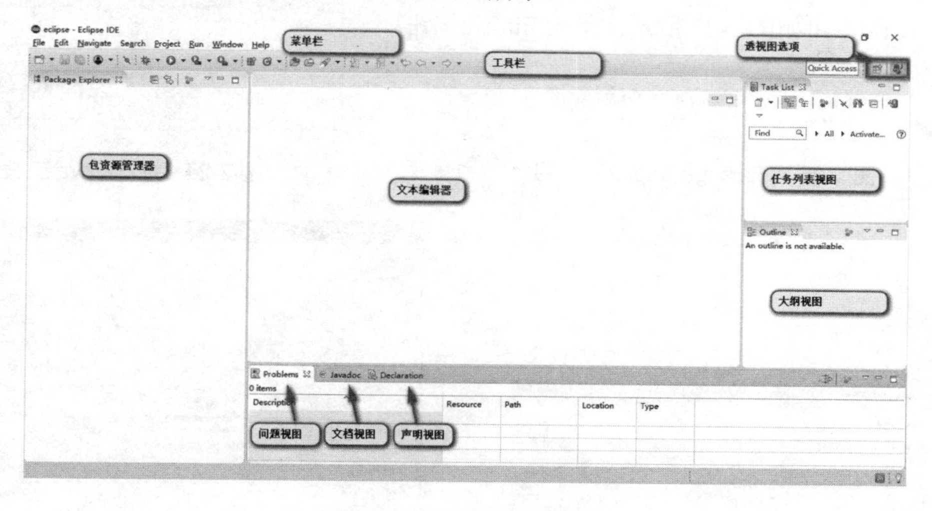

图 2-23 Eclipse 工作台

从图 2-23 中可以看出，工作台界面上有包资源管理器视图、文本编辑器视图、大纲视图等多个模块，这些视图大多用来显示信息的层次结构和实现代码编辑。下面是 Eclipse 工作台上主要视图的作用。

- Package Explorer(包资源管理器视图)：用来显示项目文件的组成结构。
- Editor(文本编辑器)：用来编写代码的区域。
- Problems(问题视图)：显示项目中的一些警告和错误。
- Console(控制台视图)：显示程序运行时的输出信息、异常和错误(如图 2-31 所示)。

关于更多 Eclipse 工作台视图的介绍，可以参考 Eclipse 官方帮助文档。

4. Eclipse 下开发 Java 应用程序

经过之前的学习，读者已经能够在命令行界面打印“Hello World!”，接下来通过 Eclipse 创建一个 Java 应用程序，并实现在控制台上打印“Hello World!”，具体步骤如下。

(1) 创建 Java 项目。

在 Eclipse 窗口中选择 File→New→Java Project 命令，或者在 Package Explorer 视图中右击，然后选择 new→Java Project 命令，会弹出 New Java Project 对话框，这里将项目命名为 project2，其余参数保持默认设置，然后单击 Finish 按钮完成项目的创建，如图 2-24 所示。

创建完成后，在 Package Explorer 视图中便会出现一个名称为 project2 的 Java 项目，如图 2-25 所示。

(2) 创建 Java 类。

在 src 目录上右击，选择 new→Package 命令，输入包名 com.nle.demo 后单击 Finish 按钮，如图 2-26 所示。

在包上右击，选择 new→Class 命令，在 New Java Class 对话框中的 Name 文本框中输入类名称“HelloWorld”，如图 2-27 所示。单击 Finish 按钮。

图 2-25　Package Explorer 项目

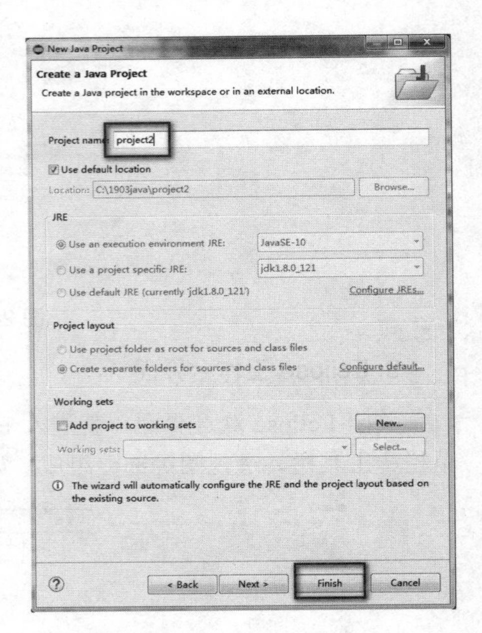

图 2-24　创建 Java 工程

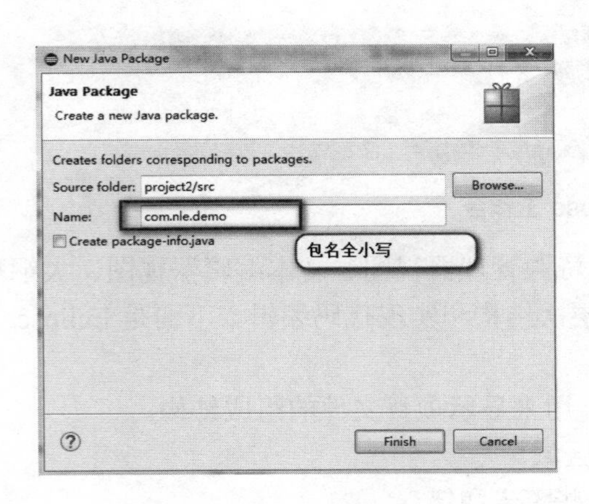

图 2-26　新建包

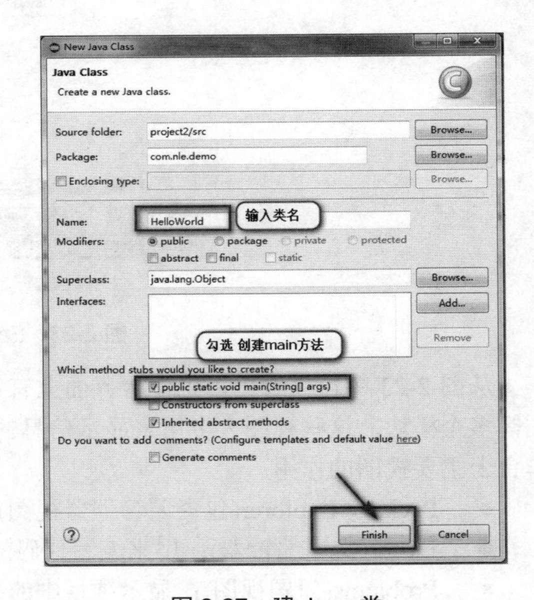

图 2-27　建 Java 类

创建好的 HelloWorld.java 文件会在编辑区域自动打开，如图 2-28 所示。

(3) 编写程序代码。

在编辑框内写入 main()方法和一条输出语句“System.out.println("Hello World!");”，如图 2-29 所示。

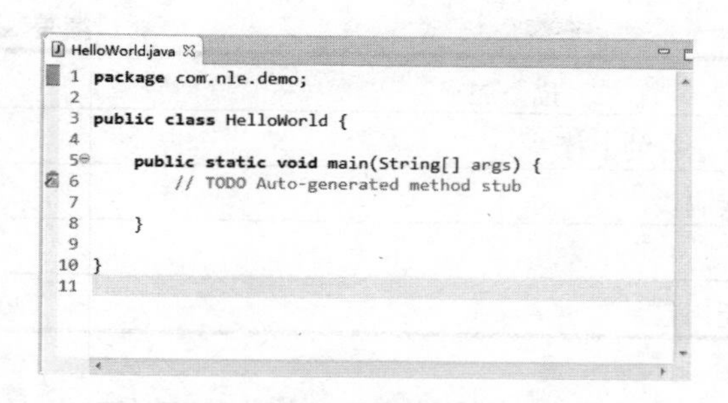

图 2-28　HelloWorld.java

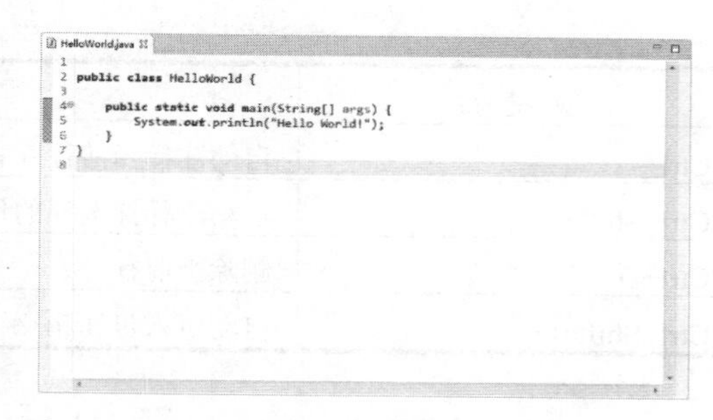

图 2-29　编写代码

(4) 运行应用程序。

在 Package Explorer 视图中选中 HelloWorld.java 文件或者在编辑框空白处右击，选择 Run As→Java Application 命令，如图 2-30 所示。

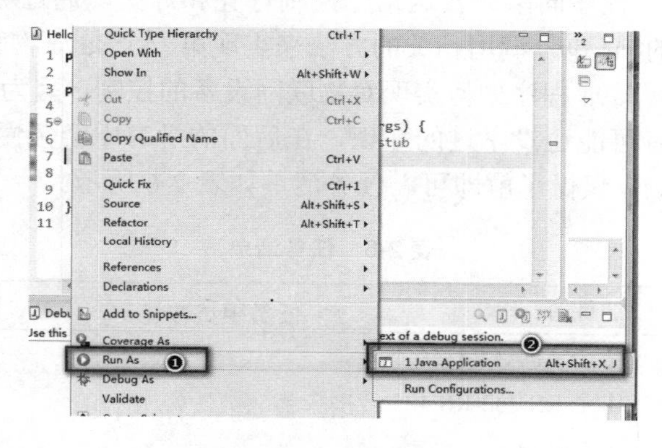

图 2-30　运行程序

也可在选中文件后，单击工具栏上的按钮运行程序。程序执行完毕后，在控制台 Console 视图中可以看到运行的结构，如图 2-31 所示。

图 2-31　运行结果

5. Eclipse 中常用快捷键

Eclipse 中常用快捷键如表 2-5 所示。

表 2-5　Eclipse 中常用快捷键

快 捷 键	描　述
Alt+? 或 Alt+/	自动补全代码或者提示代码
Ctrl+Shift+R	打开资源列表
Ctrl+Shift+F	格式化代码
Alt+方向键上下	上下行交换内容或把当前行内容往上或下移动

续表

快 捷 键	描 述
Ctrl+/	自动注释当前行或者选择的多行
Ctrl+Shift+/	自动注释掉选择的代码块
Ctrl+D	删除当前行
Ctrl+Shift+O	自动引入包和删除无用包

任务 5　第一个 Java 物联网程序(随心所欲亮灯灭灯)

【任务描述】

扫码观看视频讲解

本任务要求用 Java 程序向串口发送指令控制灯亮和灯灭，通过导入已经写好的部分功能的 jar 包，调用相关的方法来实现第一个 Java 物联网应用程序，体会并理解 Java 程序如何实现对物联网设备的控制。因为是第一个 Java 物联网程序，会用到很多读者可能还没学过的知识，在后面的学习中会慢慢展开，本任务不要求读者掌握调用到的方法，仅做了解即可。任务清单如表 2-6 所示。

表 2-6　任务清单

任务课时	2 课时	任务组员数量	建议 3 人
任务组采用设备	1 个 ADAM-4150 1 个 RS 232-RS 485 转接头 1 个继电器 1 个照明灯 1 台 PC 相关电源，导线，工具		

【拓扑图】

本任务拓扑图如图 2-32 所示。

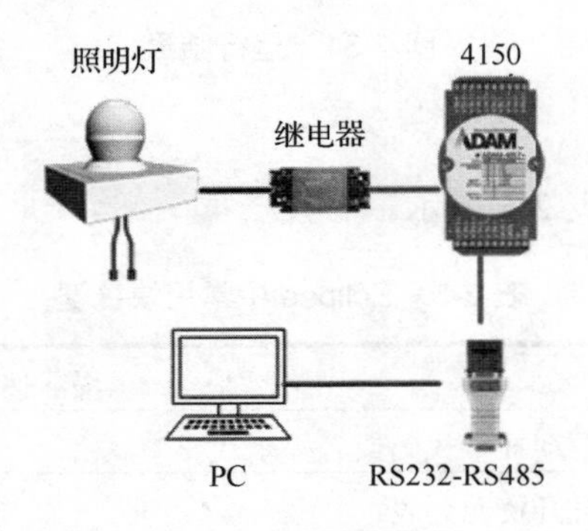

图 2-32　拓扑图

备注：本任务照明灯接 DO0 口，COM 口根据实际情况确定。

【知识解析】

以下我们来学习 ADAM-4150 数字量采集器各通道开关控制指令。

ADAM-4150 数字量采集模块应用 EIA RS-485 通信协议，它是工业上广泛使用的双向、平衡传输线标准。它使得 ADAM-4150 模块可以远距离高速传输和接收数据，可以进行数据采集和控制设备。

ADAM-4150 模块 8 通道控制设备的指令如图 2-33 所示。

ADAM-4150 开关继电器为 DO0-DO7 共 8 通道输出，对应的命令如下表：

通道号	开关继电器命令	状态
DO0	01 05 00 10 FF 00 8D FF	开
DO0	01 05 00 10 00 00 CC 0F	关
DO1	01 05 00 11 FF 00 DC 3F	开
DO1	01 05 00 11 00 00 9D CF	关
DO2	01 05 00 12 FF 00 2C 3F	开
DO2	01 05 00 12 00 00 6D CF	关
DO3	01 05 00 13 FF 00 7D FF	开
DO3	01 05 00 13 00 00 3C 0F	关
DO4	01 05 00 14 FF 00 CC 3E	开
DO4	01 05 00 14 00 00 8D CE	关
DO5	01 05 00 15 FF 00 9D FE	开
DO5	01 05 00 15 00 00 DC 0E	关
DO6	01 05 00 16 FF 00 6D FE	开
DO6	01 05 00 16 00 00 2C 0E	关
DO7	01 05 00 17 FF 00 3C 3E	开
DO7	01 05 00 17 00 00 7D CE	关

图 2-33　ADAM-4150 控制指令

【任务实施】

1. 任务分析

(1) 连接设备并检查无误。

(2) 创建工程并导入相关库文件。

(3) 打开串口发送控制指令控制灯亮灯灭。

2. 任务实施

(1) 按接线图对设备进行连接并检查无误，把 RS232-RS485 转接头接到 PC，查看使用的 COM 口号。本程序假设用的是 COM200 口。

(2) 创建工程并导入相关库文件。

创建新的工程 project2_task5，在工程的 src 下新建包 com.nle。选中工程，右击并选择 New→Folder 命令，设置文件夹名字为“libs”，把随书资料中的串口通信库 RXTXcomm.jar、串口管理工具类库 SerialPortLib.jar 和数字量设备 4150 控制库 Controller.jar 复制到 libs 下，这三个文件中有写好的可以打开和关闭串口以及通过串口发送控制指令的方法，调用方法就可以实现相关的功能。

RXTXcomm.jar 是读取串口数据用的串口通信包，SerialPortLib.jar 是基于 RXTXcomm.jar 提供的方法进行封装以提供打开串口、关闭串口、获取串口数据等的方法，Controller.jar 对获取的传感器数据进行分析计算并提供对应的方法把传感器数据返回给调用者。

将三个.jar 文件复制到 libs 文件夹后，选中这三个.jar 文件后右击，按以下操作步骤把库文件添加到工程的引用库中，这样工程才可以使用这三个.jar 库文件提供的方法，操作过程如图 2-34 中的①②③所示，操作成功应该可以看到④的结果。

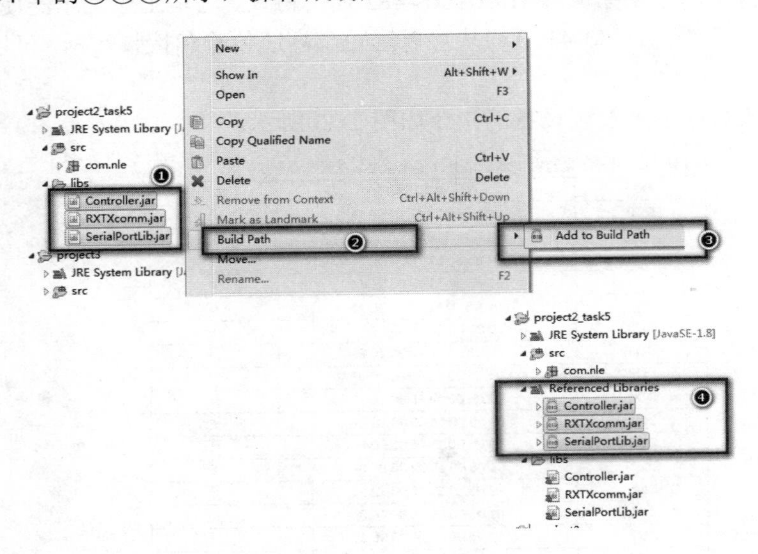

图 2-34　添加库文件到工程中

要使用串口通信库，还必须把两个.dll 文件复制到 jdk 的安装目录 bin 下，才能正常使用串口通信，添加过程如图 2-35 所示。(注意：如果添加到 jdk 目录下不能成功运行，建议也同时添加到 jre 下的 bin 目录中)

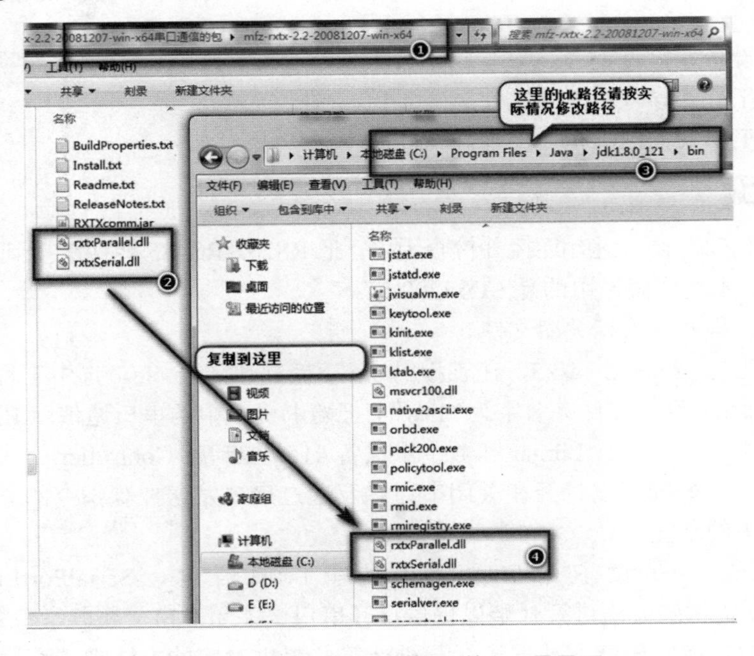

图 2-35　添加.dll 文件到 jdk 的 bin 目录下

(3) 打开串口发送控制指令控制灯亮灯灭。

要控制灯亮灯灭需要往串口中发送对应 DO 口的控制指令，因为本任务的灯接在 DO0 口，查 ADAM-4150 的指令，得知开灯指令是 01 05 00 10 FF 00 8D FF，关灯指令是 01 05 00 10 00 00 CC 0F。

在包 com.nle 下新建类 TestLed，在 main 方法中输入下面从第 11 行开始的代码，因为用到了串口，有可能出错，所以 main 后面添加 throws SerialPortException 代表相关的出错的处理。

第 11～25 行用于获取串口对象、打开串口、初始化数字量设备 4150 对象、发送控制指令和关闭串口。

当输入 12、14、16 行代码时用 Ctrl+Shift+O 自动导入包，就会产生第 3～5 行间的代码，表示程序中用到的相关类在哪个包下。这些类可以使用是因为上面已经把相关类所在的 jar 文件添加到工程中了。

```
1.  package com.nle;
2.
3.  import com.nle.demo.Adam4150Controller;
4.  import com.nle.serialport.SerialPortManager;
5.  import gnu.io.SerialPort;
6.
7.
8.  public class TestLed {
9.
10.     public static void main(String[] args) throws Exception {
11.         //1.获取串口管理对象  (SerialPortLib.jar 中提供的)
12.         SerialPortManager manager=new SerialPortManager();
13.         //2.打开串口(使用的 COM 口号要根据实际情况填写)
14.         SerialPort  serialPort=manager.openPort("COM200", 9600);
15.         //3.初始化数字量设备 4150
16.  Adam4150Controller controller  = new Adam4150Controller(manager,
    serialPort);
17.
18.         //4.灯接在 DO0 口，发控制指令开灯
19.         controller.openLed("01 05 00 10 FF 00 8D FF");
20.
21.         //5.灯接在 DO0 口，发控制指令关灯
22.     //  controller.closeLed("01 05 00 10 00 00 CC 0F");
23.
24.         //6.关闭串口
25.         manager.closePort(serialPort);
26.     }
27. }
```

3. 运行结果

(1) 运行程序，可以看到灯亮了。

(2) 把代码的 19 行用//进行注释，把 22 行前的//去掉，重新运行程序，可以看到灯灭了。

(3) 按照图 2-33 的 ADAM-4150 控制指令图，把灯接在不同的 DO 口，再发送对应的控制指令，测试是否可以随心所欲地控制灯亮灯灭。

思考与练习

1. 简述 JDK、JRE、JVM 之间的关系。
2. 简述设置 path、classpath 的作用。
3. 简述 Java 程序控制灯亮灯灭的过程。

项目 3 传感数据解析和控制指令生成

【项目描述】

本项目通过分析采集到的温湿度和光照传感器数据，并按照协议生成设备控制指令，让读者了解 Java 的基本语法，并利用 Java 的基本语法对采集到的真实物联网设备数据进行解析和对设备进行控制，从而知道 Java 物联网应用程序是怎么一步步开发出来的。具体任务列表如图 3-1 所示。

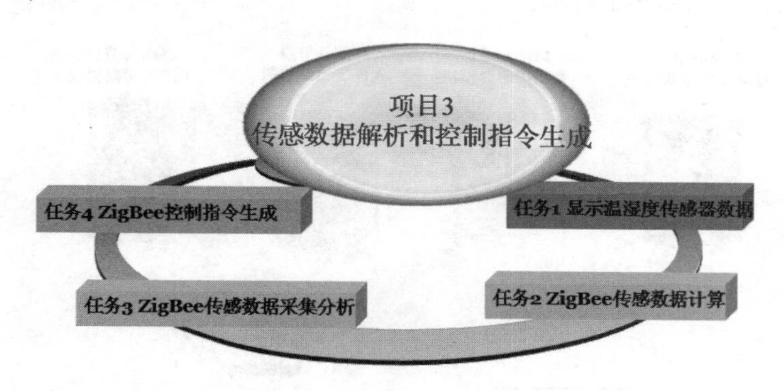

图 3-1　项目 3 任务列表

【学习目标】

会使用 Java 的注释；知道常量与变量的区别；会使用 Java 运算符；会使用流程控制中的条件与循环；会使用数组。

【技能目标】

能使用合适的数据类型显示传感数据；能使用合适的运算符计算传感数据；能使用条件和循环分析传感数据；能生成设备控制指令。

任务 1　显示温湿度传感器数据

扫码观看视频讲解

【任务描述】

本任务将 ZigBee 温湿度传感器、光照传感器、人体传感器、火焰传感器组成一个 ZigBee 无线网络，按物联网工程实训的要求对 ZigBee 进行烧写后，采集数据并实现选用合适的数据类型来保存采集到的温湿度传感器、光照传感器、人体传感器、火焰传感器的数值，并输出到控制台上。具体任务清单如表 3-1 所示。

表 3-1　任务清单

任务课时	4 课时	任务组员数量	建议 3 人
任务组采用设备	1 个 ZigBee 温湿度模块 1 个 ZigBee 光照模块 1 个 ZigBee 人体模块 1 个 ZigBee 火焰模块 1 个 ZigBee 协调器模块 1 台 PC 电源线、USB 转串口线		

【拓扑图】

本任务的拓扑图如图 3-2 所示。

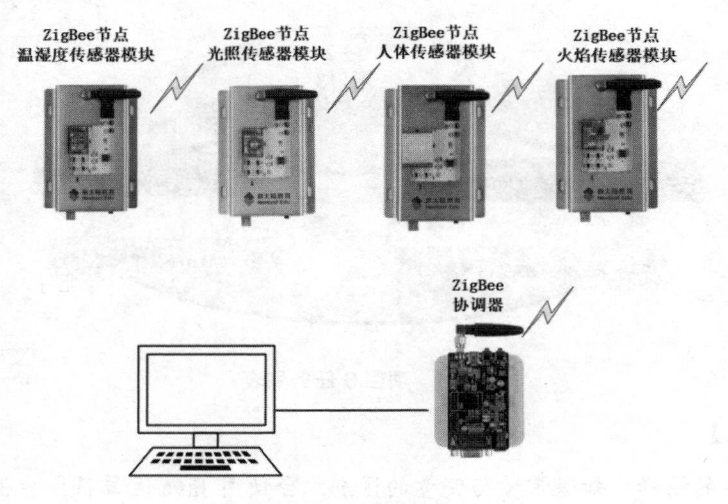

图 3-2　拓扑图

备注：本任务 ZigBee 协调器通过串口线直连 PC 机的 COM 口，COM 口号根据实际情况确定。

【知识解析】

1. 基本数据类型

在物联网的应用程序中，你会得到下面这样的数据。

温湿度传感器数据：温度 25.5℃　湿度 75%。

人体传感器数据：true，false。

温度传感器范围：-10℃～60℃。

风扇的控制：“开”“关”。

这些数据在 Java 程序中要如何表示呢？温湿度传感器数据是小数，人体传感器数据是 true 或 false，风扇的控制数据是字符。Java 定义了 8 种基本数据类型来描述这些数据，不同类型的数据在存储时需要的空间各不相同，取值范围也各不相同，具体如表 3-2 所示。

表 3-2　数据类型

类型	数据类型	占用空间	应用场合：取值范围
布尔型	boolean	8 位(1 个字节)	存储逻辑变量：true、false
字节型	byte	8 位(1 个字节)	存储字节数据：-2^{7}～$2^{7}-1$
整型	short	16 位(2 个字节)	存储短整型数据：-2^{15}～$2^{15}-1$
	int	32 位(4 个字节)	存储整型数据：-2^{31}～$2^{31}-1$
	long	64 位(8 个字节)	存储长整型数据：-2^{63}～$2^{63}-1$
浮点型	float	32 位(4 个字节)	存储单精度浮点数据： -3.4E+38～-1.4E-45，1.4E-45 ～3.4E+38
	double	64 位(8 个字节)	存储双精度浮点数据： -1.7E+308～-4.9E-324，4.9E-324～1.7E+308
字符型	char		存储一个字符

(1)　Java 中的常量。

常量即保持不变的值。常量在程序的运行过程中是不允许改变的。例如，数字 1、小数 2.3、字符'a'等。Java 中的常量包括整型常量、浮点型常量、字符型常量、字符串常量等。

①　整型常量。

整型常量有 4 种表示形式，分别为二进制、八进制、十进制和十六进制。

二进制：由数字 0 和 1 组成的数字序列。JDK1.7 以后版本可以在字面量前面加上 0b 或者 0B 来表示二进制数，如 0b00000101，0B10110111。

八进制：以 0 开头其后由数字 0～7(包含 0 和 7)整数组成的数字序列，如 056。

十进制：由数字 0～9(包含 0 和 9)的整数组成的数字序列。

十六进制：以 0x 或者 0X 开头其后由 0～9、A～F(包含 0 和 9、A 和 F，字母不区分大小写)组成的数字序列，如 0x003A。

需要注意的是，以上二进制、八进制、十六进制整数的表示是以数字 0 开头而不是以字母 o 开头。

② 浮点型常量。

浮点型常量指的就是小数，可分为单精度 float 浮点数和双精度 double 浮点数两种。其中，单精度浮点数必须以 f 或 F 结尾，而双精度浮点数可以以 d 或 D 结尾(也可省略，浮点数不加任何后缀的时候，Java 默认把该浮点数当作双精度浮点数)。具体示例如下：

```
1.23f  3.14d  3.14
```

③ 字符型常量。

字符型常量用于表示一个字符，一个字符型常量使用一对英文半角格式的单引号' '引起来，它可以是字母、数组、标点符号以及转义字符。具体示例如下：

```
'a'  '1'  '$'  '\n'  '\072'  '\u0000'
```

以上示例中，转义字符'\n'表示换行的意思。字符'\072'表示八进制数值 072 对应 Unicode 字符集中的字符，因为 Java 采用的是 Unicode 字符集。同理，字符'\u0000'表示十六进制数值 0x0 对应 Unicode 中的字符。

④ 字符串常量。

字符型常量只能用来表示一个字符，当想表示一串连续的字符时可以使用字符串常量。字符串常量使用一对英文半角格式的双引号(" ")引起来。具体示例如下：

```
"HelloWorld"  "123456"  "你好! Java"
```

⑤ 布尔型常量。

布尔型常量只有两个值 true 和 false，该常量用于表示真和假。

(2) Java 中的变量。

Java 使用变量来装载程序运行过程中可以发生变化的值，在编程中使用变量可以让程序语言表达更方便简洁。

```
1.   int num1 = 10;//存储整数 10
2.   int num2 = 20;//存储整数 20
3.   int sum = num1+num2;//存储两个整数的和 30
4.   System.out.println("sum="+sum);//输出变量 sum 中的值 30
5.   num1 = 40;//存储的值由 10 变成 40
6.   sum = num1+num2;//存储变化后的两个整数的和 40
7.   System.out.println("sum="+sum);//再次输出变量 sum 中的值 60
```

这段代码说明了存储数据和使用数据的过程。要使用变量，需遵循以下步骤。

第一步：声明变量，根据变量的数据类型在内存中申请一个空间，通过变量名可以访问这个空间，语法如下：

```
1.   数据类型  变量名 ;
```

上述代码中的第一行就是声明了一个变量 num1，用来存储一个整数。

第二步：给变量赋值，将数据存储到这个空间中。

有两种写法，一种是声明时同时赋值，如上面代码中的 1～3 行。一种是需要变量时重新赋值，如上面代码中的第 5～6 行。

第三步：使用变量，获取存储空间中的值，如上面代码中的第 5，7 行。

需要注意的是，变量的命名要符合标识符的规定，变量名首字母小写，同时变量必须先声明和赋值后才可以使用，以下是常见的使用变量时的错误：

```
1.  public static void main(String[] args) {
2.      num = 10;// 错误1：变量未声明就使用
3.      int sum;
4.      System.out.println(sum);// 错误2：局部变量使用前必须先初始化
5.
6.      {
7.          int num2 = 30;// num2可以使用的范围在这个{}内
8.      }
9.      System.out.println(num2);// 错误3：超出num2的作用范围，不能访问
10.
11.     int num3 = 10;
12.     long num4 = 40;
13.     num3 = num4;// 错误4：变量的操作必须与类型匹配
14. }
```

(3) 布尔类型变量。

布尔类型变量用来存储布尔值，变量使用关键字boolean声明，该类型的变量只有两个值true和false。具体示例如下：

```
1.  boolean flag1 =false;// 为boolean类型变量flag1赋值false
2.  boolean flag2 =true;// 为boolean类型变量flag2赋值true
```

(4) 整数类型变量。

整数类型变量用来存储整数数值。为了给不同大小范围的整数合理分配存储空间，Java的整数类型分为4种：字节型(byte)、短整型(short)、整型(int)和长整型(long)。4种整数类型变量所占存储空间的大小以及取值范围如表3-2所示。

4种整型都规定了数据所占的空间大小和取值范围。在定义变量时，不应该超出各类型规定的取值范围。需要注意的是，给long类型变量赋值时，值的后面要加上一个字母小写l或大写L，说明该值为long类型。具体示例如下：

```
1.  byte b = 127;
2.  // byte b1 = 128;//byte的取值范围为-128～127，128超出取值范围，不能直接赋值。
3.  short s = 1234;
4.  int i = 12345;
5.  long l1 = 123;// 未超出int类型的取值范围，可以不用加l或者L
6.  long l2 = 123L;// 未超出int类型取值范围，可以加l或者L
7.  long l3 = 12300000000L;// 超出int类型取值范围，必须加l或者L
```

(5) 浮点类型变量。

浮点类型变量有两种：单精度浮点数float和双精度浮点数double。double类型的浮点数可以表示的精度比float类型的浮点数高，两种浮点数所占的存储空间大小和取值范围如表3-2所示。

在表3-2所示取值范围中，E(或者小写e)表示以10为底的指数，E后面的+号表示正指数，-号表示负指数。例如-3.4E+38表示$-3.4*10^{38}$。

在Java中，小数默认为double类型的浮点数，所以在给double类型的变量赋值时，浮点数后面可以加d(或D)，也可以不加。但是给float类型的变量赋值时，值的后面必须加上

f(后者 F)，整数数字也可赋值给浮点型变量，具体示例如下：

```
1.   float f = 1.23f;// 为float类型变量赋值，值后面必须加f(或者F)
2.   double d1 = 3.14;// 小数默认为double类型，值后面可以不加d(或者D)
3.   double d2 = 3.14d;// 小数默认为double类型，值后面也可以加d(或者D)
```

(6) 字符类型变量。

字符类型变量用来存储一个字符，在 Java 中使用关键字 char 表示，字符类型变量占用 2 个字节。在给 char 类型变量赋值时，可以使用一对英文半角格式的单引号把字符括起来，如'a'、'\n'、'\072'、'\u003a'；也可以使用 0～65535(包括 0 和 65535)整数给 char 类型变量赋值，例如整数 97 对应字符'a'。这是因为 Java 虚拟机会自动把这些整数转换成 Unicode 字符集对应的字符(Java 的字符使用 Unicode 字符集)，具体示例如下：

```
1.   char c1 ='a';// 为char类型变量c1赋值字符a
2.   char c2 ='\n';// 为char类型变量c2赋值换行字符，\n表示转义字符换行符
3.   char c3 ='\072';// 为char类型变量c3赋值八进制数72对应的Unicode字符。
4.   char c4 ='\u003a';// 为char类型变量c4赋值十六进制数3a对应的Unicode字符
5.   char c5 = 97;// 为char类型变量c5赋值十进制整数97对应的Unicode字符
```

下面用一个例子来定义 8 种数据类型变量，便于读者更好地理解在代码中如何定义变量。

例 3-1：基本数据类型的使用。

```
1.   package com.nle.demo1;
2.
3.   public class DataType {
4.     public static void main(String[] args) {
5.        // 1. 整数类型变量
6.        byte b = 127;
7.        // byte b1 = 128;//byte的取值范围为-128～127，128超出取值范围，不能直接赋值。
8.        short s = 1234;
9.        int i = 12345;
10.       long l1 = 123;// 未超出int类型的取值范围，可以不用加l或者L
11.       long l2 = 123L;// 未超出int类型的取值范围，可以加l或者L
12.       long l3 = 12300000000L;// 超出int类型的取值范围，必须加l或者L
13.       // 2. 浮点型变量
14.       float f = 1.23f;// 为float类型变量赋值，值后面必须加f(或者F)
15.       double d1 = 3.14;// 小数默认为double类型，值后面可以不加d(或者D)
16.       double d2 = 3.14d;// 小数默认为double类型，值后面也可以加d(或者D)
17.       // 3. 字符型变量
18.       char c1 = 'a';// 为char类型变量c1赋值字符a
19.       char c2 = '\n';// 为char类型变量c2赋值换行字符，\n表示转义字符换行符
20.       char c3 = '\072';// 为char类型变量c3赋值八进制数72对应的Unicode字符。
21.       char c4 = '\u003a';// 为char类型变量c4赋值十六进制数3a对应的Unicode字符
22.       char c5 = 97;// 为char类型变量c5赋值十进制整数97对应的Unicode字符
23.
24.       // 4. 布尔型变量
25.       boolean flag1 = false;// 为boolean类型变量flag1赋值false
26.       boolean flag2 = true;// 为boolean类型变量flag2赋值true
27.
28.    }
29.  }
```

2. 基本数据类型的转换

在变量的赋值过程中，有可能把一种类型的值赋给另一种类型的变量，这时就需要进行数据类型的转换。数据类型的转换可分为两种：自动类型转换和强制类型转换。

(1) 自动类型转换。

自动类型转换也称为隐式类型转换，指的是两种数据类型在转换的过程中不需要进行显式声明，自动完成转换。自动类型转换要满足两个条件，第一个是两种数据类型必须彼此兼容，第二个是目标类型的取值范围要大于源类型的取值范围。具体例子代码如下：

```
1.  package com.nle.demo1;
2.  public class TypeCast {
3.      public static void main(String[] args) {
4.          byte b = 1;
5.          int i = b; // 变量b的byte类型取值范围小于int类型，可以自动转换
6.          double d = i;// 变量i的int类型取值范围小于double类型，可以自动转换
7.      }
8.
9.  }
```

上述代码中，源类型的取值范围都小于目标类型的取值范围，所以编译器自动完成类型转换，编译时不报任何错误。从取值范围由低到高可自动转换的为：byte，short，char > int > long > float > double(其中符号>表示可以自动转换的方向)。其中，byte 类型还可转换成 short 类型。

(2) 强制类型转换。

强制类型转换也称为显式类型转换，指的是两种数据类型在转换的过程中需要进行显式声明。当两种类型不兼容，或者目标类型的取值范围小于源类型时，就需要进行强制类型转换了。具体例子代码如下：

```
1.  byte b=1;
2.  int i=2;
3.  double d=3;
4.      i=(int)d;
5.      b=(byte)i;
```

从上面的代码可以看出，int 类型的取值范围小于 double 类型，byte 类型的取值范围小于 int 类型，即目标类型的取值范围小于源类型，这时需要进行强制类型转换。强制类型转换的格式如下：

```
1.  目标类型  变量名  =  (目标类型)值;
```

需要注意的是，变量在进行强制转换时，如果值的大小超出了目标类型的取值范围，会造成数据精度的丢失。接下来通过一个例子来演示数据精度丢失的情况。

```
1.  package com.nle.demo1;
2.  public class TypeCast3 {
3.      public static void main(String[] args) {
4.          byte b;
5.          int i = 129;
6.          b = (byte) i;
7.          System.out.println("i=" + i);
8.          System.out.println("b=" + b);
9.
```

```
10.         double d = 3.14;
11.         i = (int) d;
12.         System.out.println("d=" + d);
13.         System.out.println("i=" + i);
14.     }
15. }
```

运行的结果如图 3-3 所示。

```
Console ⊠  Problems  Debug Shell
<terminated> TypeCast3 [Java Application] C:\Program Files\Java\jdk-10.0.2\bin\javaw.exe
i=129
b=-127
d=3.14
i=3
```

图 3-3　运行结果

从结果中可分析得出，int 类型变量 i 的值 128，强制转换成 byte 类型时，变量 b 的值不是 129，而是-127。对此分析一下内存存储情况，int 类型变量分配的存储大小为 4 个字节，byte 类型变量分配的存储大小为 1 个字节，当变量 i 强制转换成变量 b 时，前面 3 个字节的数据将丢失，得到 byte 类型数据 10000001(二进制)。在 Java 中，数据的最高位为符号位，0 表示正数，1 表示负数。数据是以补码的形式存储的。10000001(补码)换成原码为 11111111，该数值的十进制形式为-127，如图 3-4 所示。

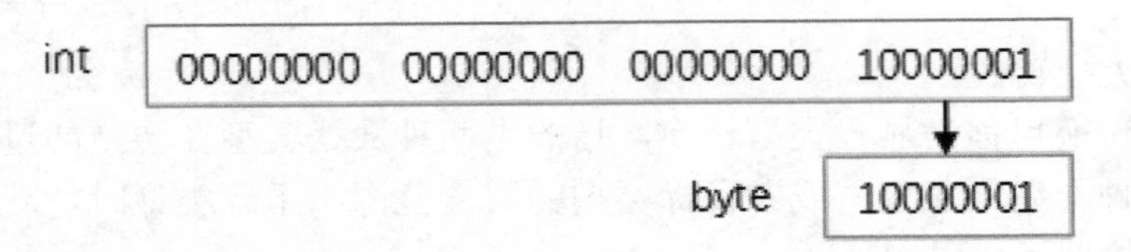

图 3-4　int 类型变量强制转换为 byte 类型

【知识拓展】

补码小知识：

计算机中的符号数有 3 种表示法：原码、反码和补码。3 种表示方法均有符号位和数值位两部分，数的最高位用来表示符号位，符号位用 0 表示“正”，用 1 表示“负”。

原码：符号位加上真值的绝对值，即用第一个二进制位表示符号(正数该位为 0，负数该位为 1)，其余位表示值。

反码：正数的反码与其原码相同；负数的反码符号位不变，其余位为其原码逐位取反。

补码：正数的补码与其原码相同；负数的补码是在其反码的基础上加 1。

举个例子如下：

```
1.   [+8] = [0000 1000]原码 = [0000 1000]反码 = [0000 1000]补码
2.   [-8] = [1000 1000]原码 = [1111 0111]反码 = [1111 1000]补码
```

计算机中为什么要使用补码？补码的本质是什么？本书不再对补码的知识做更多的拓展。有兴趣的读者可以查阅相关补码知识。

【任务实施】

1. 任务分析

本任务要使用合适的数据类型来显示温湿度传感器、光照传感器、人体传感器、火焰传感器数据。

(1) 创建工程，导入相关 jar 包。

(2) 编写传感器常量类。

(3) 定义合适的变量来存储采集到的传感器数据并输出到控制台。

2. 任务实施

(1) 搭建工程并添加 jar 包。

新建工程 project3_task1，并在 src 目录下创建包 com.nle。参照项目 2 任务 5 在工程中添加以下 jar 文件，如图 3-5 所示。

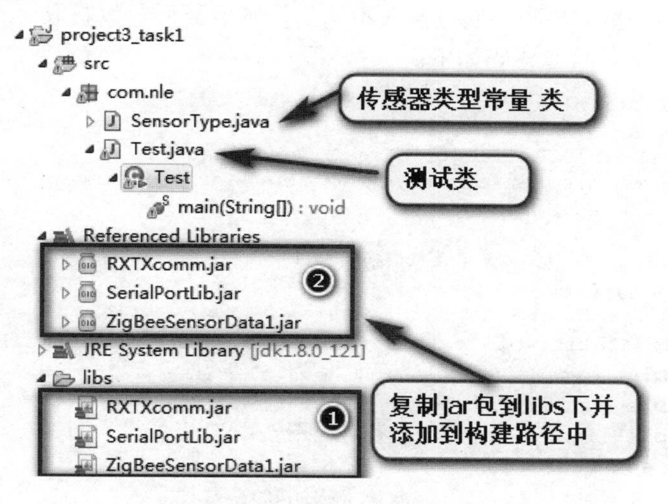

图 3-5 创建工程并导入相关的 jar 文件

其中，ZigBeeSensorData1.jar 中提供了获取 ZigBee 传感器数据的相关方法。

(2) 编写传感器常量类。

在包 com.nle 下新建类 SensorType.java，用于记录传感器的类型，这里用到了 static、final 关键字，这些关键字的用法在后面相关项目中会有介绍。一般定义常量都会用 static 和 final 做修饰，代表这些变量的值不能被改变而且它们可以通过类名直接访问。

```
1.  package com.nle;
2.
3.  public class SensorType {
4.      /** 温度传感器   */
5.      public static final String TEMPERATURE="温度";
6.      /**湿度传感器 */
7.      public static final String HUMIDITY="湿度";
8.
9.      /**光照传感器 */
10.     public static final String LIGHT="光照";
```

```
11.
12.     /**人体传感器 */
13.     public static final String PERSON="人体";
14.
15.     /**火焰传感器 */
16.     public static final String FIRE="火焰";
17.
18. }
```

(3) 定义合适的变量来存储采集到的传感器数据并输出到控制台。

在包 com.nle 下创建测试类 Test，在 main 方法里分别定义以下变量。

- 定义 double 类型变量 temperature 用来表示温度；
- 定义 double 类型变量 humidity 用来表示湿度；
- 定义 double 类型变量 light 用来表示光照；
- 定义 boolean 类型变量 hasPerson 表示有人无人和 hasFire 表示有无火警。

定义好变量后，获取串口管理对象、打开串口、通过 ZigBeeSensorData1.jar 中提供的方法 getSensorData(传感器类型)，通过传入不同的传感器类型参数获取 ZigBee 采集回来的数据并显示到控制台上。具体代码如下：

```
1.  package com.nle;
2.
3.  import com.nle.serialport.SerialPortManager;
4.  import com.nle.serialport.exception.SerialPortException;
5.  import gnu.io.SerialPort;
6.
7.  //项目 3 任务 1  显示温湿度传感器数据
8.  public class Test {
9.
10.   public static void main(String[] args) throws SerialPortException {
11.     double temperature;  // 存储温度传感器数据
12.     double humidity;     // 存储湿度传感器数据
13.     double light;        // 存储光照传感器数据
14.     boolean hasPerson;   // 存储人体传感器数据
15.     boolean hasFire;     // 存储火焰传感器数据
16.     // 1.获取串口管理对象
17.     SerialPortManager manager = new SerialPortManager();
18.     // 2.打开串口
19.     SerialPort serialPort = manager.openPort("COM200", 38400);
20.     // 3.初始化获取 ZigBee 传感器数据的对象
21.     ZigBeeSensorData sensorData = new ZigBeeSensorData(manager,serialPort);
22.     // 让程序休眠一会儿，以防止程序刚启动时未取到传感器数据产生空指针异常
23.     try {
24.         Thread.sleep(3000);
25.     } catch (InterruptedException e) {
26.         e.printStackTrace();
27.     }
28.
29.
30.     while (true) {
31.         //4.1 通过调用 getSensorData()方法获取采集回来的字符串类型的温度传感器的数据
32.         String temperature_val = ZigBeeSensorData.getSensorData(SensorType.TEMPERATURE);
33.         //4.2 把字符串类型的温度值转换成小数
34.         temperature=Double.valueOf(temperature_val);
```

```
35.          //4.3 显示带单位的温度值
36.          System.out.println("温度: " + temperature + "℃");
37.
38.          //5.获取湿度传感器的数据并显示
39.          String humidity_val = ZigBeeSensorData.getSensorData(SensorType.HUMIDITY);
40.          humidity = Double.valueOf(humidity_val);
41.          System.out.println("湿度: " + humidity + "%RH");
42.
43.          //6.获取光照传感器的数据并显示
44.          String light_val = ZigBeeSensorData.getSensorData(SensorType.LIGHT);
45.          light = Double.valueOf(light_val);
46.          System.out.println("光照: " + light + " lx");
47.
48.          //7.获取人体传感器的数据并显示
49.          String person = ZigBeeSensorData.getSensorData(SensorType.PERSON);
50.          //人体传感器有人时返回 false，无人时返回为 true，所以这里要取反
51.          hasPerson=person.equals("false")?true:false;
52.          System.out.println("人体: " + person);
53.
54.          //8.获取火焰传感器的数据并显示
55.          String fire = ZigBeeSensorData.getSensorData(SensorType.FIRE);
56.          hasFire=fire.equals("true")?true:false;
57.          System.out.println("火焰: " + fire);
58.
59.          //9.每获取一次传感器数据休眠 1s
60.          try {
61.             Thread.sleep(1000);
62.          } catch (InterruptedException e) {
63.             e.printStackTrace();
64.          }
65.       }
66.       // 5.关闭串口，因为上面的 while 循环是恒真的，所以这句代码不会执行到这里
67.       //manager.closePort(serialPort);
68.    }
69.
70. }
```

上述代码中，第 23～28 行是为了让程序延时一会儿，以免还没采集到数据发生空指针异常。第 30～68 行的 while(true)循环只是为了让程序一直获取采集到的数据并输出到控制台上，详细的 while 循环用法在任务 3 中介绍。第 60～64 行只是为了让程序每隔 1 秒采集一次。第 67 行因为上面用了循环，程序不会执行到这里，所以注释掉了。

3. 运行结果

运行程序，结果如图 3-6 所示。

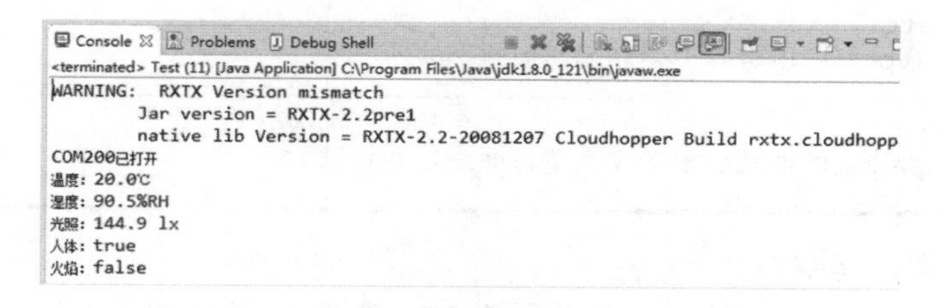

图 3-6　传感器数值显示

任务 2　ZigBee 传感数据计算

【任务描述】

将基于 ZigBee 协议的温湿度传感器、光照传感器、人体传感器、火焰传感器组成一个 ZigBee 无线网络，按物联网工程实训的要求对 ZigBee 进行烧写后，采集数据并使用合适的运算符进行计算后显示到控制台上。具体任务清单如表 3-3 所示。

表 3-3　任务清单

任务课时	4 课时	任务组员数量	建议 3 人
任务组采用设备	与任务 1 相同		

【拓扑图】

与任务 1 相同。

【知识解析】

运算符是计算机用来操作数据的特殊符号。在 Java 中，运算符可分为算术运算符、赋值运算符、比较运算符、逻辑运算符等。

1. 算术运算符

Java 中的算术运算符用来提供运算功能，如表 3-4 所示。

表 3-4　算术运算符

运算符	描　述	例子(假设 int a=2, b=3)	结　果
+	正号，放在一个操作数前面表示正数	+a	2
	加法，两个操作数进行加法运算	a+b	5
-	负号，放在一个操作数前面表示负数	-a	-2
	减法，两个操作数进行减法运算	a-b	-1
*	乘法，两个操作数进行乘法运算	a*b	6
/	除法，两个操作数进行除法运算	a/b	0
%	取余，左操作数除以右操作数的余数	a%b	2
++	自增，操作数的值增加 1(放在变量前面或后面)	a=2;b=++a;	a=3;b=3;
		a=2;b=a++;	a=3;b=2;
--	自减，操作数的值减少 1(放在变量前面或后面)	a=2;b=--a;	a=1;b=1;
		a=2;b=a--;	a=1;b=2;

(1)　普通的算术运算符。

算术运算符中的加(+)、减(−)、乘(*)、除(/)、求余(%)的使用相对比较简单，下面通过一个例子来演示它们的用法。

```
1.  package com.nle.demo2;
2.  public class ArithmeticTest {
3.      public static void main(String[] args) {
4.          int a = 1;
5.          int b = 2;
6.          int c = -3;
7.          System.out.println("a + b = " + (a + b));
8.          System.out.println("a - b = " + (a - b));
9.          System.out.println("a * b = " + (a * b));
10.         // 注意 1：整数做除法，得到的值只能是整数
11.         System.out.println("a / b = " + (a / b));
12.         System.out.println("a % b = " + (a % b));
13.         // 求余的结果的正负只与第一个数有关
14.         System.out.println("c % b = " + (c % b));
15.         System.out.println("b % c = " + (b % c));
16.
17.     }
18. }
```

运行的结果如图 3-7 所示。

```
1 package com.nle.demo2;
2 public class ArithmeticTest {
3     public static void main(String[] args) {
4         int a = 1;
5         int b = 2;
6         int c = -3;
7         System.out.println("a + b = " + (a + b));
8         System.out.println("a - b = " + (a - b));
9         System.out.println("a * b = " + (a * b));
10        // 注意1：整数做除法，得到的值只能是整数
11        System.out.println("a / b = " + (a / b));
12        System.out.println("a % b = " + (a % b));
13        // 求余的结果的正负只与第一个数有关
14        System.out.println("c % b = " + (c % b));
15        System.out.println("b % c = " + (b % c));
16
17    }
18 }
19
```

运行结果：
a + b = 3
a - b = -1
a * b = 2

a / b = 0
a % b = 1

c % b = -1
b % c = 2

图 3-7　运行结果

从运行结果中可以看出，在进行算术运算的时候需要注意如下两点。

① 在进行除法运算时，当除数与被除数都为整数时，得到的结果也是一个整数。例子中 a 的值为 1，b 的值为 2。a 除以 b，得到整数值为 0。如果除法运算中有小数参与，得到的结果将会是一个小数。例如，1.0/2 的结果为 0.5， 1/2.0 的结果也为 0.5。

② 在进行取模运算时，运算结果和被除数的符号相关，和除数符号无关，如例子中 b=2，c=-3。取模运算 c%b 即-3%2 的结果为-1。b%c 即 2%-3 的结果为 2。

(2) 自增和自减的运算符。

算术运算符中有两个非常特殊的运算符：自增运算符“++”和自减运算符“--”，它们运算时只需要与一个操作数结合，例如：

```
1.  public class ArithmeticTest2 {
2.      public static void main(String[] args) {
3.          int a = 5, b = 10;
4.          a++;
5.          System.out.println("a="+a); //结果为 6
6.          --b;
```

```
7.          System.out.println("b="+b);//结果为 9
8.      }
9.  }
```

上述代码中，第 4 行等价于 a=a+1。第 6 行等价于 b=b-1，所以经过运算，a 值为 6，b 值为 9。

自增运算符“++”和自减运算符“--”可以写在变量前，也可以写在变量后，写在前面，叫作前置，如++a，写在后面，叫作后置，如 a++。不管前置还是后置，对于变量而言，都是自增 1 或自减 1，例如上面代码中的第 4 行和第 6 行。

但是，对于有前置和后置组成的表达式的值，结果却是不一样的，例如：

```
1.  package com.nle.demo2;
2.
3.  public class ArithmeticTest3 {
4.      public static void main(String[] args) {
5.          int a = 5, b = 10, res=0;
6.          res = a++;
7.          System.out.println("a="+a+" res="+res); //结果为 a=6, res=5
8.          res=--b;
9.          System.out.println("b="+b+" res="+res);//结果为 b=9, res=9
10.     }
11. }
```

上述代码中的第 6 行，变量 a 会自增，所以 a 的值为 6，但是对于表达式 a++，因为 a 在前，所以会先用 a 的值作为表达式 a++的值赋给 res。

上述代码中的第 8 行，变量 b 会自减，所以 b 的值为 9，但是对于表达式--b，因为 b 在后，所以会用 b 减后的值作为表达式--b 的值赋给 res。

也就是说，谁在前，就先用谁的值作为表达式的值。res=a++则 res 得到 a 的值，然后 a 自增；res=--a 则 res 得到 a 自减后的值。

2. 赋值运算符

赋值运算符的作用就是将常量、变量或者表达式的值赋给某个变量，如表 3-5 所示。

表 3-5　赋值运算符

运算符	描　述	例子(假设 x=10)	结　果
=	简单赋值，将符号右边的值赋给左边的变量	x=5;	5
+=	加和赋值，将符号左边的变量与右边的值相加后再赋值给左边的变量	x+=5;等价于 x=x+5;	15
-=	减和赋值，将符号左边的变量减去右边的值后再赋值给左边的变量	x-=5;等价于 x=x-5;	5
*=	乘和赋值，将符号左边的变量乘以右边的值后再赋值给左边的变量	x*=5;等价于 x=x*5;	50
/=	除和赋值，将符号左边的变量除以右边的值后再赋值给左边的变量	x/=5;等价于 x=x/5;	2
%=	取模和赋值，将符号左边的变量除以右边的值后再将余数赋值给左边的变量	x%=5;等价于 x=x%5;	0

在赋值的过程中，运算顺序采用右结合，即将右边表达式的结果赋值给左边的变量。在 Java 中允许通过一条赋值语句对多个变量进行赋值，例如：

```
1.    int a, b, c;
2.    a = b = c = 10;//为 3 个变量同时赋值
```

但是需要注意的是，下面这种写法是不可以的：

```
1.    int a = int b = int c = 10;//写法错误
2.    int x = y = z = 10;//写法错误
```

+=、*=、/=、%=的用法如下：

```
1.    int a = 5, b = 10, res = 10;
2.    b += a + 2; //等同于 b = b+(a+2)    结果 b=17
3.    res /=a;   //等同于 res = res/a 结果 res=2
```

3. 比较运算符

比较运算符也称为关系运算符，用于对两个数值或变量进行比较，得到一个布尔值 true 或 false。关系运算符> 、>=、<、<=的优先级一样，比==、!=的优先级高，具体如表 3-6 所示。

表 3-6 比较运算符

运 算 符	描 述	例 子	结 果
>	大于	2>1	true
<	小于	2<1	false
>=	大于等于	2>=1	true
<=	小于等于	2<=1	false
==	等于	2==1	false
!=	不等于	2!=1	true

需要注意的是，运算符>、<、>=、<=，符号两边只能比较数值，布尔值不能参与比较。运算符==、!=，符号两边可以比较数值，也可以比较布尔值。

```
1.   package com.nle.demo2;
2.   public class Test4 {
3.       public static void main(String[] args) {
4.           int a = 5, b = 7;
5.           boolean flag;
6.           flag = a>b;  //结果为 false
7.           flag = a>b == false ;//true
8.           flag = a!=b ==true;// true
9.       }
10.  }
```

上述代码的第 7 行，“>”的优先级比“==”高，先做 a>b 的结果为 false，false==false 结果为 true。

第 8 行，! =和==的优先级一样，关系运算符从左边做起，先做 a!=b 的结果为 true，true==true 结果为 true。

4. 逻辑运算符

逻辑运算符用于对布尔型的数据进行操作，其结果仍是一个布尔值，如表 3-7 所示。

表 3-7 逻辑运算符

运算符	描 述	例 子	结 果
&&	当且仅当两个运算量的值都为“真”时，运算结果为“真”，否则为“假”	true&&true	true
		true&&false	false
		false&&true	false
		false&&false	false
||	当且仅当两个运算量的值都为“假”时，运算结果为“假”，否则为“真”	true||true	true
		true||false	true
		false||true	true
		false||false	false
!	当运算量的值为“真”时，运算结果为“假”；当运算量的值为“假”时，运算结果为“真”	!true	false
		!false	true

下面通过一个例子来深入了解逻辑运算符的使用。

```
1.   package com.nle.demo2;
2.   public class LogicalTest {
3.       public static void main(String[] args) {
4.           // 1.与&&的使用例子
5.           int x = 3;
6.           int y = 1;
7.           // 当且仅当两个运算量的值都为“真”时，运算结果为“真”
8.           // 3>0 为 true，y-->0 为 true，所以结果为 true，y 的值自减 1 为 0
9.           boolean and1 = x > 0 && y-- > 0;
10.          System.out.println("and1=" + and1 + ", y=" + y);
11.          // 有一个运算量的值为“假”时，运算结果为“假”
12.          // 3<0 为 false，与结果即为 false；但是与符号后面的表达式被忽略，没有参与运算，
               y 的值还为 0
13.          boolean and2 = x < 0 && y-- > 0;
14.          System.out.println("and2=" + and2 + ", y=" + y);
15.          // 2.或||的使用例子
16.          // 当且仅当两个运算量的值都为“假”时，运算结果为“假”
17.          // 3<0 为 false，y-->100 也为 false，或结果为 false；y 自减值为-1
18.          boolean or1 = x < 0 || y-- > 100;
19.          System.out.println("or1=" + or1 + ", y=" + y);
20.          // 有一个运算量的值为“真”时，运算结果为“真”
21.          // 3>0 为 true，或结果即为 true，或符号后面的表达式忽略；y 没有参与运算，其值
               还是-1
22.          boolean or2 = x > 0 || y-- > 100;
23.          System.out.println("or2=" + or2 + ", y=" + y);
24.          // 3.非!的使用例子
25.          System.out.println("!true=" + !true);
26.          System.out.println("!false=" + !false);
```

```
27.       }
28.   }
```

运行结果如图 3-8 所示。

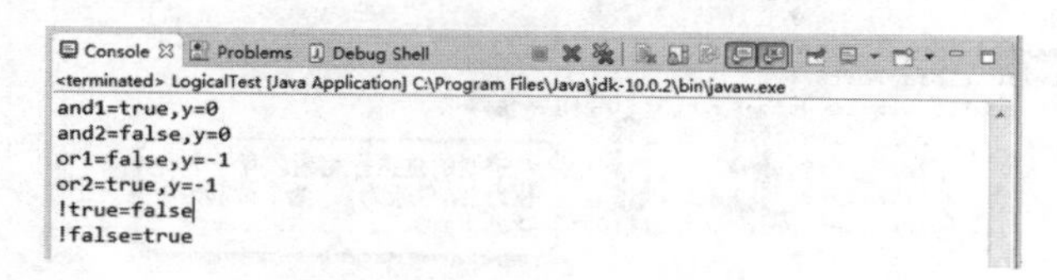

图 3-8 运行结果

例子中 x 的初始值为 3，y 的初始值为 1。

(1) x > 0 && y-- > 0，“&&”符号两边表达式的值都为 true，所以结果为 true，y 的值为自减后的值 0。

(2) x < 0 && y-- > 0，“&&”符号左边的表达式为 false，所以与的结果即为 false，此时“&&”符号运算会忽略右边表达式(即右表达式没有参与运算)，y 的值没有自减，还是 0。所以“&&”符号又被称为短路与。

(3) 同理或“||”符号运算 x < 0 || y-- > 100，两边表达式结果同时为 false，或结果才为 false，y 的值自减 1 为 0。

(4) x > 0 || y-- > 100，或符号左边的表达式为 true，即或运算结果为 true，或符号右边表达式忽略，y 的值还是-1。所以“||”符号又被称为短路或。

(5) 非运算比较简单，在布尔值或布尔变量前面加上!，结果取反。

除了符号“&&”可以进行与运算外，符号“&”也可以进行与运算。符号“&”又称为非短路与。除了符号“||”可以进行或运算外，符号“|”也可以进行或运算。符号“|”又被称为非短路或。至于非短路与和短路与，非短路或和短路或的区别，请读者自行进行验证。

5. 位运算符

Java 支持的位运算符有：&位与、｜位或、～位非、^位异或、<<左移、>>右移、>>>无符号右移。

逻辑运算符的&&、||主要针对两个关系运算符来进行逻辑运算，而位运算中的&、｜主要针对二进制数进行逻辑运算，&和|没有短路现象。

- 按位与&运算符：两个操作数的位都为 1，结果为 1，否则为 0，如图 3-9 所示。

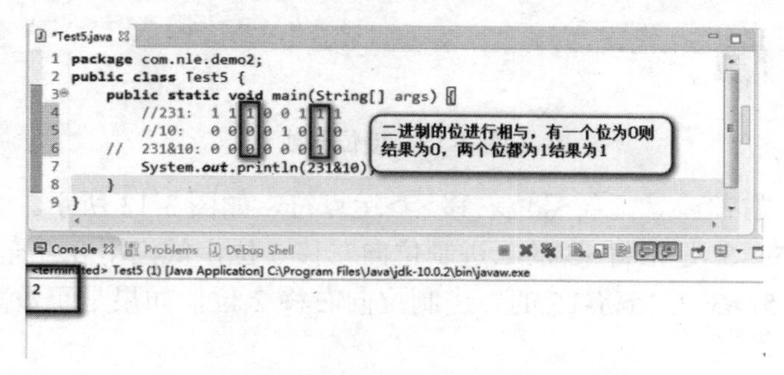

图 3-9 按位与

- 按位或 | 运算符：两个操作数中有一位为 1，结果就为 1，两位都为 0 则结果为 0，如图 3-10 所示。

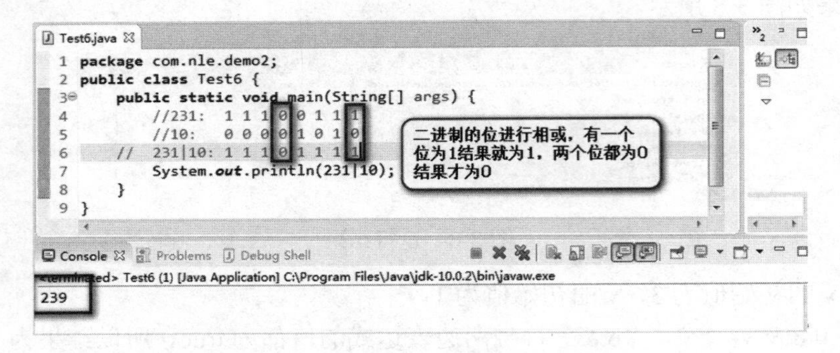

图 3-10　按位或

- 按位取反～运算符：如果位为 0，结果为 1；如果位为 1，结果为 0，如图 3-11 所示。

```
Test7.java
1 package com.nle.demo2;
2 public class Test7 {
3     public static void main(String[] args) {
4         //10:    0 0 0 0 1 0 1 0
5     //  ~10:   1 1 1 1 0 1 0 1
6         System.out.println(~10);
7     }
8 }
9
```

二进制的位取反~，位为1结果为0，位为0结果为1

高位为0代表负数

Console Problems Debug Shell

<terminated> Test6 (1) [Java Application] C:\Program Files\Java\jdk-10.0.2\bin\javaw.exe

-11

图 3-11　按位取反

- 按位异或^ 运算符：两个操作数的位同号为 0，异号为 1，如图 3-12 所示。

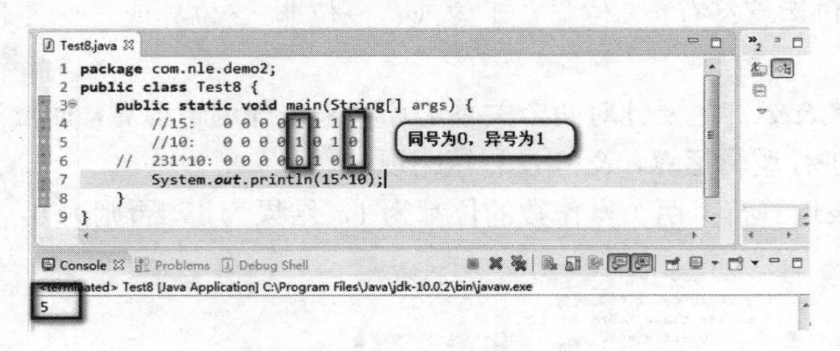

图 3-12　按位异或

- 左移<<、右移>>、无符号位右移>>>运算符，如图 3-13 所示。

左移<< ：　15<<3 是指将 15 的二进制位向左移 3 位，低位用 0 补充。

右移<< ：　-15>>3 是指将-15 的二进制位向右移 2 位，如果是正数高位用 0 补充，负数高位用 1 补充。

无符号位右移>>> ：　只针对正数的右移，高位用 0 补充。

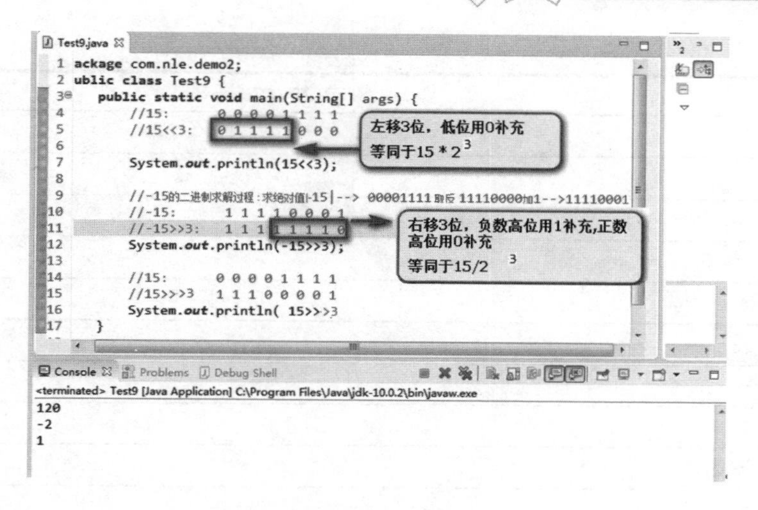

图 3-13　移位

6. 三目运算符

三目运算符又称条件运算符，语法是：

```
条件？结果 1：结果 2
```

运算规则是：如果条件表达式的值是真的，则结果 1 的值是整个表达式的值，否则结果 2 的值是整个表达式的值，如图 3-14 所示。

7. 运算符的优先级

在对一些复杂的表达式运算时，要明确表达式中所有运算符参与运算的先后顺序，这种顺序一般被称作运算符的优先级。接下来通过一张表说明 Java 中运算符的优先级，数字越小优先级越高，如表 3-8 所示。

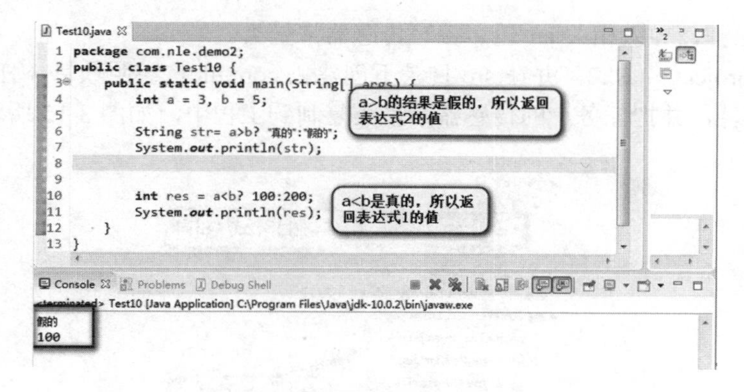

图 3-14　三目运算符

表 3-8　运算符的优先级

优先级	运 算 符
1	.　[]　()
2	++　--　～　!

续表

优先级	运 算 符
3	* / %
4	+ -
5	<< >> >>>
6	< > <= >=
7	== !=
8	&
9	^
10	|
11	&&
12	||
13	?:
14	= *= /= %= += -= <<= >>= >>>= &= ^= |=

【任务实施】

1. 任务分析

本任务要使用合适的运算符来计算温湿度传感器、光照传感器、人体传感器、火焰传感器数据。

(1) 创建工程，导入相关 jar 包、编写传感器常量类。

(2) 按 ZigBee 协议对采集回来的传感器数据做计算并显示。

2. 任务实施

(1) 创建工程，导入相关 jar 包、编写传感器常量类。

新建工程 project3_task2，并在 src 目录下创建包 com.nle。参照项目 3 任务 1 在工程中添加以下 jar 文件，并把任务 1 的传感器常量类复制到工程中，如图 3-15 所示。

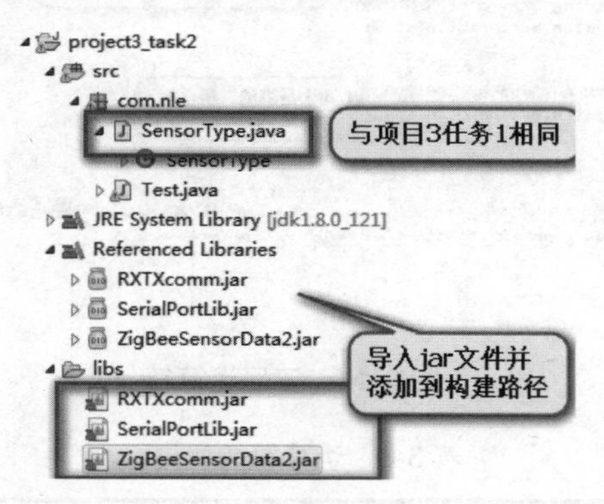

图 3-15 创建工程并导入相关 jar 文件

(2) 按 ZigBee 协议对采集回来的传感器数据做计算。

编写测试类 Test，在 main 方法里，获取串口管理对象，打开串口，通过 ZigBeeSensor

Data2.jar 里提供的方法 getSensorData(传感器类型)传入不同的传感器类型参数获取 ZigBee 采集回来的数据，按照 ZigBee 协议进行计算分析并显示到控制台上。

大部分代码与任务 1 相同，这里不再做解释，只针对传感器真实数据计算部分做分析。

ZigBeeSensorData2.jar 里返回的传感器数据是经过截取的十六进制的字符串，要取到真正的传感器数值，就要将十六进制的字符串换算成十进制，这个换算后的十进制值就是采集到的传感器数据模拟量。不同类型的传感器，有不同的转换值，把模拟量值除以转换值得到的就是真实的数据。

以下为传感器对应的转换值。

温度：10

湿度：10

光照：100

空气质量：100

可燃气体：100

火焰：100

酒精：100

比如，获取的温度 temp 值为 0070，它是十六进制的字符串，要用 Integer.parseInt(tempk,16) 转换成整数，这个就是采集到的模拟量数据，再按照转换值进行转换：double restemp = Integer.parseInt(temp,16) / 10.0。

这样算出来的值才是真正的温度值。

```
1.   package com.nle;
2.
3.   import com.nle.serialport.SerialPortManager;
4.   import com.nle.serialport.exception.SerialPortException;
5.
6.   import gnu.io.SerialPort;
7.
8.   public class Test {
9.       public static void main(String[] args) throws SerialPortException {
10.          double temperature; // 存储温度传感器数据
11.          double humidity; // 存储湿度传感器数据
12.          double light; // 存储光照传感器数据
13.          // 1.获取串口管理对象
14.          SerialPortManager manager = new SerialPortManager();
15.          // 2.打开串口
16.          SerialPort serialPort = manager.openPort("COM200", 38400);
17.          // 3.初始化获取 ZigBee 传感器数据的对象
18.          ZigBeeSensorData sensorData = new ZigBeeSensorData(manager,
               serialPort);
19.
20.          // 让程序休眠一会儿，以防止程序刚启动时未取到传感器数据产生空指针异常
21.          try {
22.             Thread.sleep(3000);
23.          } catch (InterruptedException e) {
24.             e.printStackTrace();
25.          }
```

```
26.
27.         while (true) {
28.             //4.1 通过调用 getSensorData()方法获取采集回来的十六进制的字符串类型的
                温度传感器数据
29.             String temperature_val = ZigBeeSensorData.getSensorData
                (SensorType.TEMPERATURE);
30.             //4.2 把十六进制的温度传感器数据转换成十进制的值得到模拟量，除以温度传感
                器的转换值 10 得到真实的温度
31.             temperature = Integer.parseInt(temperature_val, 16) / 10.0;
32.             //4.3 通过工具类的 m1()方法 只保留一位小数的温度值
33.             String nowtemperature = Tools.m1(temperature);
34.             System.out.println("温度: " + nowtemperature + "°C");
35.
36.             // 5.获取湿度并计算和输出
37.             String humidity_val = ZigBeeSensorData.getSensorData
                (SensorType.HUMIDITY);
38.             humidity = Integer.parseInt(humidity_val, 16) / 10.0;
39.             String nowhumidity = Tools.m1(humidity);
40.             System.out.println("湿度: " + nowhumidity + "%RH");
41.
42.             //6. 获取光照并计算和输出
43.             String light_val = ZigBeeSensorData.getSensorData
                SensorType.LIGHT);
44.             light = Integer.parseInt(humidity_val, 16) / 100.0;
45.             System.out.println("光照: " + light + "lx");
46.
47.             //7. 获取人体并计算和输出 ,getSensorData()方法中当 ZigBee 人体传感器当
                有人时返回的是 0100 ，因为低位在前高位在后，0100 实际上是 0010
48.             String person = ZigBeeSensorData.getSensorData
                (SensorType.PERSON);
49.             String hasPerson = person.equals("0100") ? "无人" : "有人";
50.             System.out.println("人体: " + hasPerson);
51.
52.             //8. 获取火焰并计算和输出 ,getSensorData()方法中当 ZigBee 火焰传感器有
                火时返回的是 true
53.             String fire = ZigBeeSensorData.getSensorData(SensorType.FIRE);
54.             String hasFire = fire.equals("true") ? "报警" : "正常";
55.             System.out.println("火焰: " + hasFire);
56.
57.             try {
58.                 Thread.sleep(1000);
59.             } catch (InterruptedException e) {
60.                 e.printStackTrace();
61.             }
62.         }
63.         // 关闭串口
64.         // manager.closePort(serialPort);
65.     }
66. }
```

3. 运行结果

运行程序，结果如图 3-16 所示。

```
Console ⌧ Problems Debug Shell
<terminated> Test (9) [Java Application] C:\Program Files\Java\jdk1.8.0_121\bin\javaw.exe
WARNING:  RXTX Version mismatch
        Jar version = RXTX-2.2pre1
        native lib Version = RXTX-2.2-20081207 Cloudhopper Build rxtx.cloudhopper.net
COM200已打开
温度: 25.0°C
湿度: 89.8%RH
光照: 8.98lx
人体: 有人
火焰: 报警
```

图 3-16　程序运行结果

任务 3　ZigBee 传感数据采集分析

【任务描述】

用 ZigBee 温湿度传感器、光照传感器、人体传感器、火焰传感器组成一个 ZigBee 无线网络，按物联网工程实训的要求对 ZigBee 进行烧写后，采集数据并按 ZigBee 协议对数据进行分析，并显示到控制台上。任务清单如表 3-9 所示。

表 3-9　任务清单

任务课时	4 课时	任务组员数量	建议 3 人
任务组采用设备	与任务 1 相同		

【拓扑图】

与任务 1 相同。

【知识解析】

Java 的流程控制包括顺序控制、条件控制和循环控制。顺序控制，即程序从头到尾依次执行语言。条件控制，基于条件选择执行语句。循环控制，根据循环初始条件和终结要求，执行循环体内的操作。顺序控制很简单，前面介绍的任务用的就是顺序控制。本任务知识重点介绍条件控制和循环控制的使用。

1. 条件控制

Java 中的条件控制使用两种条件结构：if 条件语句与 switch 条件语句。

if 条件语句又可细分为 3 种语法格式，分别为 if 语句，if…else 语句，if…else if…else 语句。

(1) if 语句。

if 语句指的是满足某种判断条件，就进行某种处理，语法格式如下。

```
1.   if(判断条件){
2.        代码块
```

```
3.    }
```

当室内温度大于 30℃时，自动打开空调，这段话用 if 语句来表示如下所示。

```
1.    if(室内温度大于 30℃){
2.        自动打开空调
3.    }
```

if 语句把条件放在 if 关键字后面的()中，这个判断条件得到的是一个布尔值，当这个判断条件为 true 时，执行 if 语句结构中{}内的语句。if 语句的执行流程如图 3-17 所示。

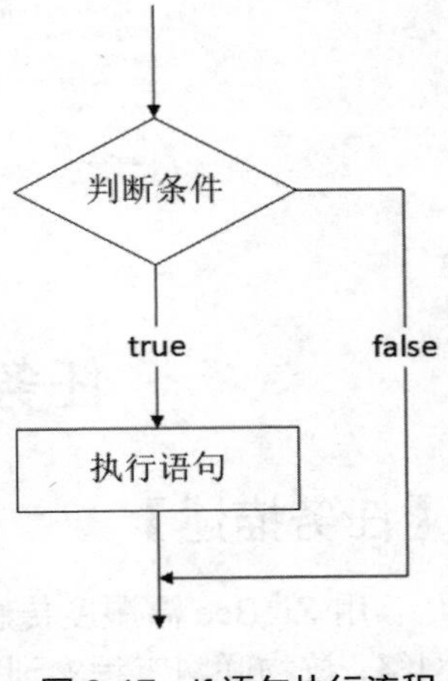

图 3-17　if 语句执行流程

接下来通过一个例子来学习 if 语句的用法，具体代码如下所示。

```
1.    package com.nle.demo3;
2.    public class TestIf01 {
3.        public static void main(String[] args) {
4.            double temp = 38.0;
5.            if(temp>=30) {
6.                System.out.println("打开空调");
7.            }
8.        }
9.    }
```

运行结果如图 3-18 所示。

图 3-18　运行结果

上面的程序例子中，定义了一个变量 temp 用来表示温度。在 if 语句中判断 temp>=30 为 true 时，{}中的语句将会被执行。temp 的初始值为 38.0>30，所以{}内语句被执行，控制台界面显示“打开空调”。

(2)　if…else 语句。

if…else 语句是指如果满足某种条件，就进行某种处理，否则就进行另一种处理。例如，判断一个数的奇偶性，如果该数能够被 2 整除，则是一个偶数，否则该数字就是一个奇数。if…else 语句的具体语法如下：

```
1.    if(判断条件){
2.        代码块 1
3.    }else{
4.        代码块 2
5.    }
```

上述语法格式中，判断条件得到一个布尔值。当这个布尔值为 true 时，执行代码块 1，当这个布尔值为 false 时，执行代码块 2。if…else 语句的执行流程如图 3-19 所示。

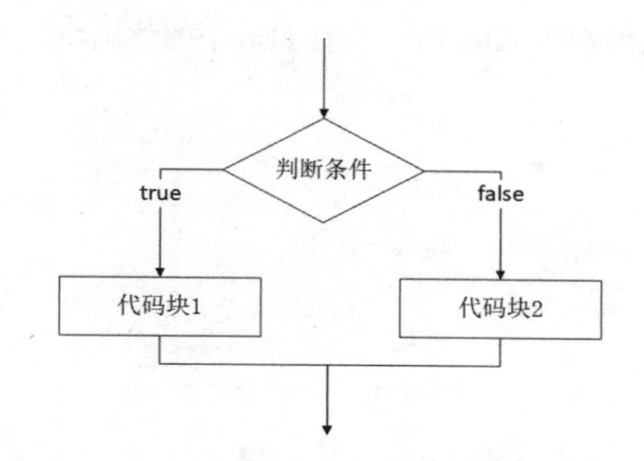

图 3-19 if...else 语句执行流程

```
1.  package com.nle.demo3;
2.  public class TestIf02 {
3.      public static void main(String[] args) {
4.          int number = 7;
5.          if(number%2 == 0) {
6.              //number%2==0 为 true 时，执行该代码段
7.              System.out.println("number 是一个偶数");
8.          }else {
9.              //number%2==0 为 false 时，执行该代码段
10.             System.out.println("number 是一个奇数");
11.         }
12.     }
13. }
```

运行结果如图 3-20 所示。

图 3-20 运行结果

(3) if…else if…else 语句。

if…else if…else 语句用于对多个条件进行判断，进行多种不同的处理。例如，对一个学生的成绩进行分级，如果分数大于或等于 80 分等级为优；如果分数大于或等于 70 分，等级为良；如果分数大于或等于 60 分，等级为及格；否则，等级为差。if…else if…else 的具体语法格式如下：

```
1.  if(判断条件 1) {
2.      代码块 1
3.  }else if(判断条件 2) {
4.      代码块 2
5.  }else if(判断条件 n) {
6.      代码块 n
7.  }else {
8.      代码块 n+1
9.  }
```

上述格式中，当判断条件 1 为 true 时，执行代码块 1 的代码。否则，执行下一个判断条件 2。当判断条件 2 为 true 时，执行代码块 2 的代码。否则，执行下一个判断条件。依次

类推，如果所有的判断条件都为 false 时，执行 else 中的代码块。if…else if…else 语句执行的流程如图 3-21 所示。

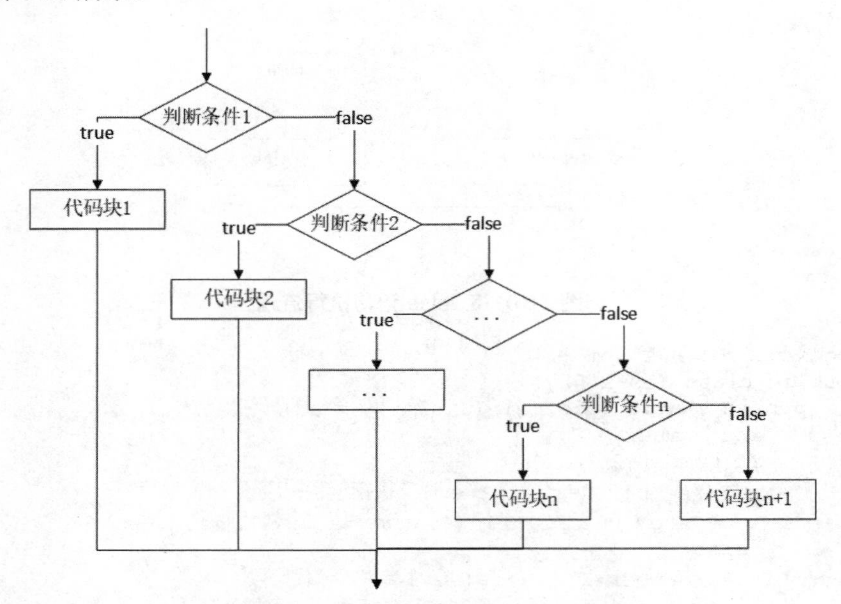

图 3-21　if...else if...else 语句执行流程

```
1.   package com.nle.demo3;
2.
3.   public class TestIf03 {
4.       public static void main(String[] args) {
5.           int score = 67;// 学生的成绩
6.           if (score >= 80) {
7.               // 满足条件 score>=80
8.               System.out.println("该学生成绩等级为优");
9.           } else if (score >= 70) {
10.              // 不满足条件 score>=80，满足条件 score>=70，即分数区间为 70<=score<80
11.              System.out.println("该学生成绩等级为良");
12.          } else if (score >= 60) {
13.              // 不满足条件 score>=70，满足条件 score>=60，即分数区间为 60<=score<70
14.              System.out.println("该学生成绩等级为及格");
15.          } else {
16.              // 不满足条件 score>=60
17.              System.out.println("该学生成绩等级为不及格");
18.          }
19.
20.      }
21.  }
```

运行结果如图 3-22 所示。

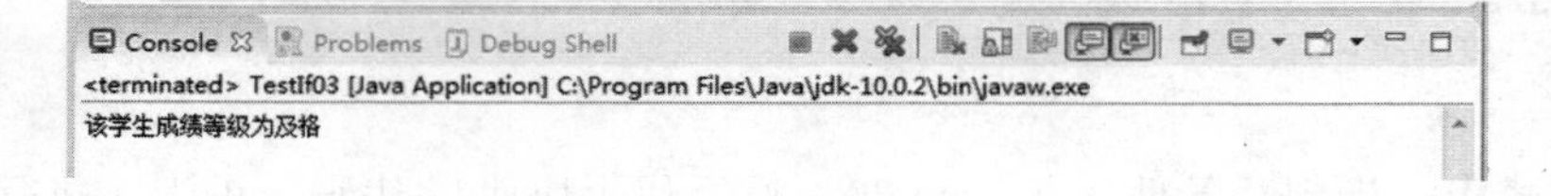

图 3-22　运行结果

该学生的成绩 score 为 67，不满足第一个条件 score>=80，也不满足第二个条件 score>=70，接下来判断满足第三个条件 score>=60，执行该条件下的代码块，输出“该学生成绩等级为及格”。需要注意的是，在该例子中，学生的成绩 score 默认在 0～100 之间。当出现该范围以外的数值时，会出现什么情况？请读者思考，提出解决方法。

对于 if 语句，在确认满足条件只执行一行代码的时候，可以省略{}，如下所示。

```
1.   if (a > b)
2.       max = a;
3.   else
4.       max = b;
```

这种结构不适用满足条件时执行多条语句的情况。一般情况下也不建议使用这种结构，建议读者在使用 if 语句时都把执行代码放入{}内。

(4) switch 条件语句。

switch 条件语句也是一种常见的选择语句，与 if 不同的是，它使用的是值匹配的形式，即根据值的不同，从而决定执行哪一段代码。switch 条件语句的具体格式如下：

```
1.   switch(表达式) {
2.       case 值1:
3.           代码块1
4.           break;
5.       case 值2:
6.           代码块2
7.           break;
8.       ...
9.       case 值n:
10.          代码块n
11.          break;
12.      default:
13.          代码块n+1
14.          break;
15.  }
```

上面的格式中，switch 关键字后面的()内放入一个表达式，根据这个表达式的值去匹配 case，匹配成功即执行 case 后面的代码块，执行完 break 的时候，switch 语句结束；如果没有 case 匹配，则执行 default 后面的代码块。

在使用 switch 时需要注意以下几点。

case 后面的值必须是常量，且类型只能是 byte、short、char、int、enum(枚举类型)、String(JDK1.7 及以上可用)。

case 执行代码块后面的 break 可省略，省略后程序将继续执行下一个 case 代码块，直到遇到 break 时结束 switch 语句。

default 在 case 都没有匹配项的情况下才执行。default 块可省略，省略后表示未匹配 case 的情况下没有其他处理。

```
1.   package com.nle.demo3;
2.
3.   public class TestIf04 {
4.       public static void main(String[] args) {
5.           int dayOfWeek = 3;
6.           switch (dayOfWeek) {
```

```
7.          case 1:
8.              System.out.println("星期一");
9.              break;
10.         case 2:
11.             System.out.println("星期二");
12.             break;
13.         case 3:
14.             System.out.println("星期三");
15.             break;
16.         case 4:
17.             System.out.println("星期四");
18.             break;
19.         case 5:
20.             System.out.println("星期五");
21.             break;
22.         case 6:
23.             System.out.println("星期六");
24.             break;
25.         case 0:
26.             System.out.println("星期天");
27.             break;
28.         default:
29.             System.out.println("输入的数字不正确！");
30.             break;
31.         }
32.
33.         System.out.println("-------------");
34.
35.         switch (dayOfWeek) {
36.         case 1:
37.         case 2:
38.         case 3:
39.         case 4:
40.         case 5:
41.             // dayOfWeek 为 1, 2, 3, 4, 5 时，都执行该代码块
42.             System.out.println("工作日");
43.             break;
44.         case 6:
45.         case 0:
46.             // dayOfWeek 为 6, 0 时，执行该代码块
47.             System.out.println("休息日");
48.             break;
49.         default:
50.             System.out.println("输入的数字不正确！");
51.             break;
52.         }
53.     }
54. }
```

以上程序例子的第一个 switch 语句表示，变量 dayOfWeek 为 1～6 时分别输出“星期一”到“星期六”，为 0 时输出“星期天”。当 dayOfWeek 的值不是 0～6 时，输出“输入的数字不正确！”。

程序的第二个 switch 语句表示，变量 dayOfWeek 为 1～5 时，输出“工作日”，为 6 或 0 时，输出“休息日”。

运行结果如图 3-23 所示。

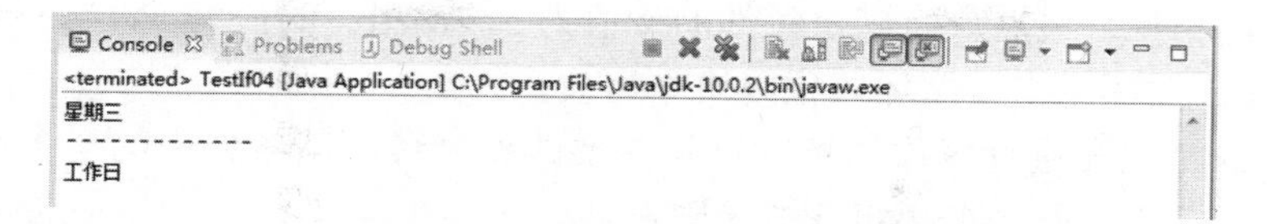

图 3-23　运行结果

2. 循环控制

循环控制指的是满足某种条件下重复执行某段代码。Java 的循环控制分为 3 种语句结构，下面对这 3 种结构进行详细的讲解。

(1) while 循环语句。

while 循环语句，根据循环条件来判断是否执行大括号内的代码块，如果循环条件结果为真，则重复执行代码块，直到条件不满足，while 循环才结束。while 循环语句的语法结构如下：

```
1.  while(循环条件){
2.      代码块
3.  }
```

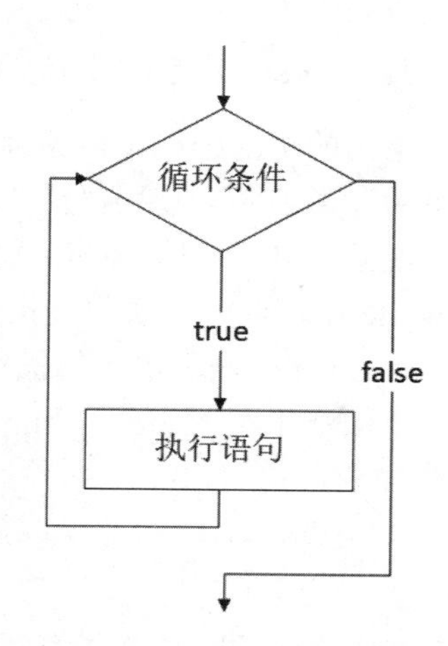

图 3-24　while 循环语句执行流程

while 循环语句的执行流程如图 3-24 所示。

下面通过代码演示利用 while 循环语句实现输出 1～5 的整数。

```
1.  package com.nle.demo3;
2.
3.  public class TestWhile {
4.      public static void main(String[] args) {
5.          int x = 1;
6.          while (x <= 5) {// 循环判断条件
7.              System.out.println("x = " + x);// 输出 x 的值
8.              x++;// x 自增 1
9.          }
10.     }
11. }
```

运行结果如图 3-25 所示。

```
Console  Problems  Debug Shell
<terminated> TestWhile [Java Application] C:\Program Files\Java\jdk-10.0.2\bin\javaw.exe
x = 1
x = 2
x = 3
x = 4
x = 5
```

图 3-25　运行结果

上述代码中，整型变量 x 的初始值为 1，循环条件为 x<=5，第一次 x 满足条件，执行循环体代码块，在代码块内部对 x 的值增加 1。第二次判断 x<=5 满足条件，继续执行循环

体代码块。依次类推，最后一次满足条件的 x 值为 5，输出 x 的值为 5。当 x=6 时，已经不满足条件，while 循环退出，循环体代码块不再被执行。

需要注意的是，在进行每次循环的过程中要改变循环判断条件中的变量值，以保证最终可以退出循环，否则整个循环将变成无限循环。

(2) do...while 循环语句。

do…while 循环语句的语法结构如下：

```
1.   do{
2.       代码块
3.   }while(循环条件);
```

上面的语法结构中，关键字 do 后面的大括号中放入循环体代码块。循环条件放在大括号后面，最后必须使用“;”号结束。

do…while 循环与 while 循环的区别在于，do…while 循环至少执行一次循环体代码块，而 while 循环可能一次都不执行循环体代码块。

do…while 循环语句的执行流程如图 3-26 所示。

下面演示利用 do…while 循环语句实现输出 1～5 的整数。

```
1.   package com.nle.demo3;
2.
3.   public class TestdoWhile {
4.       public static void main(String[] args)
     {
5.           int x = 1;
6.           do {
7.               System.out.println("x = " + x);
8.               x++;
9.           } while (x <= 5);
10.
11.      }
12.  }
```

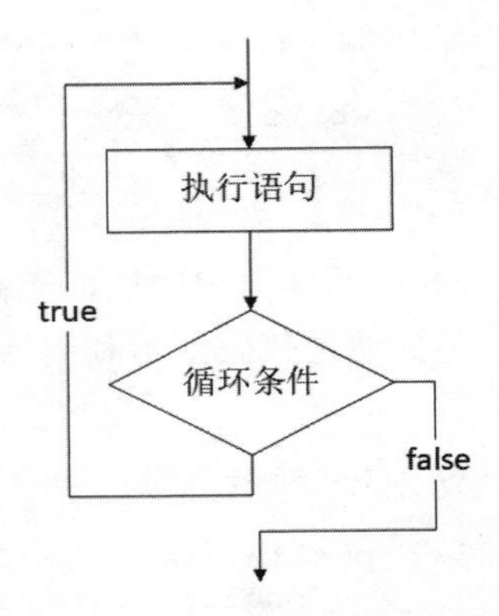

图 3-26　do...while 循环语句执行流程

运行结果如图 3-27 所示。

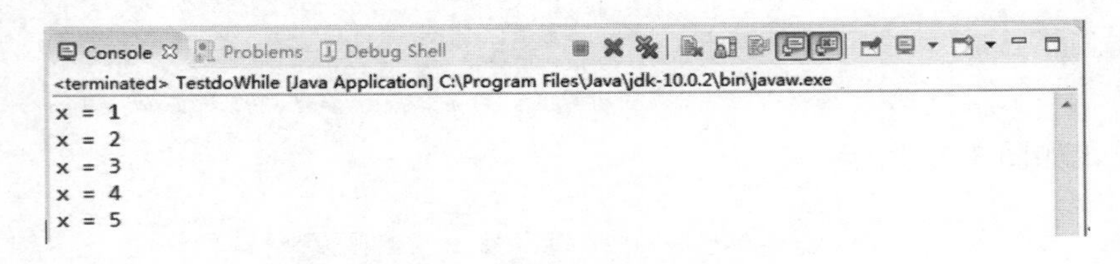

图 3-27　运行结果

从该例子结果中可以看到，do…while 循环和 while 循环的运行结果一致，但是如果循环条件改为 x<0，那么 do…while 循环还是会输出第一次 x 的值 x=0，而 while 循环什么都没输出。

(3) for 循环语句。

for 循环与 while 循环和 do…while 循环相比较，一般用在循环次数已知的情况下。for 循环语句的语法格式如下：

```
1.   for(①初始化表达式; ②循环条件; ③操作表达式){
2.       ④代码块
3.   }
```

在上面的for循环结构中，共分为4部分内容。for关键字后面的()中包括3部分内容：①初始化表达式，②循环条件，③操作表达式，它们之间使用“;”号隔开。大括号内为④循环体代码块。for循环的执行流程如下。

第一步，执行①。

第二步，执行②，如果条件判断结果为true，执行第三步；如果条件判断结果为false，执行第五步。

第三步，执行④。

第四步，执行③，然后重复执行第二步。

第五步，退出循环。

下面利用for循环实现数字1～5的累加。

```
1.   package com.nle.demo3;
2.
3.   public class TestFor {
4.       public static void main(String[] args) {
5.           int sum = 0;// 定义求和变量
6.           for (int i = 1; i <= 5; i++) {
7.               sum += i;// sum会进行对i的累加
8.           }
9.           System.out.println("sum = " + sum);//输入sum的值
10.      }
11.  }
```

运行结果如图3-28所示。

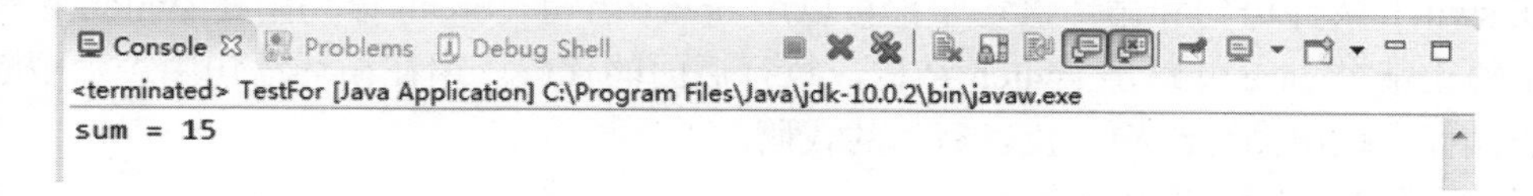

图3-28　运行结果

(4) 嵌套循环。

嵌套循环，即在一个循环语句的循环体内再定义一个循环语句。while、do…while、for循环语句都可以互相嵌套。例如，for循环嵌套for循环的格式如下：

```
1.   for(初始表达式;循环条件;操作表达式){
2.       …
3.       for(初始表达式;循环条件;操作表达式){
4.           …
5.       }
6.       …
7.   }
```

下面利用for循环嵌套实现输出直角三角形。

```
1.   package com.nle.demo3;
2.
3.   public class TestFor2 {
```

```
4.      public static void main(String[] args) {
5.          for (int i = 1; i <= 9; i++) {// 外层循环，控制打印多少行
6.              for (int j = 1; j <= i; j++) {// 内层循环，控制每行打印多少个*
7.                  System.out.print("*");// 输出*，不换行
8.              }
9.              System.out.println();// 输出一行*后，再进行换行
10.         }
11.     }
12. }
```

运行结果如图 3-29 所示。

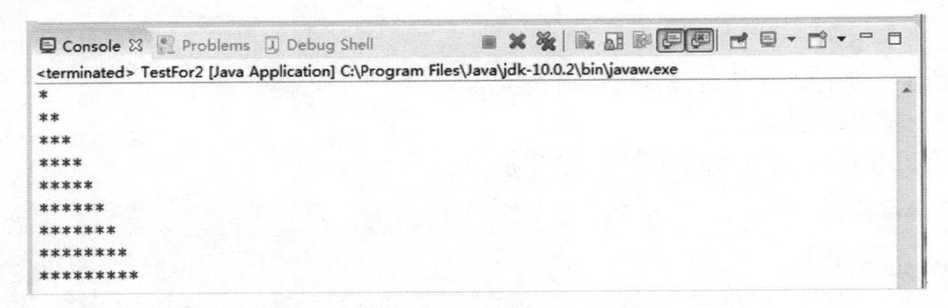

图 3-29 运行结果

在上面代码中定义了两层的嵌套循环，外层循环用来控制直角三角形的行数，内层循环用来控制每行“*”的数量。

(5) break 语句与 continue 语句。

在循环执行的过程中，经常用到 break、continue 来控制循环流程的跳转。

① break 语句。

在 switch 条件语句和循环语句中都可以用 break 语句。break 语句用在 switch 条件语句中表示终止某个 case 并跳出 switch 结构，执行后面的代码。用在循环语句中表示中断循环语句的执行，跳出循环结构，执行后面的代码。

下面利用 break 实现中断循环。

```
1.  package com.nle.demo3;
2.
3.  public class TestBreak {
4.
5.      public static void main(String[] args) {
6.          int x = 1;
7.          while (x <= 5) {// 循环判断条件
8.              System.out.println("x = " + x);// 输出 x 的值
9.              if (x == 3) {
10.                 break;
11.             }
12.             x++;// x 自增 1
13.         }
14.     }
15. }
```

运行结果如图 3-30 所示。

Console ⊠ Problems Debug Shell
<terminated> TestBreak [Java Application] C:\Program Files\Java\jdk-10.0.2\bin\javaw.exe
x = 1
x = 2
x = 3

图 3-30　运行结果

上面的代码循环体中，当 x 的值为 3 时，使用 break 语句跳出循环结构，在控制台可以看到运行结果只到“x=3”。

当 break 用在嵌套循环中的时候，它只能跳出 break 语句所在那层的循环，如果想指定跳出外层循环，可以为外层循环添加标记。

下面通过代码演示用 break 和中断标志退出循环。

```
1.   package com.nle.demo3;
2.
3.   public class TestBreak {
4.
5.       public static void main(String[] args) {
6.           outer:for (int i = 1; i <= 9; i++) {// 外层循环，控制打印多少行
7.               for (int j = 1; j <= i; j++) {// 内层循环，控制每行打印多少个*
8.                   if(i==4) {
9.                       break outer;//指定跳出 outer 标签循环
10.                  }
11.                  System.out.print("*");// 输出*，不换行
12.              }
13.              System.out.println();// 输出一行*后，再进行换行
14.          }
15.      }
16.  }
```

运行结果如图 3-31 所示。

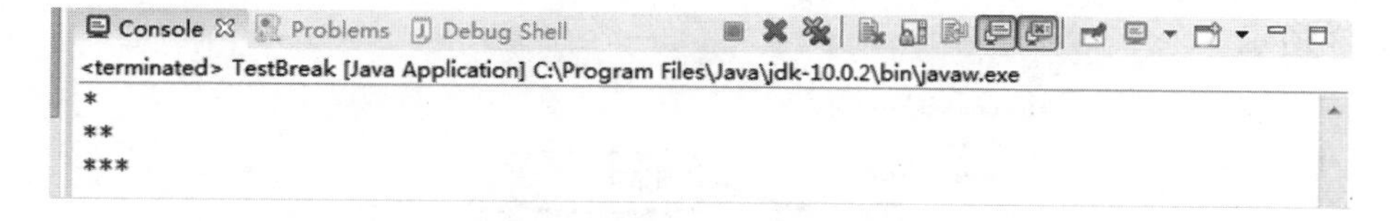

图 3-31　运行结果

② continue 语句。

continue 用在循环语句中，作用是提前终止本次循环，执行下一次循环。

下面通过代码演示 continue 的使用。

```
1.   package com.nle.demo3;
2.
3.   public class TestContinue {
4.       public static void main(String[] args) {
5.           for (int i = 1; i <=5; i++) {
6.               if(i==3) {
7.                   continue;
8.               }
```

```
9.             System.out.println("i = " + i);
10.        }
11.    }
12. }
```

运行结果如图 3-32 所示。

```
Console  Problems  Debug Shell
<terminated> TestContinue [Java Application] C:\Program Files\Java\jdk-10.0.2\bin\javaw.exe
i = 1
i = 2
i = 4
i = 5
```

图 3-32　运行结果

上面程序第 6 行加了一个条件判断，当 i 的值为 3 时，使用 continue 语句，此时会跳过本次循环，忽略循环体的剩余语句，所以“i = 3”没有打印。

【任务实施】

1. 任务分析

本任务要对采集回来的 ZigBee 数据按 ZigBee 协议进行解析。

(1) 创建工程，导入相关 jar 包。

(2) 按 ZigBee 协议对采集回来的传感器数据做解析。

(3) 编码实现对 ZigBee 协议数据做解析。

2. 任务实施

(1) 创建工程，导入相关 jar 包。

新建工程 project3_task3，并在 src 目录下创建包 com.nle，在工程中添加相关 jar 文件，如图 3-33 所示。

图 3-33　创建工程并导入相关 jar 包

(2) 按 ZigBee 协议对采集回来的传感器数据做分析。

连接好设备并检查无误后通电，把协调器用 USB 转串口线接到 PC 机上(假设接的是 COM200 口)，先用串口助手采集 ZigBee 协调器发到串口的数据，如图 3-34 所示。

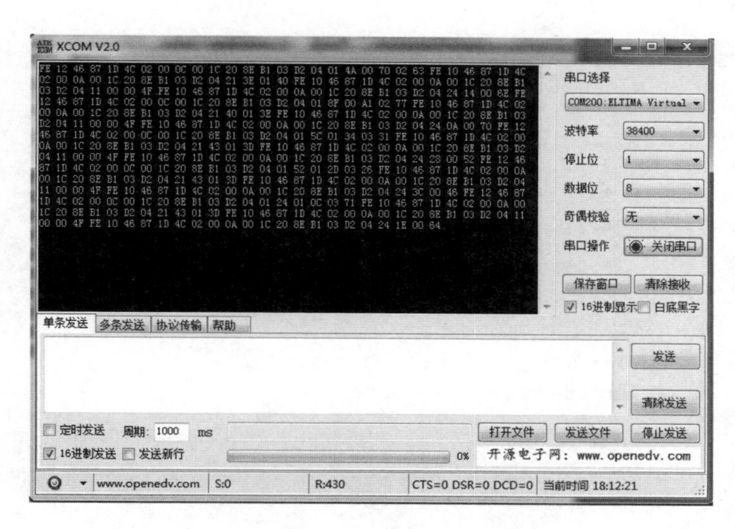

图 3-34　用串口助手接收数据

协调器通过串口往上位机发送的数据，格式如下：

```
HEAD + LEN + CMD0 + CMD1 + ADRL + ADRH + DTYPEL+ DTYPEH+DLEN+ REV +[ SDATA] + CHK
```

其中，HEAD：数据头，固定为 0xfe。

CHK：校验码，从 LEN 开始到 CHK 前一个字节的所有字节依次按字节异或的值。其他字段这里先不做解释，有兴趣的读者可以自行查阅相关文档。

需要真正关心的传感器数据在 SDATA 中。

把窗口中接收到的数据复制一份出来，并按 ZigBee 协议数据头 FE 进行分隔：

```
FE 12 46 87 1D 4C 02 00 0C 00 1C 20 8E B1 03 D2 04 01 70 00 8C 02 A5
FE 10 46 87 1D 4C 02 00 0A 00 1C 20 8E B1 03 D2 04 21 40 01 3E
FE 10 46 87 1D 4C 02 00 0A 00 1C 20 8E B1 03 D2 04 11 00 00 4F
FE 10 46 87 1D 4C 02 00 0A 00 1C 20 8E B1 03 D2 04 24 3C 00 46
```

其中，第一个数字是传感器类型，后面是对应采集到的传感器数据。

01 代表温湿度传感器，70 00 是温度的低位和高位，8C 02 代表湿度的低位和高位。

21 代表光照传感器，40 01 是光照的低位和高位。

11 代表人体传感器，00 00 是人体的低位和高位。

24 代表火焰传感器，3C 00 是光照的低位和高位。

计算时需要把高低位的值进行互换，比如温度的 70 00，实际值是 00 70，并且它是十六进制的字符串，要用 Integer.parseInt(temp,16)转换成整数，这个就是采集到的模拟量数据。

再按照任务 2 中说明的转换值进行转换：

```
double restemp = Integer.parseInt(temp, 16) / 10.0;
```

这样算出来的值才是真正的温度值。

(3) 编码实现对 ZigBee 协议数据做解析。

按上面的分析，用代码的形式解析数据，编写 Test 类，具体代码如下：

```
1.   package com.nle;
2.
3.   import com.nle.serialport.SerialPortManager;
4.   import com.nle.serialport.exception.SerialPortException;
5.
6.   import gnu.io.SerialPort;
7.
8.   public class Test {
9.      static SerialPortManager manager;
10.     static SerialPort serialPort;
11.
12.     public static void main(String[] args) {
13.        // 1.获取串口管理工具类对象
14.        manager = new SerialPortManager();
15.        // 2.打开串口
16.        try {
17.           serialPort = manager.openPort("COM200", 38400);
18.        } catch (SerialPortException e1) {
19.           e1.printStackTrace();
20.        }
21.        //3.ZigBee 协调器在采集数据并发送到串口
22.        ZigBeeSensorData sensorData = new ZigBeeSensorData(manager,
             serialPort);
23.
24.        while (true) {
25.           //4.取出从串口采集到的数据
26.           String strHexData = sensorData.getStrData();
27.           System.out.println("从串口采集到的数据是：="+strHexData);
28.
29.           //5.按 ZigBee 协议分析传感器数据
30.           getSensorData(strHexData);
31.
32.
33.           //6.采集一次休息一秒
34.           try {
35.              Thread.sleep(1000);
36.           } catch (InterruptedException e) {
37.              e.printStackTrace();
38.           }
39.        }
40.
41.     }
42.
43.     /**
44.      * 获取 ZigBee 传感器数据
45.      *
46.      * @param strData com 口采集回来的数据
47.      */
48.     public static void getSensorData(String strData) {
49.        if (strData == null || strData.length() == 0) {
50.           System.out.println("数据还在采集中，请稍等.....");
51.           return;
52.        }
53.        // ZigBee 协议数据头，固定为 0xfe，按协议数据头 FE 做字符串的分割
54.        String[] datas = strData.split("FE");
55.        int count = 0;
```

```
56.         String temp = null, humi = null, light = null, people = null, hire
              = null, tempStr;
57.
58.         for (int i = 0; i < datas.length; i++) {
59.
60.             if (datas[i].length() != 0) {// 如果第一行分割出来的字符串长度不为0
61.                 count++;
62.                 tempStr = datas[i];// 把每一行数据复制一份做处理
63.                 String s = tempStr.substring(32, 34);// 截取传感器的类型
64.
65.                 if (s.equals("01"))// 温湿度
66.                 {
67.                     // 截取温度数据
68.                     String s1 = tempStr.substring(34, 38);// FA00
69.                     // 高低位交换
70.                     temp = s1.substring(2, 4) + s1.substring(0, 2);
71.                     // 计算温度
72.                     double restemp = Integer.parseInt(temp, 16) / 10.0;
73.                     System.out.println("温度: " + restemp + "℃");
74.
75.                     // 截取湿度数据
76.                     s1 = tempStr.substring(38, 42);
77.                     humi=s1.substring(2, 4)+s1.substring(0, 2);// 高低位互换
78.                     // 计算湿度
79.                     double reshumi = Integer.parseInt(humi, 16) / 10.0;
80.                     System.out.println("湿度: " + reshumi + "%RH");
81.
82.                 } else if (s.equals("21"))// 光照 //42 01
83.                 {
84.                     String s1 = tempStr.substring(34, 38);
85.                     light = s1.substring(2, 4) + s1.substring(0, 2);
                          // 高低位互换 0142
86.
87.                     double reslight = Integer.parseInt(light, 16) / 100.0;
88.                     double val = Math.pow(10, ((1.78 - Math.log10(33 / reslight
                          - 10)) / 0.6));
89.                     // System.out.println("光照: " + val + " lx");
90.                     String sval = String.valueOf(Tools.m1(val));
                          // 小数点后面留一位
91.                     System.out.println("光照: " + sval + " lx");
92.                 } else if (s.equals("11"))// 人体
93.                 {
94.                     people = tempStr.substring(34, 38);
95.                     // 值为0100，因为低位在前高位在后，所以实际上值是0001，代表无人
96.                     if (people.equals("0100")) {
97.                         System.out.println("人体: 无人");
98.                     } else {
99.                         System.out.println("人体: 有人");
100.                    }
101.                } else if (s.equals("24"))// 火焰，返回来的是模拟量
102.                {
103.                    hire = tempStr.substring(34, 38);
104.                    // 高低位互换
105.                    String h = hire.substring(2, 4) + hire.substring(0, 2);
106.
```

```
107.                    int reshire = Integer.parseInt(h, 16) / 100;
108.                    if (reshire == 0)// 值为0 就是正常，大于0要报警
109.                    {
110.                        System.out.println("火焰：正常");
111.                    } else {
112.                        System.out.println("火焰：有火警发生");
113.                    }
114.                }
115.                // 本任务只用到温湿度传感器、光照传感器、人体传感器、火焰传感器四个
                        传感器数据，所以一次只采集 4 条数据
116.                if (count == 4) {
117.                    break;
118.                }
119.            }
120.        }
121.    }
122. }
```

代码第 1～7 行为包名和用 Ctrl+Shift+O 快捷键自动导入的包。第 8～22 行在之前的任务中做过说明，24～41 行用了一个循环，不停地取出从串口中采集到的数据进行处理，并且每隔 1s 处理一次。

getSensorData(String strData)用于处理采集到的数据，详细的说明已经写在代码处。第 54 行用了字符串的分割方法，把采集回来的十六进制的字符串按 ZigBee 协议数据头 EF 进行分割并存放到数组中。第 60 行是从数组中取出一条传感器数据进行分析。关于数组的知识在下一个任务中介绍，读者在这里先照着用。

3. 运行结果

运行程序，结果如图 3-35 所示。

图 3-35　传感数据分析结果

任务4 ZigBee控制指令的生成

【任务描述】

在任务3的基础上只保留温度的采集数据，并增加一个ZigBee单联继电器，风扇接在继电器上(模拟新风系统)，在采集到的温度传感器数据的基础上，判断当温度大于等于35度时，风扇启动，否则风扇停止工作。任务清单如表3-10所示。

表3-10 任务清单

任务课时	4课时	任务组员数量	建议1人
任务组采用设备	1个ZigBee温湿度传感器(只采集温度) 1个ZigBee协调器 1个ZigBee双联继电器(只使用其中一联) 1个风扇 1台PC 电源、USB转串口线		

【拓扑图】

本任务拓扑图如图3-36所示。

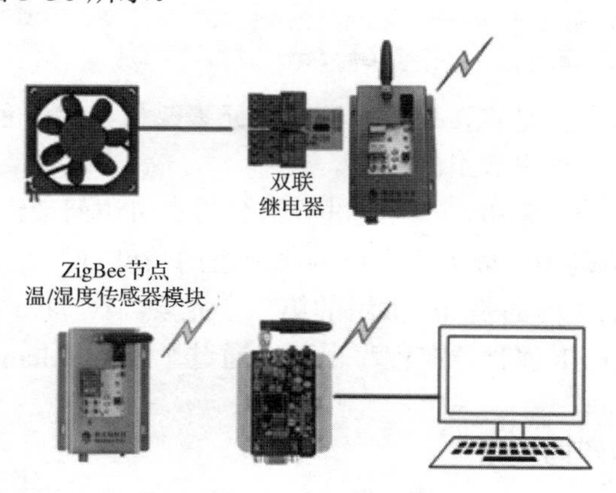

图3-36 拓扑图

备注：本任务风扇接COM1口，COM口根据实际情况确定。

【知识解析】

数组是有序数据的集合，数组中的每个元素具有相同的数据类型，可以用一个统一的数组名和下标来唯一地确定数组中的元素。数组有一维数组和多维数组。在Java中，数组是一种引用数据类型。

1. 一维数组

(1) 一维数组的定义。

在 Java 中，一维数组的定义格式如下：

```
datatype[]  arrayRefVar;// 建议使用
```

或

```
datatype  arrayRefVar[];// 不建议使用
```

上面的 datatype 表示 Java 中的数据类型，可以是基本数据类型和引用数据类型。arrayRefVar 表示数组名称。数组符号[]可以放在数据类型的后面，也可以放在数组名称的后面。建议数组的符号放在数据类型的后面。

(2) 一维数组的初始化。

在 Java 中，数组的定义并没有为数组的元素分配内存。所以定义完数组后还需要为数组分配内存，这样数组变量才可以存储元素。创建数组的过程也被称为数组的初始化。

数组的初始化可以分为静态初始化和动态初始化。数组的动态初始化，使用关键字 new 创建数组对象，具体示例如下：

```
1.   int[] arr = new int[3];
```

静态初始化，即在定义数组的同时，使用大括号直接给数组元素进行赋值。具体示例如下：

```
1.   静态初始化示例 1: int[] arr = { 1, 2, 3 };
2.   静态初始化示例 2: int[] arr = new int[]{1, 2, 3};
```

静态初始化有两种，一种是直接使用{}给数组元素赋值。另一种是使用 new 接上数组符号，数组符号后面使用{}给出数组的元素。

数组的每个元素都有一个索引，可以通过数组名与索引访问数组中的元素，例如 arr[0]、arr[1]。数组的最小索引值是 0，最大索引值为“数组的长度-1”。所以 arr[0]访问的是 arr 数组的第一个元素，arr[1]访问的是 arr 数组的第二个元素。在 Java 中，为了方便地获得数组的长度，提供了一个 length 属性，在程序中可以通过“数组名.length”的方式来获得数组的长度，即元素的个数。

下面通过代码演示数组的使用。

```
1.   package com.nle.demo4;
2.
3.   public class TestArray1 {
4.
5.       public static void main(String[] args) {
6.           // 数组的动态初始化:
7.           int[] arr1 = new int[3];
8.           // 数组的静态初始化方式一
9.           int[] arr2 = { 1, 2, 3 };
10.          // 数组的静态初始化方式二:
11.          int[] arr3 = new int[] { 1, 2, 3 };
12.
13.          // 访问数组的元素
14.          System.out.println("arr1[0] = " + arr1[0]);
```

```
15.         System.out.println("arr2[1] = " + arr2[1]);
16.         System.out.println("arr3[2] = " + arr3[2]);
17.
18.         // 对数组的元素重新赋值
19.         arr1[1] = 1;
20.         System.out.println("重新赋值后: arr1[1] = " + arr1[1]);
21.
22.         // 数组的长度
23.         System.out.println("数组的长度是: " + arr2.length);
24.     }
25. }
```

运行结果如图 3-37 所示。

```
Console ⊠  Problems  Debug Shell
<terminated> TestArray1 [Java Application] C:\Program Files\Java\jdk-10.0.2\bin\javaw.exe
arr1[0] = 0
arr2[1] = 2
arr3[2] = 3
重新赋值后: arr1[1] = 1
数组的长度是: 3
```

图 3-37 运行结果

上面程序列出了数组的几种初始化方式，并且动态初始化数组后，输出数组的每个元素值都为 0，这是因为数组在动态初始化后，数组中的元素会自动获得一个默认值，根据元素类型的不同，默认值也不一样，具体如表 3-11 所示。

表 3-11 数组元素的默认值

数据类型	默 认 值
byte、short、int、long	0
float、double	0.0
char	一个空字符，即'\u0000'
boolean	false
引用数据类型	null

需要注意的是，在使用数组索引访问数组元素时，索引必须是 0～数组长度-1 范围内的整数，使用范围外的索引会出现数组索引越界异常 ArrayIndexOutOfBounds。代码如下：

```
1.  package com.nle.demo4;
2.
3.  public class TestArray2 {
4.
5.      public static void main(String[] args) {
6.          int[] arr = new int[2];
7.          arr[3] = 3;//对索引为 3 的元素进行赋值，索引超出边界会产生
              ArrayIndexOutOfBounds 异常
8.      }
9.  }
```

运行结果如图 3-38 所示。

从运行结果可以看出，程序在 main 方法中的第 7 行代码中出现了

java.lang.ArrayIndexOutOfBoundsException 异常，异常信息为 3，即索引 3 超出边界了。有关异常的详细介绍和解决方法，在后面的任务中会详细讲解。

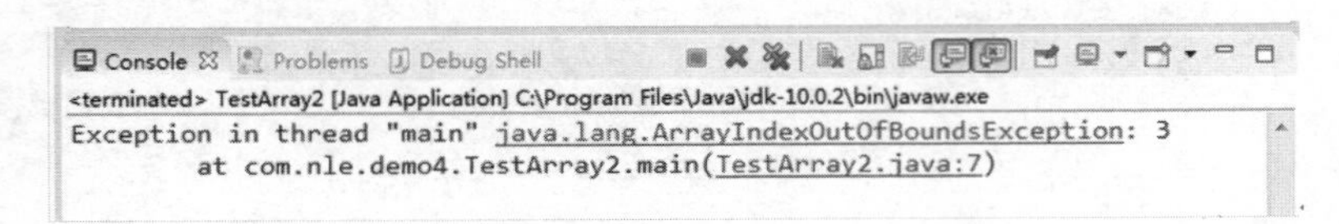

图 3-38　运行结果

(3)　一维数组的遍历。

在操作数组的元素时，如果需要访问数组中的每个元素，在数组元素比较多的情况下，一个一个地访问数组的元素是比较烦琐的，甚至是不切实际的。这时可以使用循环依次访问数组中的每个元素，这被称为数组的遍历。

下面通过代码演示循环遍历数组。

```
1.  package com.nle.demo4;
2.  public class TestArray3 {
3.      public static void main(String[] args) {
4.          int[] arr = { 1, 2, 3, 4, 5, 6 };
5.          // 数组索引从 0 开始，最大值为长度-1，循环遍历每次自增 1
6.          for (int i = 0; i < arr.length; i++) {
7.              System.out.println("arr[" + i + "] = " + arr[i]);
8.          }
9.      }
10. }
```

运行结果如图 3-39 所示。

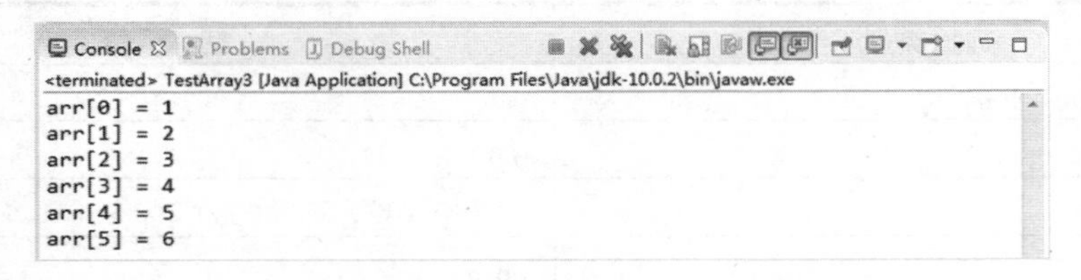

图 3-39　运行结果

(4)　数组的内存解析。

数组在创建的过程中，数组变量和数组元素在内存中是如何分配的呢？下面列出一个例子，请分析数组 arr1 与数组 arr2 是否相等。

```
1.  int[] arr1 = new int[2];
2.  int[] arr2 = new int[2];
3.  //arr1 == arr2 的结果是 true 还是 false 呢?
```

在 Java 中，虚拟机的内存可分为 3 个区：栈空间(stack)、堆空间(heap)和方法区(method)。

- 栈空间：连续的存储空间，遵循先进后出的原则，用于存放局部变量。
- 堆空间：不连续的存储空间，用于存放出来的对象，即类的实例。
- 方法区：方法区在堆空间内，用于存放类的代码信息、静态变量和方法、常量池(字符串常量)。

上例的数组 arr1 与 arr2 在内存中的存储情况，如图 3-40 所示。

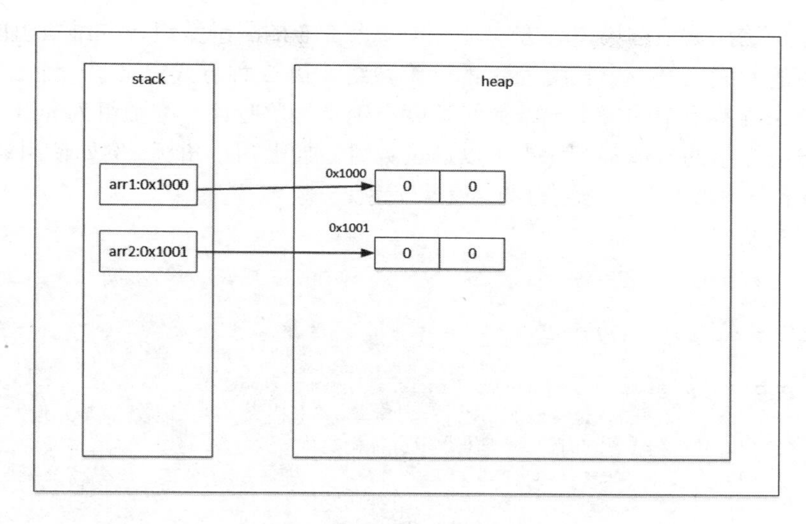

图 3-40　内存状态图

从图 3-40 可以看出，数组变量 arr1 与数组变量 arr2 存放于栈空间中，而 arr1 与 arr2 分别使用 new 创建数组，在堆空间分别有两个数组对象，arr1 指向其中一块内存，arr2 指向另外一块内存。栈中的 arr1 与 arr2 存储的是堆中数组对象的地址，而数组元素存放在堆中。所以 arr1==arr2 比较的是堆中的地址，故 arr1==arr2 为 false。

请思考，若 arr1 = arr2，即把 arr2 的引用地址赋值给 arr1，在内存中是如何变化的。这时 arr1==arr2 的结果是 true 还是 false?

2. 多维数组

多维数组指的就是在数组中嵌套数组。在程序中比较常见的就是二维数组，即一维数组中嵌套一维数组。

二维数组的定义有如下几种方式(以 int 类型为例)。

```
1.    第 1 种方式：int[][] arr;
2.    第 2 种方式：int[] arr[];
3.    第 3 种方式：int arr[][];
```

建议使用第一种方法定义二维数组。二维数组的初始化与一维数组类似，也分为静态初始化和动态初始化两种。

静态初始化示例如下：

```
1.    第 1 种方式：int[][] arr = {{1, 2}, {3, 4, 5}, {6}};
2.    第 2 种方式：int[][] arr = new int[][]{{1, 2}, {3, 4, 5}, {6}};
```

上面的这两种静态初始化的二维数组可看成一个特殊的一维数组，这个一维数组的长度是 3，且这个一维数组的第 1 个元素为一维数组{1，2}，第 2 个元素为一维数组{3，4，5}，第 3 个元素为一维数组{6}。

动态初始化示例如下：

```
1.    第 1 种方式：int[][] arr = new int[2][3];
2.    第 2 种方式：int[][] arr = new int[2][];
```

上面动态初始化数组的第 1 种方式定义了一个 2*3 的二维数组，二维数组的长度为 2，每个二维数组的元素又是一个长度为 3 的一维数组。第 2 种方式定义了一个长度为 2 的二维数组，但是二维数组中的每个一维数组长度不确定，此时该一维数组为 null。

访问二维数组中的具体某个元素可以通过数组名加上两个角标，例如使用 arr[0][1]是访问二维数组 arr 中的第 1 个一维数组的第 2 个元素。

二维数组初始化示例如下：

```
1.   package com.nle.demo4;
2.
3.   public class TestArray4 {
4.
5.      public static void main(String[] args) {
6.         int[][] arr = new int[2][];// 定义一个长度为 2 的二维数组，默认值为{null，null}
7.         // 给二维数组的第 1 个一维数组进行初始化
8.         arr[0] = new int[] { 1, 2 };
9.         // 给二维数组的第 2 个一维数组进行初始化
10.        arr[1] = new int[] { 3, 4, 5 };
11.
12.        // 遍历二维数组的元素(使用嵌套循环)
13.        for (int i = 0; i < arr.length; i++) {
14.           for (int j = 0; j < arr[i].length; j++) {
15.              // 打印二维数组中第 i 个一维数组的第 j 个元素
16.              System.out.print(arr[i][j] + "\t");
17.           }
18.           // 换行
19.           System.out.println();
20.        }
21.     }
22.
23.  }
```

上面已经创建了一个长度为 2 的二维数组，但是该二维数组中的每个一维数组长度未定义，接着对二维数组的每个一维数组进行初始化，最后使用嵌套循环遍历二维数组的所有元素，并打印到控制台上显示，如图 3-41 所示。

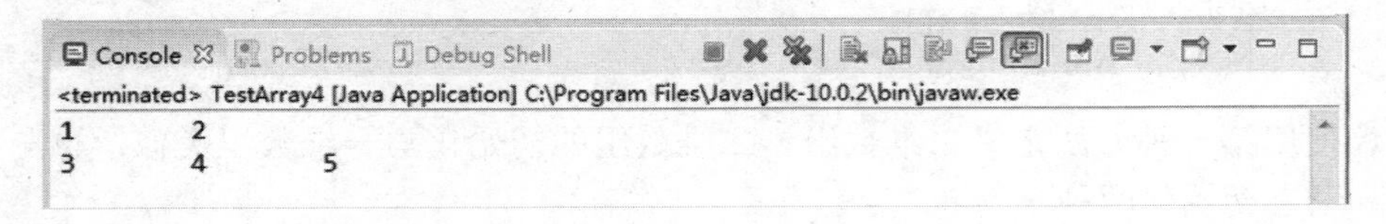

图 3-41　运行结果

【任务实施】

1. 任务分析

本任务要完成当采集到的温度超过预设的 35 度时，开启风扇通风，否则关闭风扇。

(1) 在任务 3 的基础上搭建工程，只采集温度数据。

(2) 编写生成继电器的控制指令。

(3) 编写业务代码，实现当温度大于等于 35 度时，开风扇，反之关风扇。

2. 任务实施

(1) 在任务 3 的基础上搭建工程，只采集温度数据。

第一步： 复制任务 3 的工程，并重命名为 project3_task4，jar 包与任务 3 相同，如图 3-42 所示。

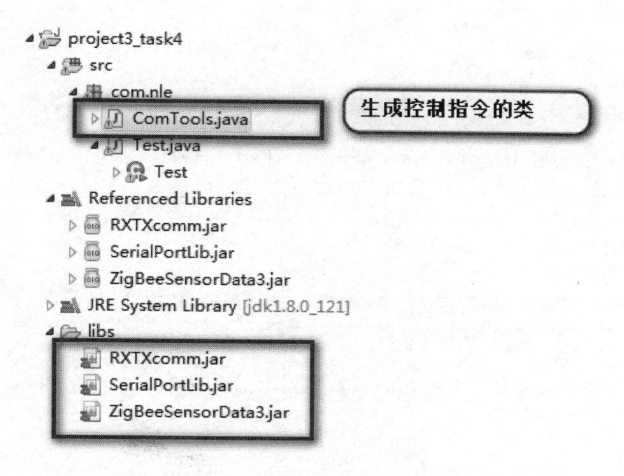

图 3-42　创建工程并添加 jar 包

第二步： 在 Test 类里只保留温度数据的处理部分：

```
1.   public class Test {
2.       static SerialPortManager manager;
3.       static SerialPort serialPort;
4.   //   static boolean isOpenFlag = false;
5.
6.       public static void main(String[] args) {
7.           // 1.获取串口管理工具类对象
8.           manager = new SerialPortManager();
9.           // 2.打开串口
10.          try {
11.              serialPort = manager.openPort("COM200", 38400);
12.          } catch (SerialPortException e1) {
13.              e1.printStackTrace();
14.          }
15.          // 3.ZigBee 协调器采集数据并发送到串口
16.          ZigBeeSensorData sensorData = new ZigBeeSensorData(manager,
             serialPort);
17.
18.          while (true) {
19.              // 4.取出从串口采集到的数据
20.              String strHexData = sensorData.getStrData();
21.          //  System.out.println("从串口采集到的数据是：=" + strHexData);
22.
23.              // 5.按 ZigBee 协议分析传感器数据
24.              getSensorData(strHexData);
25.
26.              // 6.采集一次休息一秒
27.              try {
28.                  Thread.sleep(1000);
```

```
29.             } catch (InterruptedException e) {
30.                 e.printStackTrace();
31.             }
32.         }
33. 
34.     }
35. 
36.     /**
37.      * 获取 ZigBee 传感器数据
38.      *
39.      * @param strData com 采集回来的数据
40.      */
41.     public static void getSensorData(String strData) {
42.         if (strData == null || strData.length() == 0) {
43.             System.out.println("数据还在采集中，请稍等.....");
44.             return;
45.         }
46.         // ZigBee 协议数据头，固定为 0xfe，按协议数据头 FE 做字符串的分割
47.         String[] datas = strData.split("FE");
48.         int count = 0;
49.         String temp = null, humi = null, light = null, people = null, hire
         = null, tempStr;
50. 
51.         for (int i = 0; i < datas.length; i++) {
52. 
53.             if (datas[i].length() != 0&& datas[i].length() >=38) {// 如果第
                 一行分割出来的字符串长度不为 0
54.                 count++;
55.                 tempStr = datas[i];// 把每一行数据复制一份做处理
56.                 String s = tempStr.substring(32, 34);// 截取传感器的类型
57. 
58.                 if (s.equals("01"))// 温湿度
59.                 {
60.                     // 截取温度数据
61.                     String s1 = tempStr.substring(34, 38);// FA00
62.                     // 高低位交换
63.                     temp = s1.substring(2, 4) + s1.substring(0, 2);
64.                     // 计算温度
65.                     double restemp = Integer.parseInt(temp, 16) / 10.0;
66.                     System.out.println("采集到的温度值是: " + temp + " 实际温度: "
                     + restemp + "℃");
67.                   //此处要添加根据温度控制风扇的代码
68.                  //一次只要采集到一条温度数据就退出
69.                     break;
70.                 }
71.             }
72.         }
73. }
```

源代码与任务 3 相同，此处不再展开。

(2) 编写生成继电器的控制指令。

由 ZigBee 协议得知(相关协议请参考物联网工程实训相关文档)，ZigBee 协调器控制指令如下：

```
FF F5 05 02 01 00 00 01 03====序列号 0001，开
FF F5 05 02 01 00 00 02 02====序列号 0001，关
```

```
FF F5 05 02 02 00 00 01 02====序列号 0002，开
FF F5 05 02 02 00 00 02 01====序列号 0002，关

FF F5 05 02 03 00 00 01 01====序列号 0003，开
FF F5 05 02 03 00 00 02 00====序列号 0003，关
```

由分析得知继电器控制指令的结构如图 3-43 所示。

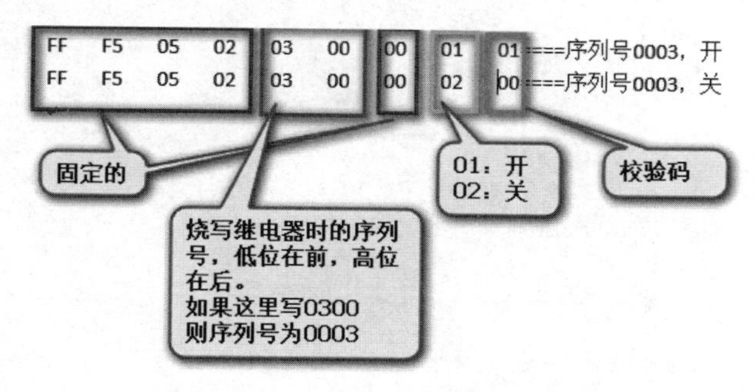

图 3-43　ZigBee 控制指令结构分析

在包内新建工具类 ComTools.java，添加 getLrc()方法用于计算继电器控制指令的校验码，计算规则是图 3-43 中除了校验码部分以外的数据全部累加后进行取反加 1。

```
1.  package com.nle;
2.
3.  import java.nio.ByteBuffer;
4.  import java.util.Locale;
5.
6.  public class ComTools {
7.
8.      /**
9.       * 计算控制继电器动作的控制指令的校验码
10.      *
11.      * @param pSendBuf
12.      * @param nEnd
13.      * @return
14.      */
15.     private static byte getLrc(byte[] pSendBuf, int nEnd) {
16.         byte byLrc = 0;
17.         for (int i = 0; i < nEnd; i++) {
18.             byLrc += pSendBuf[i];
19.         }
20.         byLrc = (byte) ~byLrc;
21.         byLrc = (byte) (byLrc + 1);
22.         return byLrc;
23.     }
24.
25.     /**
26.      * 函数名称：hexStr2Byte</br>
27.      * 功能描述：String 转数组
28.      *
29.      * @param hex
30.      * @return
31.      *
32.      */
33.     public static byte[] hexStrToByte(String hex) {
```

```
34.         // 移除字符串中的空格
35.         hex = hex.replace(" ", "");
36.
37.         ByteBuffer bf = ByteBuffer.allocate(hex.length() / 2);
38.         for (int i = 0; i < hex.length(); i++) {
39.             String hexStr = hex.charAt(i) + "";
40.             i++;
41.             hexStr += hex.charAt(i);
42.             byte b = (byte) Integer.parseInt(hexStr, 16);
43.             bf.put(b);
44.         }
45.         return bf.array();
46.     }
47.
48.     /**
49.      * 通过序列号获取 ZigBee 继电器的控制指令
50.      *
51.      * @param seq  烧写 ZigBee 继电器时的序列号
52.      * @param flag true 开, flase 关
53.      * @return 组装好的控制指令
54.      */
55.     public static byte[] getCom(String seq, boolean flag) // 0001 的序列号
      要变成 01 00
56.     {
57.
58.         byte[] com = { (byte) 0xFF, (byte) 0xF5, 0x05, 0x02, 0x00, 0x00, 0x00,
          0x01 };// 校验码最后一位未添加
59.         if (seq.length() == 4) {
60.             // 序列号高低位互换
61.             // 把十六进制的字符串的序列号转换成字节数据
62.             byte[] buf = hexStrToByte(seq);
63.             com[4] = buf[1];
64.             com[5] = buf[0];
65.         }
66.         if (flag)// 为 true 代表开
67.         {
68.             com[7] = 0x01;
69.
70.         } else {
71.             com[7] = 0x02;
72.         }
73.
74.         byte crc = getLrc(com, 8);
75.         byte[] newcom = new byte[9];
76.         for (int i = 0; i < 8; i++) {
77.             newcom[i] = com[i];
78.         }
79.         newcom[8] = crc;
80.
81.         return newcom;
82.     }
83. }
```

第 33 行 hexStrToByte(String hex)方法用于把字符串的序列号转换成字节数据，在 63、64 行把转换成字节数据的高低位互换后存放到控制指令的数组中。

第 55 行 getCom(String seqk, boolean flag)用于按序列号生成控制指令，在该方法中的 74 行计算完校验码后，79 行把校验码存放到数组中。

(3) 编写业务代码，实现当温度大于等于 35 度时，开风扇，反之关风扇。

启用 Test.java 类里的第 4 行代码，代码第 4 行定义了一个风扇开停的标志，用于判断当风扇已经开启时，温度达到了预设值，风扇也不会再次开启。风扇关的道理是一样的。

在 Test 类的第 67 行添加以下代码：根据温度高低判断开关风扇。第 4 行获取到控制风扇的指令后，第 5 行把控制指令通过串口发出去，从而达到控制风扇开的目的。关的道理相同。

```
1.   if (restemp >= 35) {
2.       if (!isOpenFlag)// 如果没有开启过，就开启
3.       {
4.           // 继电器的序列号为 0001
5.           byte[] contrlcom = ComTools.getCom("0001", true);
6.           manager.sendToPort(serialPort, contrlcom);
7.           isOpenFlag = true;// 开启标志
8.           System.out.println("风扇开了");
9.       }
10.  } else {
11.      if (isOpenFlag)// 如果有开启过才关闭
12.      {
13.          // 继电器的序列号为 0001
14.          byte[] contrlcom = ComTools.getCom("0001", false);
15.          manager.sendToPort(serialPort, contrlcom);
16.          isOpenFlag = false;
17.          System.out.println("风扇关了");
18.      }
19.  }
```

3. 运行结果

运行程序，结果如图 3-44 所示。

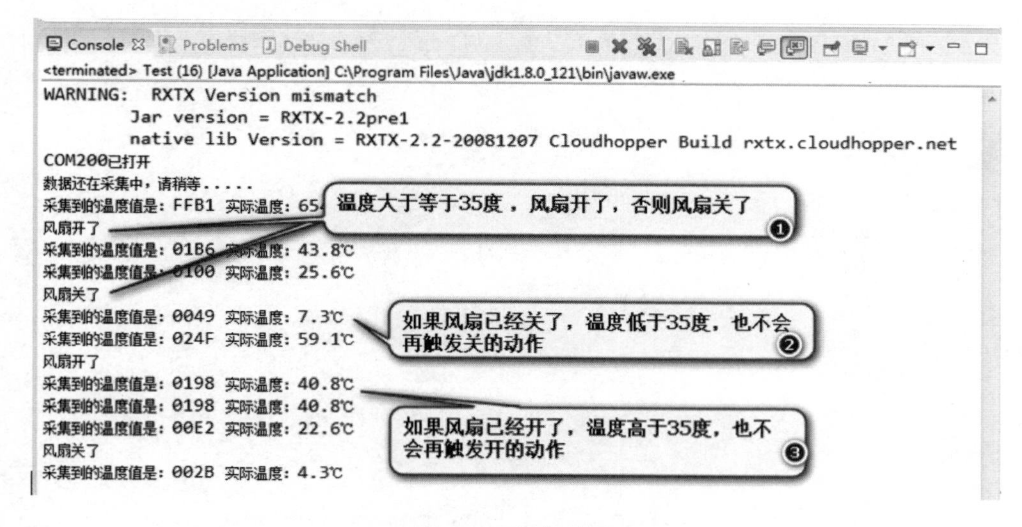

图 3-44 运行结果

思考与练习

1. 简述 8 大基本数据，它们各占多大内存空间。
2. 简述常用的运算符。
3. 简述单分支、两分支、多重分支条件的语法结构。
4. 简述 while 、do…while、for 循环的区别。
5. 简述一维数组初始化的方式。

【项目描述】

对 Java 的基础语法有了基本的了解以后，便可以通过 Java 实现具体的功能了。本项目通过完成串口管理类的封装，并通过编程实现从串口获取设备数据，让读者了解面向对象的编程思想，掌握如何使用 Java 面向对象编程实现这些功能。串口管理类主要是将常用的打开串口、关闭串口、串口添加监听、串口移除监听、获取串口数据、向串口发送指令等操作封装到一个类中，实现复用。在串口管理工具类的基础上，读者可以进一步实现通过串口获取 4150 采集器的数据。本项目的任务列表如图 4-1 所示。

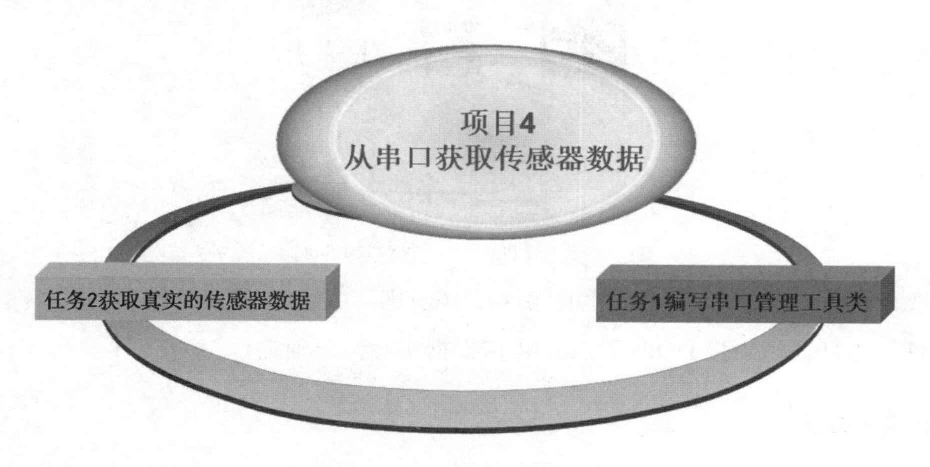

图 4-1　项目 4 的任务列表

【学习目标】

知识目标： 理解面向对象的编程思想；理解类与对象的概念；理解 Java 中包的概念；理解构造方法的作用；理解方法重载的概念。

技能目标： 能运用类创建对象；能使用 this 关键字；能灵活使用 Java 的几种访问权限；能根据需求进行类的封装。

任务 1　编写串口管理工具类

扫码观看视频讲解

【任务描述】

在开发过程中，必须经常调用 RXTX 的一些方法对串口进行操作，如打开串口、关闭串口、为串口添加监听、移除串口监听、获取串口数据、发送数据到串口等。本任务要求读者对这些方法进行封装，以便在之后的编码过程中实现复用。任务清单如表 4-1 所示。

表 4-1　任务清单

任务课时	4 课时	任务组员数量	建议 3 人
任务组采用设备	1 个 ADAM-4150 1 个 RS232-RS485 转接头 1 个继电器 1 个风扇 1 台 PC 相关电源，导线，工具		

【拓扑图】

本任务的拓扑图如图 4-2 所示。

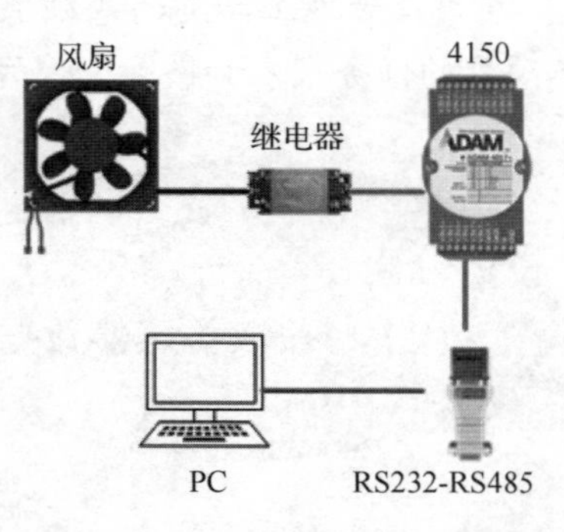

图 4-2　拓扑图

备注：本任务风扇接 DO0 口，COM 口根据实际情况确定。

【知识解析】

1. 面向对象的概念

现实生活中的各种不同的事物都可以看成对象，而面向对象就是把事物映射成对象，使用对象来表示事物之间的关系。

在程序设计中，除了面向对象外，还有一种叫作面向过程。面向过程就是把问题分解成步骤，然后用函数依次调用解决。而面向对象则是把问题分解成多个独立的对象，然后通过调用对象来解决。面向对象的编程思想也更符合人类的思维方式。面向对象的三大特征是封装性、继承性、多态性。

2. 类与对象

在面向对象思想中有两个概念，即类与对象。其中，类是对某一类事物的抽象描述，而对象是表示现实中该类事物的具体个体。例如，可以将车这个概念看作一个类，而把具体的每一辆车看成是对象，这些具体的车与概念上的车就是对象与类之间的关系。类与对象之间的关系，可以简单地理解为类是对象的模板，而对象是类具体的实现。

(1) 类的定义。

类是对象的抽象，用于描述一组对象的共同特征与行为。类中可以定义成员变量和成员方法，其中成员变量用于描述对象的特征，也被称为属性；成员方法用于描述对象的行为，也被称为方法。类定义的语法规范如下所示。

```
1.   [类的修饰符] class 类名称 [extends 父类名称][implements 接口名称列表]{
2.         变量定义及初始化;
3.         方法定义及方法体;
4.   }
```

上面类的语法规范具体说明如下。

- 类的修饰符：类的修饰符有访问权限修饰符、最终修饰符 final 等，这些修饰符在后面的学习中会逐步介绍，不是必需的。
- class：class 关键字是用来声明类的，必须有。
- extends 父类名称：继承父类，不是必需的。在后面的学习中再介绍。
- implements 接口名称列表：实现接口，如果有多个接口名称，用“,”隔开，不是必需的。在后面的学习中再介绍。

下面通过例子来学习如何定义一个类，具体代码如下所示。

```
1.  public class Person {
2.      // 定义成员变量(属性)
3.      public String name;
4.      public int age;
5.      // 定义成员方法(行为)
6.      public void speak() {
7.          System.out.println("您好！我是" + name + "，今年" + age + "岁了！");
8.      }
9.      public void sleep() {
10.         String name = "小明";
11.         System.out.println(name + "睡觉了...");
12.     }
13. }
```

上面的代码定义了一个 Person 类，类中有成员变量 name、age，成员方法 speak()、sleep()。在成员方法中可以直接访问成员变量 name 和 age。

需要注意的是，定义在类体中的变量称为成员变量，类中的方法都可访问，还可以通过对象的引用来访问。而定义在代码段中的变量称为局部变量，例如定义在方法中、条件语句块、循环语句块中的变量都是局部变量，局部变量只有在定义的代码段中才可以使用。在 Person 类的代码中，sleep()方法内部定义了一个局部变量 name，这是可以的，用的时候优先使用最近定义的变量，所以 sleep()方法中打印语句中使用的 name 为局部变量，且该局部变量只在 sleep()方法中才能被访问。

(2) 对象的创建与使用。

程序想要完成具体的功能，仅仅定义类是不够的，还需要通过类创建实例对象。在程序中使用 new 关键字来创建对象，具体格式如下。

```
1.   类名 对象名 = new 类名();
```

例如，创建一个 Person 类的实例对象，代码如下。

```
1.   Person p = new Person();
```

上面代码中，赋值符号左边“Person p”为声明一个 Person 类型的变量 p，右边“new Person()”才是在堆空间中创建一个对象。整条语句的意思是创建一个 Person 类型的对象，并将对象的引用地址赋值给 Person 类型的变量 p，这时这个变量 p 也可以称为对象引用。变量 p 和对象之间的关系如图 4-3 所示。

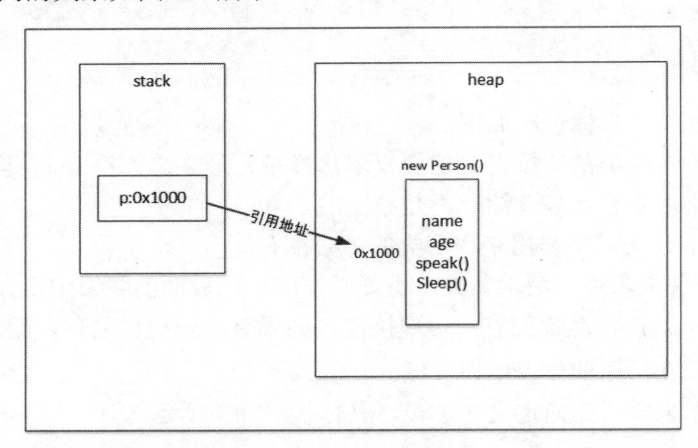

图 4-3 变量与对象的关系

在创建对象后，可以通过对象的引用来访问对象的所有成员，具体格式如下。

```
1.   对象引用.对象成员
```

接下来通过一个例子来学习如何访问对象的成员，具体代码如下。

```
1.   public class Test01 {
2.       public static void main(String[] args) {
3.           Person p1 = new Person();// 创建第一个 Person 对象
4.           Person p2 = new Person();// 创建第二个 Person 对象
5.           p1.name = "小明";// 给对象 p1 的成员变量 name 赋值
6.           p1.age = 18;// 给对象 p1 的成员变量 age 赋值
7.           p1.speak();// 调用对象 p1 的成员方法 speak()
8.           p2.speak();// 调用对象 p2 的成员方法 speak()
9.       }
```

```
10. }
```

运行结果如图 4-4 所示。

```
Problems  Javadoc  Declaration  Console
<terminated> Test01 [Java Application] D:\java\bin\javaw.exe
您好！我是小明，今年18岁了！
您好！我是null，今年0岁了！
```

图 4-4　运行结果

上面代码中，p1、p2 分别引用 Person 类的两个实例对象(简称 p1 对象，p2 对象)。从运行结果可以看出来，p1 对象调用 speak()方法打印的结果与 p2 对象调用 speak()方法打印的结果不同。这是因为 p1 对象和 p2 对象是两个独立的个体，对 p1 对象的属性值修改不会影响 p2 对象的属性值。从运行结果还可以看出，p2 对象的成员变量 name 和 age 也有值，name 的值为 null，age 的值为 0。这是因为在实例化对象时，Java 虚拟机会自动对成员变量进行初始化，针对不同类型的成员变量会赋予不同的初始值，如表 4-2 所示。

表 4-2　成员变量的初始值

成员变量类型	初 始 值
byte	0
short	0
int	0
long	0L
float	0.0F
double	0.0D
char	空字符，'\u0000'
boolean	false
引用数据类型	null

对象在创建后，可以通过对象的引用来调用对象内的成员。但是大部分初学者容易遇到一个问题，即对象的引用为 null 的情况，这时使用该引用访问对象的成员会出现错误。这种错误在 Java 中被称为空指针异常，即 NullPointerException。接下来通过一个例子来演示空指针异常的情况，具体代码如下所示。

```
1.  public class Test02NullPointExce {
2.      public static void main(String[] args) {
3.          Person p = new Person();
4.          p = null;
5.          p.speak();
6.      }
7.  }
```

运行结果如图 4-5 所示。

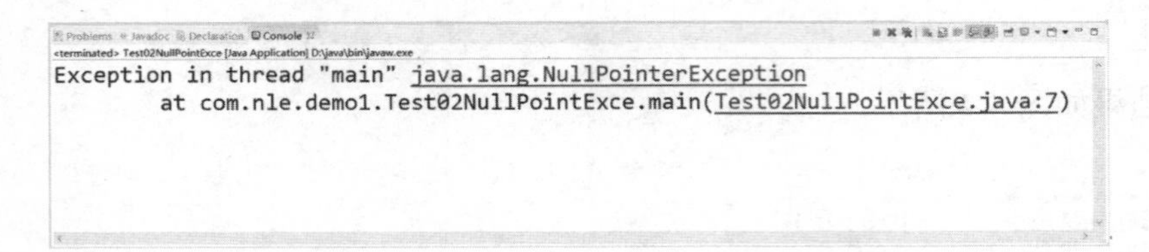

图 4-5　运行结果

从运行结果中可以看出，在类 Test02NullPointExce 的 main()方法的第 5 行代码出现了空指针异常，原因是 p 对象的引用为 null。因为在程序中把 null 赋值给了变量 p，变量 p 没有指向具体的对象，所以 p 调用对象中的 speak()方法出现了空指针异常。

3. 成员变量

在项目 3 的内容中，读者已经学习了变量的定义与使用，大家是否有注意在 main()方法中定义的变量，这种在代码块中定义的变量称为局部变量。

局部变量是在代码块中定义的变量，它的作用域只是在定义的代码块中。区别于局部变量，成员变量是在类范围内定义的变量，它的作用域是整个类，成员变量可以分为实例变量与静态变量两种。接下来，分别介绍实例变量与静态变量，并介绍它们之间的区别。

(1) 实例变量。

实例变量定义在类范围内，在对象被实例化时创建，对象销毁时消亡。对象创建完成后，可以通过对象访问实例变量，为实例变量赋值或者获取实例变量的值。接下来通过一个例子学习实例变量的使用，具体代码如下所示。

```
1.  public class Employee1 {
2.      //定义实例变量，员工姓名与薪水
3.      String name;
4.      double salary;
5.  
6.      public static void main(String[] args) {
7.          //创建员工对象
8.          Employee1 emp  = new Employee1();
9.          //为该对象的实例变量赋值
10.         emp.name = "小新";
11.         emp.salary = 3000;
12.         //在控制台上打印对象实例变量的字符串值
13.         System.out.println("名字："+emp.name+"\n 薪水："+emp.salary);
14.     }
15. }
```

运行结果如图 4-6 所示。

图 4-6　运行结果

(2) 静态变量。

静态变量也定义在类范围内，使用 static 关键字修饰，也被称为类变量。静态变量使用类名访问，不建议使用对象访问静态变量。接下来通过类 Employee2 学习静态变量的使用，具体代码如下所示。

```
1.  public class Employee2 {
2.      //定义静态变量，员工所在城市
3.      static String city;
4.
5.      public static void main(String[] args) {
6.          Employee.city = "福州";
7.          System.out.println("员工所在的城市是："+Employee2.city);
8.      }
9.  }
```

(3) 实例变量与静态变量的区别。

在学习了如何简单使用实例变量与静态变量之后，下面通过类 Employee3 演示它们之间的区别，具体代码如下所示。

```
1.  public class Employee3 {
2.      // 定义实例变量，员工姓名与薪水
3.      String name;
4.      double salary;
5.      // 定义静态变量，员工所在城市
6.      static String city;
7.
8.      public static void main(String[] args) {
9.          Employee3.city = "福州";
10.
11.         //创建对象
12.         Employee3 emp1 = new Employee3();
13.         emp1.name = "小新";
14.         emp1.salary = 3000;
15.         Employee3 emp2 = new Employee3();
16.         emp2.name = "小丽";
17.         emp2.salary = 10000;
18.
19.         System.out.println("=====修改小新的工资前=======");
20.         System.out.println(emp1.name+"的薪水是："+emp1.salary);
21.         System.out.println(emp2.name+"的薪水是："+emp2.salary);
22.         //修改小新的工资
23.         emp1.salary = 6000;
24.         System.out.println("=====修改小新的工资后=======");
25.         System.out.println(emp1.name+"的薪水是："+emp1.salary);
26.         System.out.println(emp2.name+"的薪水是："+emp2.salary);
27.
28.         System.out.println("=====修改员工城市前=======");
29.         System.out.println(emp1.name+"的城市是："+emp1.city);
30.         System.out.println(emp2.name+"的城市是："+emp2.city);
31.         //修改员工城市
32.         emp1.city = "北京";
33.         System.out.println("=====修改员工城市后=======");
34.         System.out.println(emp1.name+"的城市是："+emp1.city);
35.         System.out.println(emp2.name+"的城市是："+emp2.city);
```

```
36.     }
37. }
```

运行结果如图 4-7 所示。

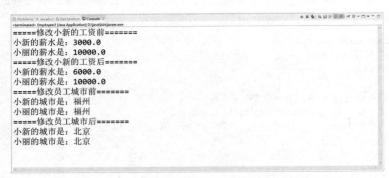

图 4-7　运行结果

从运行结果可以看出，实例变量是对象独有的，修改对象的实例变量值不会影响其他对象的实例变量值，而静态变量是所有对象共享的，所有对象使用的是同一个静态变量的值。

为了更方便读者区分实例变量与静态变量，下面列出实例变量与静态变量的区别。

- 共享性：静态变量被所有对象共享，即同一类的所有对象共享此类的静态变量；实例变量为每个对象独有，操作自己的实例变量不会影响其他对象。
- 初始化时间：类变量在类加载时分配内存，实例变量在创建对象时分配内存。
- 消亡时间：类变量在程序退出时释放内存，对象不存在，实例变量就不存在了。
- 调用形式：类变量一般通过类名调用，实例变量通过对象名调用。
- 声明的形式：类变量用 static 声明，实例变量没有 static 声明。

4. 方法

成员方法简称方法，定义在类范围内。有些地方会把方法称为函数，两者没有区别，只是名称不一样。

在 Java 中，声明一个方法具体的格式如下所示。

```
1.  [方法的修饰符] 返回值类型 方法名 ([参数类型 参数名1,参数类型 参数名2,…]){
2.        执行语句
3.        …
4.        return 返回值;
5.  }
```

对于上面的语法格式具体说明如下。

- 方法的修饰符：方法的修饰符有访问权限修饰符、静态修饰符 static、最终状态修饰符 final 等，这些修饰符在后面的学习过程中会逐步介绍，不是必需的。
- 返回值类型：用于限定方法返回值的数据类型，必须有(除了构造方法)。
- 参数类型：用于限定调用方法时传入参数的数据类型，不是必需的。例如不带参数的方法。
- 参数名：是一个变量，用于介绍调用方法时传入的数据。
- return 返回值：当方法有具体返回值时，用 return 返回，不是必需的。

在方法的语法格式中，方法的修饰符不是必需的。返回值类型必须有(除构造方法外)，无返回值的情况使用 void 修饰。方法中的参数列表可以声明 0 到多个参数，参数之间使用“,”隔开。在有具体返回值类型的情况下，必须使用 return 返回方法的返回值，如果返回值类型用 void 修饰，return 可以省略。

方法可以分为如下几类。

- 按返回值可以分为：有返回值和无返回值。
- 按参数可以分为：有参数和无参数。
- 按功能可以分为：实例方法、静态方法、构造方法。

接下来通过一个例子来学习方法的声明与调用，具体代码如下所示。

```
1.  public class Test06function {
2.      int var;
3.
4.      // 1.方法按返回值分为：有返回值与无返回值
5.      // 1.1 有返回值方法
6.      int getVar() {
7.          System.out.println("方法getVar()被调用执行,并返回一个成员变量var的值:"
             + var);
8.          return var;
9.      }
10.
11.     // 1.2 无返回值方法
12.     void print() {
13.         System.out.println("方法print()被调用执行，该方法没有返回值");
14.     }
15.
16.     // 2.方法按参数可以分为：有参数和无参数
17.     // 2.1 有参数方法
18.     void setVar(int arg0) {// 形式参数
19.         System.out.println("方法setVar(int arg0)被调用执行，该方法为有参数方法，
             并且把传进来的参数值赋值给成员变量var");
20.         var = arg0;
21.     }
22.
23.     // 2.2 无参数方法
24.     void noArgFun() {
25.         System.out.println("方法noArgFun()被调用执行，该方法为无参数方法");
26.     }
27.
28.     public static void main(String[] args) {
29.         Test06function obj = new Test06function();
30.         int var = obj.getVar();
31.         System.out.println("调用getVar方法，得到返回值：" + var);
32.         System.out.println("---------");
33.         obj.print();
34.         System.out.println("---------");
35.         obj.setVar(100); // 传入实际参数
36.         System.out.println("---------");
37.         obj.noArgFun();
38.     }
39. }
```

运行结果如图 4-8 所示。

```
方法getVar()被调用执行，并返回一个成员变量var的值:0
调用getVar方法，得到返回值：0
---------
方法print()被调用执行，该方法没有返回值
---------
方法setVar(int arg0)被调用执行，该方法为有参数方法，并且把传进来的参数值赋值给成员变量var
---------
方法noArgFun()被调用执行，该方法为无参数方法
```

图 4-8　运行结果

需要注意的是，在定义有参数的方法时，方法声明处的参数为形式参数(形参)，具体的参数值为方法调用时传入的实际参数(实参)。

5. 构造方法

(1) 构造方法的定义。

对象创建后，如果想为这个对象的属性赋初始值，则需要通过对象访问属性值或者通过方法设置属性值。如果需要在创建对象时就能为对象的属性赋值，可以通过构造方法来实现。

构造方法的特点如下。

- 方法名与类名同名。
- 在方法名的前面没有返回值类型。
- 在方法中不能使用 return 语句返回值，但是可以使用 return 语句结束方法。

接下来通过一个例子来演示如何在类中定义构造方法，具体代码如下。

```
1.  public class Test07constructor {
2.      public Test07constructor() {
3.          System.out.println("构造方法 Test07constructor()被调用了...");
4.      }
5.
6.      public static void main(String[] args) {
7.          //创建 Test03function 对象
8.          Test07constructor obj = new Test07constructor();
9.      }
10. }
```

运行结果如图 4-9 所示。

图 4-9　运行结果

从运行结果可以看出，在创建对象时构造方法会被执行。可以这样理解，创建对象“new Test07constructor()”语句中 new 后面接的就是构造方法，如果类中没有定义构造方法，Java 会默认定义一个无参构造方法。

在一个类中除了可以定义无参的构造方法外，还可以定义有参的构造方法。接下来通过一个例子来学习有参构造方法的定义，具体代码如下所示。

```
1.  public class Test07constructor2 {
2.      int value;
3.      //定义有参的构造方法
4.      public Test07constructor2(int arg) {
5.          value = arg;
6.      }
7.      public static void main(String[] args) {
8.          //通过有参的构造方法创建对象
9.          Test07constructor2 obj = new Test07constructor2(100);
10.         System.out.println("obj.value = " + obj.value);
11.     }
12. }
```

运行结果如图 4-10 所示。

图 4-10　运行结果

从运行结果可以看出，有参的构造方法在创建对象的时候被执行了。那么这时是否还可以通过无参的构造方法创建对象呢？答案是不可以，当一个类中已经定义了构造方法的情况下，默认的无参构造方法就不存在了，所以不能通过无参的构造方法创建对象。

(2) 构造方法的重载。

在上面的例子中可以看出，一个类中可以定义有参的构造方法，也可以定义无参的构造方法。那么在一个类中能不能同时存在无参的构造方法和有参的构造方法呢？答案是可以，一个类中可以定义多个构造方法，只要构造方法的参数列表不同即可(参数类型、参数个数、参数顺序有一个不同)。一个类中定义多个构造方法称为构造方法的重载。接下来通过一个例子来学习构造方法的重载，具体代码如下所示。

```
1.  public class Test07constructor3 {
2.      String name;
3.      int age;
4.      public Test07constructor3() {
5.          System.out.println("构造方法 Test07Constructor()被执行了！");
6.      }
7.      public Test07constructor3(String argName) {
8.          System.out.println("构造方法 Test07Constructor(String argName)被执行
            了！name=" + argName);
9.          name = argName;
10.     }
11.     public Test07constructor3(String argName, int argAge) {
12.         System.out.println("构造方法 Test07Constructor(String argName, int
            argAge)被执行了！name=" + argName + ", age=" + argAge);
13.         name = argName;
14.         age = argAge;
```

```
15.     }
16.     public static void main(String[] args) {
17.         Test07constructor3 obj1 = new Test07constructor3();
18.         Test07constructor3 obj2 = new Test07constructor3("王小二");
19.         Test07constructor3 obj3 = new Test07constructor3("李星星", 18);
20.     }
21. }
```

运行结果如图 4-11 所示。

```
构造方法Test07Constructor()被执行了！
构造方法Test07Constructor(String argName)被执行了！ name=王小二
构造方法Test07Constructor(String argName,int argAge)被执行了！ name=李星星,age=18
```

图 4-11　运行结果

从上面的结果可以看出，一个类中可以定义多个构造方法，可以根据不同的情况，选择其中的一个构造方法创建对象。

除了构造方法外，其他的方法也可以进行重载。方法的重载指的是一个类中有多个同名的方法，且方法的参数列表不同。接下来通过一个例子来学习方法的重载，具体代码如下所示。

```
1.  public class Test08overload {
2.      void speak() {
3.          System.out.println("你好！...");
4.      }
5.      void speak(String name) {
6.          System.out.println("你好！我是" + name);
7.      }
8.      void speak(String name, int age) {
9.          System.out.println("你好！我是" + name + ", 今年" + age + "岁了！");
10.     }
11.     public static void main(String[] args) {
12.         Test08overload obj = new Test08overload();
13.         obj.speak();
14.         obj.speak("宝强");
15.         obj.speak("小花", 18);
16.     }
17. }
```

运行结果如图 4-12 所示。

```
你好！...
你好！我是宝强
你好！我是小花,今年18岁了！
```

图 4-12　运行结果

上面的例子代码中定义了三个 speak 方法，可以通过调用 speak 方法，分别传入对应的参数或者不传参数来执行相对应的 speak 方法。

在 Test08overload 类中再加入一些 speak 方法，请判断是否是重载的方法。

```
1.  void speak(int age, String name) //该方法是重载的方法
2.  int speak() //该方法不是重载的方法
```

上面的例子中，第一个方法是重载的方法，原因是参数的顺序不同。第二个不是方法的重载，原因是方法的重载要求方法的参数不同，与方法的返回值类型无关。

6. this 关键字

在 Java 中，this 指向当前对象的引用，可以使用 this 访问对象的成员。接下来为读者介绍 this 关键字的几种用法。

(1) 通过 this 关键字访问成员变量。

用于解决与局部变量同名的冲突问题，具体代码如下所示。

```
1.  public class Test09this1 {
2.      String name;
3.
4.      void setName(String name) {
5.          // this 区分成员变量和局部变量
6.          this.name = name;
7.      }
8.  }
```

在上面的代码中，setName()方法定义了一个 String 类型参数，参数名 name 与成员变量 name 同名，故在方法内使用 name 时，根据就近原则总是使用参数 name，所以可以在 setName()方法中使用 this 来区分成员变量与局部变量(参数)，this 引用的变量为成员变量。

(2) 通过 this 关键字调用成员方法。

具体代码如下所示。

```
1.  public class Test09this2 {
2.
3.      public void testFun1() {
4.
5.      }
6.
7.      public void testFun2() {
8.          this.testFun1();
9.      }
10. }
```

在上面的代码中，testFun2()方法中使用 this 关键字调用 testFun1()方法，一般调用本类中的方法时，this 关键字是可以省略的。在调用方法的写法上，“this.testFun1()”与“testFun1()”是一样的。

(3) 通过 this 关键字调用构造方法。

在构造方法中可以使用 this([参数])调用其他构造方法，具体代码如下所示。

```
1.  public class Test09this3 {
2.      public Test09this3() {
3.          System.out.println("无参的构造方法被调用了...");
4.      }
```

```
5.
6.      public Test09this3(String content) {
7.          this();//调用无参的构造方法
8.          System.out.println("有参的构造方法被调用了...");
9.      }
10.
11.     public static void main(String[] args) {
12.         Test09this3 obj = new Test09this3("测试");
13.     }
14. }
```

在上面的代码中，使用有参的构造方法创建对象时，调用有参的构造方法，又因有参的构造方法中使用 this()调用了无参的构造方法，所以运行结果中两个构造方法都被调用了。需要注意的是，在使用 this 关键字调用其他构造方法时，只能在构造方法中的第一行代码调用，并且只能出现一次。

7. RXTX 串口通信工具

RXTX 是一个提供串口和并口通信的开源 Java 类库，由该项目发布的文档均遵循 LGPL 协议。

针对 x86 体系结构的 Linux 操作系统平台，RXTX 的部署包括下面几个文档。

- RXTXcomm.jar：RXTX 自己的实现。
- librxtxSerial.so：底层串口库文档。
- librxtxParallel.so：底层并口库文档。

配置方法：

(1) 由官网下载最新的包，得到 RXTX 包后，将其中的 rxtxParallel.dll、rxtxSerial.dll 放到 JAVA_HOME\jre\bin\目录下。

(2) 将 RXTXcomm.jar 复制一份放到工程目录新建 libs 下，如果没有 libs，就先创建 libs 目录。

【任务实施】

1. 任务分析

本任务要完成串口管理工具类。

(1) 搭建串口开发环境。

(2) 创建串口管理工具类 SerialPortManager。

(3) 工具类中包含方法：打开串口、关闭串口、串口添加监听、串口移除监听、读取串口数据、发送指令等，为了方便调用将方法都定义为静态。

2. 任务实施

(1) 将教学资源目录下的 RXTX 包解压，将 rxtxParallel.dll、rxtxSerial.dll 复制到 JAVA_HOME\jre\bin 目录下。

(2) 新建工程 project4_task1。

(3) 在工程根目录下创建 libs 文件夹，添加 RXTX 包下的 RXTXcomm.jar 文件，在文件上右击，在弹出的快捷菜单中选择 Build Path 后再选择 Add to Build Path 命令。

(4) 在工程 src 目录下新建 com.nle.serialport.manage 包，如图 4-13 所示。

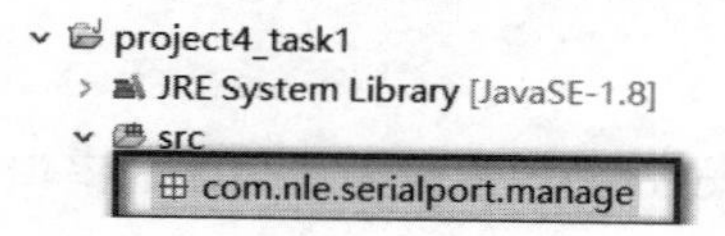

图 4-13 新建包

(5) 在 com.nle.serialport.manage 包下创建 SerialPortManager 类，如图 4-14 所示。

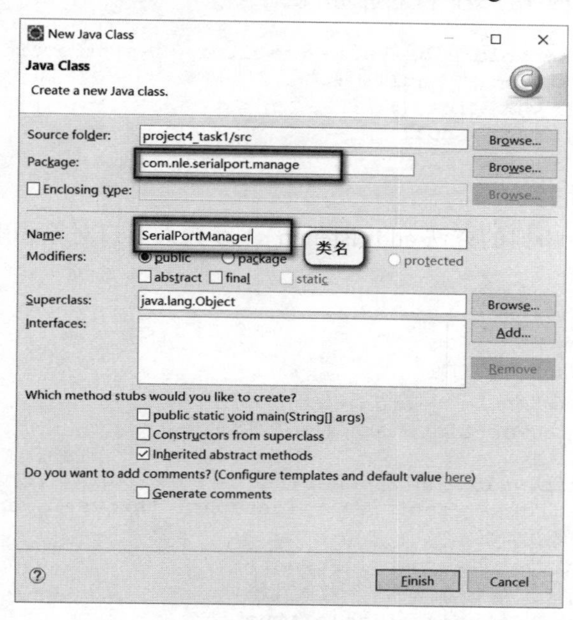

图 4-14 创建 SerialPortManager 类

(6) 封装打开串口的方法 openPort()，方法的参数为串口名称和波特率，返回值为串口对象。方法中有关异常采用声明的方式处理，异常的相关知识点会在后面相关内容中再详细说明。

```
1.  /**
2.   * 打开串口
3.   *
4.   * @param portName 端口名称
5.   * @param baudrate 波特率
6.   * @return 串口对象
7.   */
8.  public static final SerialPort openPort(String portName, int baudrate)
    throws Exception {
9.      // 通过端口名识别端口
10.     CommPortIdentifier portIdentifier = CommPortIdentifier.
          getPortIdentifier(portName);
11.     // 打开端口，并给出端口名字和一个timeout(打开操作的超时时间)
12.     CommPort commPort = portIdentifier.open(portName, 2000);
13.     // 判断是不是串口
14.     SerialPort serialPort = (SerialPort) commPort;
15.     // 设置串口的波特率等参数
```

```
16.     serialPort.setSerialPortParams(baudrate, SerialPort.DATABITS_8,
          SerialPort.STOPBITS_1, SerialPort.PARITY_NONE);
17.     return serialPort;
18. }
```

(7) 封装关闭串口的方法 closePort()，参数为待关闭的串口对象。

```
1.   /**
2.    * 关闭串口
3.    *
4.    * @param serialport 待关闭的串口对象
5.    */
6.   public static void closePort(SerialPort serialPort) {
7.      if (serialPort != null) {
8.         serialPort.close();
9.         serialPort = null;
10.     }
11. }
```

(8) 封装添加监听串口的方法 addListener()，参数为串口对象和串口监听器对象。

```
1.   /**
2.    * 添加监听器
3.    *
4.    * @param port 串口对象
5.    * @param listener 串口监听器
6.    * @throws TooManyListeners 监听类对象过多
7.    */
8.   public static void addListener(SerialPort port,
9.          SerialPortEventListener listener) throws Exception {
10.     // 给串口添加监听器
11.     port.addEventListener(listener);
12.     // 设置当有数据到达时唤醒监听接收线程
13.     port.notifyOnDataAvailable(true);
14.     // 设置当通信中断时唤醒中断线程
15.     port.notifyOnBreakInterrupt(true);
16.
17. }
```

(9) 封装串口监听移除方法 removeListener()，参数为待移除监听串口。

```
1.   /**
2.    * 移除监听器
3.    * @param port
4.    */
5.   public static void removeListener(SerialPort port) {
6.      port.removeEventListener();
7.   }
```

(10) 封装往串口发送数据的方法 sendToPort()，参数为串口对象和要发送的数据 byte 数组。获取串口的输出流，将传入的字节数组写出到输出流中。

```
1.   /**
2.    * 往串口发送数据
3.    *
4.    * @param serialPort 串口对象
5.    * @param order 待发送数据
6.    */
7.   public static void sendToPort(SerialPort serialPort, byte[] order)
```

```
8.          throws Exception {
9.      OutputStream out = serialPort.getOutputStream();
10.     out.write(order);
11.     out.flush();
12. }
```

(11) 封装从串口读取数据的方法 readFromPort()，参数为串口对象。通过串口对象的输入流每次读取一个字节数据，并将数据添加到预先定义好的 bytes 数组中。

```
1.   /**
2.    * 从串口读取数据
3.    *
4.    * @param serialPort 当前已建立连接的 SerialPort 对象
5.    * @return 读取到的数据
6.    */
7.  public static byte[] readFromPort(SerialPort serialPort) throws Exception{
8.      InputStream in = serialPort.getInputStream();
9.      byte[] bytes = {};
10.     // 缓冲区大小为一个字节
11.     byte[] readBuffer = new byte[1];
12.     int bytesNum = in.read(readBuffer);
13.     while (bytesNum > 0) {
14.        byte[] bytesNew = new byte[bytes.length+1];
15.        //复制第一个数组
16.        System.arraycopy(bytes, 0, bytesNew, 0, bytes.length);
17.        //复制刚读到的数据到前一个数组数据后
18.        System.arraycopy(readBuffer, 0, bytesNew, bytes.length,
19.              readBuffer.length);
20.        bytes = bytesNew;
21.        bytesNum = in.read(readBuffer);
22.     }
23.     return bytes;
24. }
```

(12) 编写测试类 MainApp，测试串口管理工具类，控制照明灯的开和关。

```
1.   /**
2.    * 测试串口管理工具类
3.    * @author admin
4.    *
5.    */
6.  public class MainApp {
7.
8.     public static void main(String[] args) {
9.        try {
10.          //打开串口
11.          SerialPort port = SerialPortManager.openPort("COM200", 9600);
12.          //发送指令控制风扇开关
13.          //打开风扇
14.          SerialPortManager.sendToPort(port, new byte[] {0x01, 0x05, 0x00,
               0x10, (byte) 0xFF, 0x00, (byte) 0x8D, (byte) 0xFF});
15.          //关闭风扇
16.          //SerialPortManager.sendToPort(port, new byte[] {0x01, 0x05,
               0x00, 0x10, 0x00, 0x00, (byte) 0xCC, (byte) 0x0F});
17.       } catch (Exception e) {
18.          e.printStackTrace();
19.       }
20.    }
```

```
21.
22. }
```

3. 运行结果

(1) 运行程序，可以看到风扇打开了。

(2) 把代码的第 14 行用//进行注释，把 16 行前的//去掉，重新运行程序，可以看到风扇关闭了。

任务 2　获取真实的传感器数据

扫码观看视频讲解

【任务描述】

在应用开发过程中，应用层需要获取感知层的数据进行分析、监测，并根据感知层获取的数据对设备进行控制。本任务要求读者通过串口管理工具获取真实的传感器数据，并通过代码将数据转换为可用数据。利用任务 1 封装好的串口管理工具，就可以轻松地对串口进行监听，并进行 I/O 操作。任务清单如表 4-3 所示。

表 4-3　任务清单

任务课时	4 课时	任务组员数量	建议 3 人
任务组采用设备	1 个 ZigBee 光照传感器 1 个 ZigBee 协调器 1 台 PC 相关电源，导线，工具		

【拓扑图】

本任务拓扑图如图 4-15 所示。

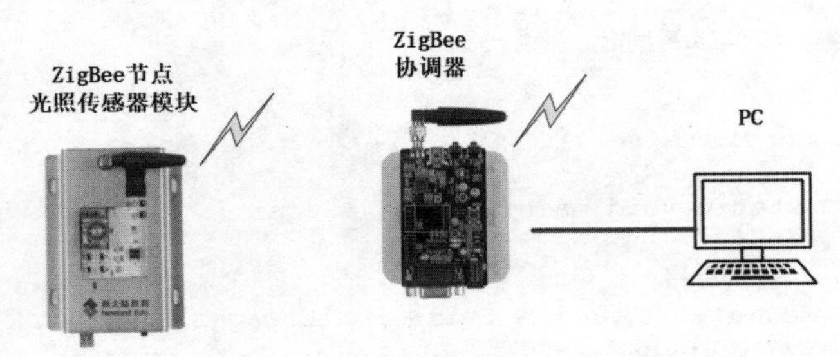

图 4-15　拓扑图

备注：本任务 COM 口根据实际情况确定。

【知识解析】

1. Java 常见代码块

代码块指的是在类中使用花括号{}包围的代码。在 Java 中常见的代码块有：局部代码块、构造代码块、静态代码块、同步代码块等。这里只介绍构造代码块和静态代码块。

(1) 构造代码块(也称对象块)，定义在类中，使用花括号{}包围，例如：

```
1.  public void Test{
2.      {
3.          System.out.println("构造代码块");
4.      }
5.  }
```

构造代码块的作用和构造方法类似，可用于对象的初始化。一个类中可以定义多个构造代码块，构造代码块的执行顺序按定义的顺序，构造代码块在构造方法前执行。对象都具有的功能可以放在构造代码块中，在对象创建时，就会实现该功能，从而减少代码的冗余度，提高代码的复用性。

(2) 静态代码块，在定义时使用 static 关键字加上花括号{}包围，例如：

```
1.  public void Test{
2.      static {
3.          System.out.println("静态代码块");
4.      }
5.  }
```

当类加载的时候，静态代码块会被执行，由于类只加载一次，因此静态代码块也只执行一次。所以若代码必须在项目启动的时候执行的话，需要使用静态代码块。

2. Java 垃圾回收机制

Java 提供一种垃圾回收机制，用来管理内存中没有被引用的对象。有了这种机制，开发人员将不用手动管理内存，Java 虚拟机会调用垃圾回收器自动清理垃圾对象，从而使程序获得更多可用的内存。开发人员也可以通过 System.gc()方法通知 Java 虚拟机立即进行垃圾回收。当对象被回收释放内存空间时，它的 finalize()方法会被自动调用，所以可以在 finalize()方法中定义对象释放时的代码。

接下来通过一个例子演示 Java 虚拟机回收垃圾对象的过程，具体代码如下所示。

```
1.  public class Test10finalize {
2.      public void finalize() {
3.          System.out.println(this+"对象被回收了！");
4.      }
5.  
6.      public static void main(String[] args) {
7.          //创建对象
8.          Test10finalize obj1 = new Test10finalize();
9.          Test10finalize obj2 = new Test10finalize();
10.         System.out.println("对象 obj1="+obj1);
11.         System.out.println("对象 obj2="+obj2);
12.         //将变量引用赋 null，使对象成为垃圾对象
13.         obj1 = null;
14.         obj2 = null;
15.         //调用方法手动进行垃圾回收
16.         System.gc();
17.     }
18. }
```

运行结果如图 4-16 所示。

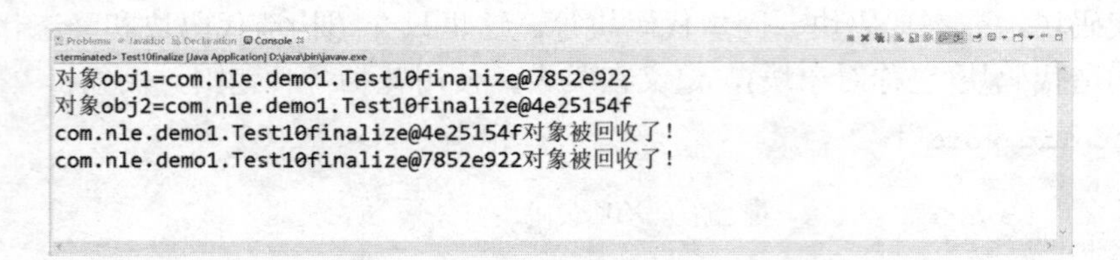

图 4-16 运行结果

从运行结果可以看出，对象被回收时，对象的 finalize()方法被调用了。

3. 包与访问权限

(1) 包的定义。

由于 Java 编译器为每个类生成一个字节码文件，且文件名与类名相同，因此同名的类有可能发生冲突。为了解决这一问题，Java 提供包来管理类名空间。包语句的语法格式如下。

```
1.    package pkg1[.pkg2[.pkg3…]];
```

例如一个 Person.java 文件，它的内容如下。

```
1.    package com.nle.demo2;
2.    public class Person{
3.        …
4.    }
```

这个文件保存在路径 com/nle/demo2 下。package 的作用就是声明该类所在的包路径。需要注意的是，package 声明包路径语句必须放在程序的第一行(除注释外)。包的名称有多个层级时，使用符号“.”隔开。包名建议使用公司的域名来命名，并且全部使用小写英文。在 Eclipse 的 Java 项目下，可以在 src 目录下右击，选择 new→Package 命令，此时会弹出一个对话框，在 Name 文本框中写入相应的包名，单击 Finish 按钮即可创建包，如图 4-17 所示。

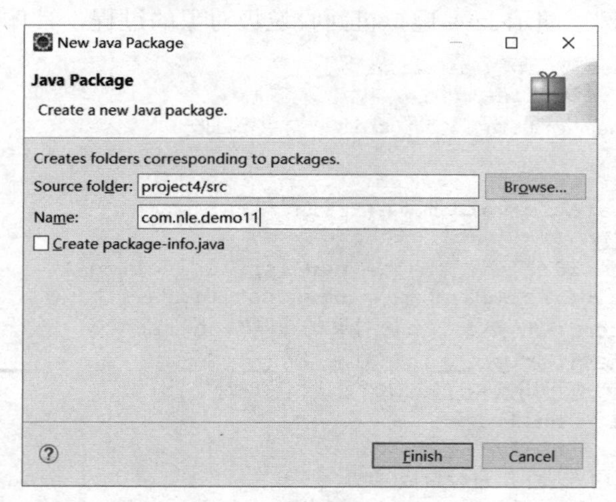

图 4-17 创建包

在之前的例子中，创建 Java 类时都没有填入具体的包名，Eclipse 会把创建的 Java 文件放在默认包下[default packge]，在今后的例子中，创建的 Java 文件都会指定具体的包名。

(2) import 关键字。

在 Java 中，为了能够在一个类中使用其他包中的类，Java 使用 import 语句来完成此功能。import 语句应位于 package 语句之后，在类定义之前，可以没有，也可以有多条。它的语法格式如下。

```
1.  import pkg1[.pkg2[.pkg3…].( classname|*);
```

接下来通过一个例子来学习 import 的用法，具体代码如下。

```
1.  import java.util.Scanner;
2.
3.  public class Test11 {
4.      public static void main(String[] args) {
5.          Scanner sc = new Scanner(System.in);
6.      }
7.  }
```

上面的代码在程序入口方法 main()中，使用 JDK 中 API 提供的 Scanner 类创建了一个对象。由于 Scanner 类在包路径 java.util 下，所以要想在 Test11 类中使用 Scanner 需要使用 import 声明该类的路径。当然也可以使用 import 声明 java.util.*，即 java.util 路径下的所有类，但是一般情况下不使用*声明，而是用到哪个类声明哪个类的路径。如果不想使用 import 声明类路径，可以使用类全名的形式声明类，即类名前面加上包名。例如，没有使用 import 关键字声明 Scanner 的包路径情况下，创建对象时可写成“java.util.Scanner sc = new java.util.Scanner(System.in);”。使用类全名来声明类会使得程序代码变长，不美观，所以一般情况下还是使用 import 来声明类路径。

在 Java 中，使用同包名下的类时，不需要声明类的路径。使用 java.lang 包下的类时也不需要声明类的路径。因为程序默认装载当前包下的类，并且默认装载 java.lang 包的类，如上例代码中的 System 类就在 java.lang 包下，不需要声明类路径。

(3) 访问权限。

在 Java 中，提供四种访问权限用来修饰类、成员变量以及成员方法。通过访问权限可以限制访问的范围。四种访问权限的访问范围如表 4-4 所示。

表 4-4　四种访问权限

访问权限修饰符	同一个类中	同包不同类	不同包的子类中	不同包非子类中
private	√			
默认不写	√	√		
protected	√	√	√	
public	√	√	√	√

从表 4-4 中可以看出，Java 中提供四种访问权限，但只有三个访问权限修饰符，还有一个是没有访问权限修饰符的情况。表中的单元格打钩说明在该范围可以访问。

- private：只在当前类中可访问，其他类不能够访问。
- 默认不写：在同一个类中、同包不同类中都可访问。
- protected：在同一个类中、同包不同类中、不同包的子类中可访问。
- public：所有的类都可访问。

4. 类的封装

将数据连同函数捆绑在一起，形成新的数据类型，这被称为封装。在 Java 语言中，对象就是一组变量和相关方法的封装，其中变量声明了对象的状态，方法表明了对象具有的行为。通过对象的封装，实现了模块化与信息隐藏。通过对类的成员施以一定的访问权限，实现了类中成员的隐藏，避免了对象的滥用。

一般类在封装时，类中的属性私有化，即使用 private 关键字修饰。这时私有属性只能在它所在的类中被访问，如果外界想要访问私有属性，需要提供一些公有方法，其中包含获取属性值的 getXxx 方法和设置属性值的 setXxx 方法。有一个比较特殊的是，如果属性是布尔类型成员变量，getXxx 方法将改为 isXxx 方法。

5. 单例模式

单例模式是一种创建型模式，目的是保证一个类仅有一个实例，并提供一个访问它的全局访问点。优点是一个单例模式的类在内存中只有一个实例，减少了内存的开销，尤其是频繁地创建和销毁实例。缺点是不能继承，违反了单一职责原则。常用的单例模式实现有饿汉式和懒汉式两种。

(1) 饿汉式单例模式。

```
1.   /**
2.    *饿汉式单例模式(线程安全)
3.    **/
4.   public class Singleton{
5.      private static Singleton instance = new Singleton();
6.      //私有构造方法
7.      private Singleton(){}
8.      //静态方法为调用者提供单例对象
9.      public static Singleton getInstance(){
10.        return instance;
11.     }
12.     public void test() {
13.        System.out.println("饿汉式单例");
14.     }
15.
16.     public static void main(String[] args) {
17.        Singleton.getInstance().test();
18.     }
19.  }
```

运行结果如图 4-18 所示。

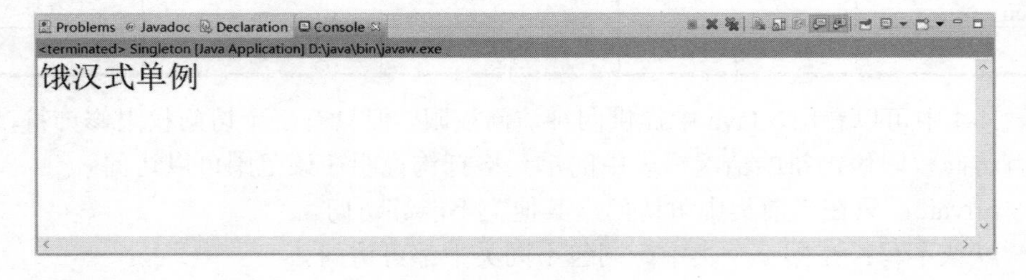

图 4-18 运行结果

(2) 懒汉式单例模式。

```
1.   /**
2.    *懒汉式单例模式
3.    **/
4.  public class Singleton2 {
5.      // 私有构造
6.      private Singleton2() {
7.      }
8.
9.      private static Singleton2 single = null;
10.
11.     public static Singleton2 getInstance() {
12.         if (single == null) {
13.             single = new Singleton2();
14.         }
15.         return single;
16.     }
17.
18.     public void test() {
19.         System.out.println("懒汉式单例");
20.     }
21.
22.     public static void main(String[] args) {
23.         Singleton2.getInstance().test();
24.     }
25. }
```

运行结果如图 4-19 所示。

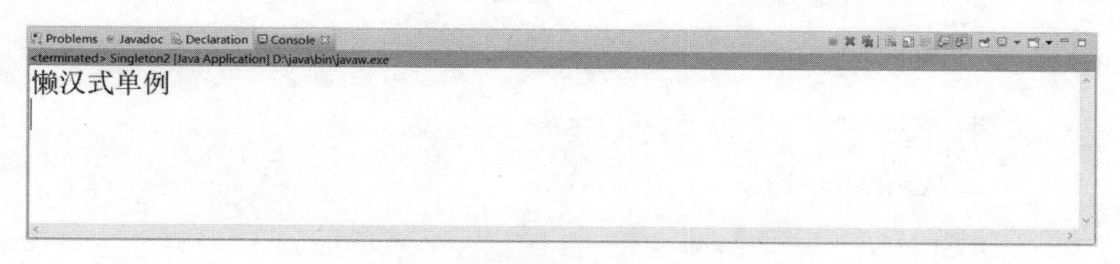

图 4-19　运行结果

6. 枚举

Enum 枚举类型是一种特殊的数据类型，能够为一个变量定义一组预定义的常量。变量必须等于为其预定义的值之一，一般用来表示一组相同类型的常量，如性别、日期、月份、颜色等。对这些属性值，用常量的好处是显而易见的，不仅可以保证单例，且在比较的时候可以用“==”来替换 equals。

(1) 枚举的基本使用。

```
1.  public enum Season {
2.
3.      SPRING,SUMMER,FALL,WINTER;
4.
5.      public static void main(String[] args) {
6.          Season season = Season.SUMMER;
7.          switch (season) {
8.          case SPRING:
9.              System.out.println("春天");
```

```
10.                 break;
11.             case SUMMER:
12.                 System.out.println("夏天");
13.                 break;
14.             case FALL:
15.                 System.out.println("秋天");
16.                 break;
17.             case WINTER:
18.                 System.out.println("冬天");
19.                 break;
20.         }
21.
22.     }
23. }
```

(2) 自定义枚举属性和方法。

```
1.  /**
2.   * 模拟量传感器枚举类型
3.   */
4.  public enum SENSOR_TYPE {
5.      /**
6.       * 光照传感器
7.       */
8.      LIGHT("光照", 0, 20000, true),
9.      /**
10.      * 温度(温湿度传感器)
11.      */
12.     TEMPERATURE("温度", -10, 60, true),
13.     /**
14.      * 湿度(温湿度传感器)
15.      */
16.     HUMIDITY("湿度", 50, 100, true),
17.     /**
18.      * 风速传感器
19.      */
20.     WIND("", 0, 70, true),
21.     /**
22.      * 大气压传感器
23.      */
24.     AIRPRESSURE("大气压", 0, 110, false),
25.     /**
26.      * 二氧化碳传感器
27.      */
28.     CO2("二氧化碳", 0, 5000, true),
29.     /**
30.      * PM2.5(PM2.5 传感器)
31.      */
32.     PM2POINT5("PM2.5", 0, 300, true),
33.     /**
34.      * 氧气传感器
35.      */
36.     O2("O2", 0, 100, true),
37.     /**
38.      * 空气质量(空气质量传感器)
39.      */
40.     AIRQUALITY("空气质量", -1, -1, false);
```

```
41.
42.     private String name;
43.     private int maxRange;
44.     private int minRange;
45.     private boolean decimalFlag;//是否格式化
46.
47.     // 最小量程，最大量程，小数点格式
48.     private SENSOR_TYPE(String name, int minRange, int maxRange, boolean
          decimalFlag) {
49.         this.name = name;
50.         this.minRange = minRange;
51.         this.maxRange = maxRange;
52.         this.decimalFlag = decimalFlag;
53.     }
54.
55.     public int getMaxRange() {
56.         return maxRange;
57.     }
58.
59.     public int getMinRange() {
60.         return minRange;
61.     }
62.
63.     @Override
64.     public String toString() {
65.         return name;
66.     }
67.
68.     public static void main(String[] args) {
69.         SENSOR_TYPE type = SENSOR_TYPE.TEMPERATURE;
70.         System.out.println(type);
71.         System.out.println(type.getMinRange());
72.         System.out.println(type.getMaxRange());
73.     }
74. }
```

运行结果如图 4-20 所示。

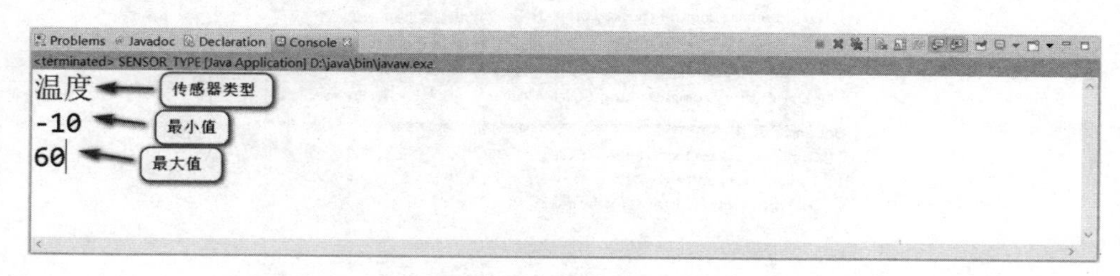

图 4-20 运行结果

7. 导出 jar 依赖包

对于封装好的工具类，若需要在其他工程上实现复用，可以将编写好的 Java 文件导出为 jar 包，然后导入其他工程即可调用。

(1) 在需要导出的文件上右击，选择 Export 命令，如图 4-21 所示。

(2) 在弹出的对话框中选择 Java 下的 JAR file 类型，如图 4-22 所示。

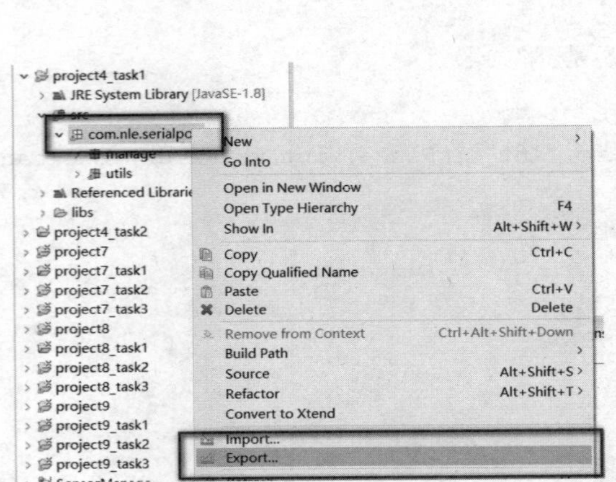

图 4-21　导出依赖包

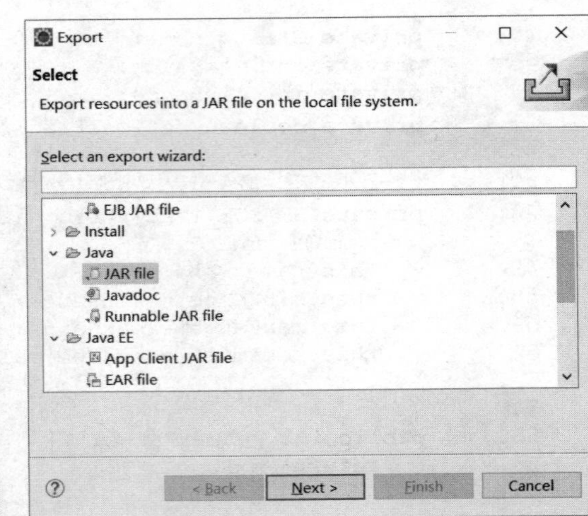

图 4-22　选择导出类型

(3) 设置需要导出的文件与导出文件的存放路径，单击 Finish 按钮，如图 4-23 所示。

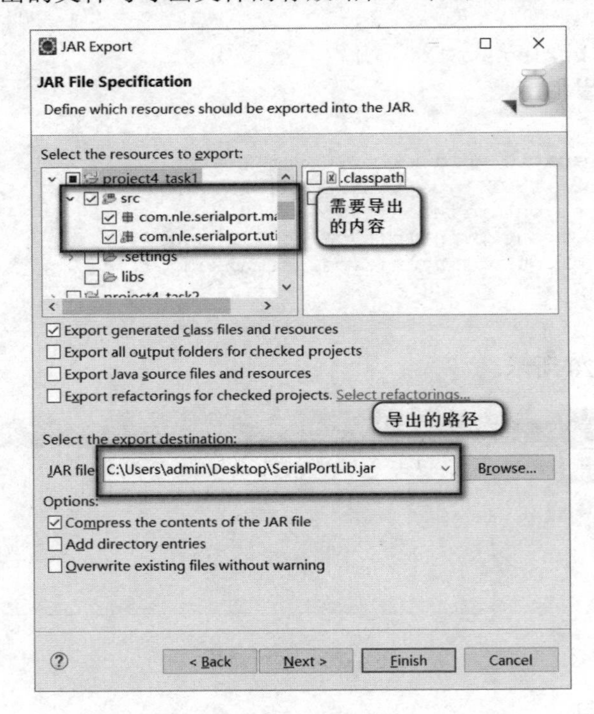

图 4-23　设置导出路径

【任务实施】

1. 任务分析

本任务要求获取 ZigBee 采集的光照传感器的真实数据。

(1) 添加 SerialPortLib.jar、RXTX.jar 两个依赖包。

(2) 通过串口管理工具打开 ZigBee 协调器所连接的串口。

(3) 通过串口管理工具为串口添加监听。

(4) 分析传感器数据。

2. 任务实施

(1) 新建 Java 工程 project4_task2，并在 src 下创建 com.nle.task2 包。

(2) 在根目录下创建 libs 文件夹，将 RXTXcomm.jar、SerialPortLib.jar 包放入文件夹中，并构建路径，如图 4-24 所示。

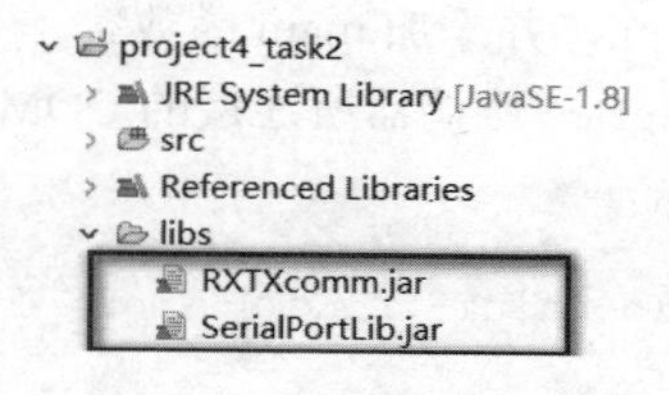

图 4-24 导入依赖包

(3) 创建 ZigBee 串口监听器类 ZigBeeListener 实现 implements SerialPortEventListener 接口，接口相关内容将在后续课程中详细说明。此处读者只需将数据分析代码写在如图的 serialEvent 方法的 switch 代码块中即可。

```
1.   /**
2.    * ZigBee 串口监听类
3.    * @author admin
4.    */
5.   public class ZigBeeListener implements SerialPortEventListener{
6.
7.      private SerialPort port;
8.
9.      public ZigBeeListener(SerialPort port) {
10.        this.port = port;
11.     }
12.
13.     @Override
14.     public void serialEvent(SerialPortEvent arg0) {
15.        try {
16.           switch(arg0.getEventType())
17.           {
18.           case SerialPortEvent.DATA_AVAILABLE:
19.              if(port!=null) {
20.                    //读取串口数据
21.                 byte[] res = SerialPortManager.readFromPort(port);
22.                    //根据解析协议判断发送的数据为光照传感器数据
23.                 if(res[17]==33) {
24.                    String lightStr = ByteUtils.byteToHex(res[19])+
                        ByteUtils.byteToHex(res[18]);
25.                    double reslight = Integer.parseInt(lightStr,
                        16)/100.00;
26.                    double val = Math.pow(10, ((1.78 - Math.log10(33 /
                        reslight - 10)) / 0.6));
27.                    System.out.println("光照:"+new DecimalFormat
                        ("#.00").format(val));
28.                 }
```

```
29.                 }
30.
31.                 break;
32.             }
33.         } catch (Exception e) {
34.             e.printStackTrace();
35.         }
36.
37.     }
38.
39. }
```

(4) 创建程序主入口 MainApp，并添加 main 函数。

(5) 通过串口工具类打开 ZigBee 协调器所连接的 COM 口，并添加监听。

```
1.  public class MainApp {
2.
3.      public static void main(String[] args) {
4.          try {
5.              //打开 ZigBee 协调器连接的串口
6.              SerialPort port = SerialPortManager.openPort("COM6", 38400);
7.              //通过串口管理工具为串口添加监听
8.              SerialPortManager.addListener(port, new ZigBeeListener(port));
9.
10.         } catch (Exception e) {
11.             e.printStackTrace();
12.         }
13.
14.     }
15.
16. }
```

(6) 运行时可将手电筒对准光照传感器查看数值变化。

3. 运行结果

运行程序，结果如图 4-25 所示。

图 4-25 运行结果

思考与练习

1. 简述类与对象的区别与联系。
2. 简述单例设计模式的实现步骤。
3. 简述如何利用面向对象的思想实现石头剪刀布小游戏。

项目5

采集传感数据的API的构建

【项目描述】

模块化的设计可以提高代码的复用性，提升开发人员的工作效率，所以将一些常用的代码进行封装打包是非常必要的。在物联网开发过程中，数字量采集器和ZigBee模块的数据通信是最常用的功能，串口通信所抛出的异常也多种多样，因此，可以将这些代码与异常进行封装，方便在后期开发中直接调用。本项目将演示如何对数字量采集器、ZigBee四输入模块、自定义异常等模块的API进行封装和构建。项目任务列表如图5-1所示。

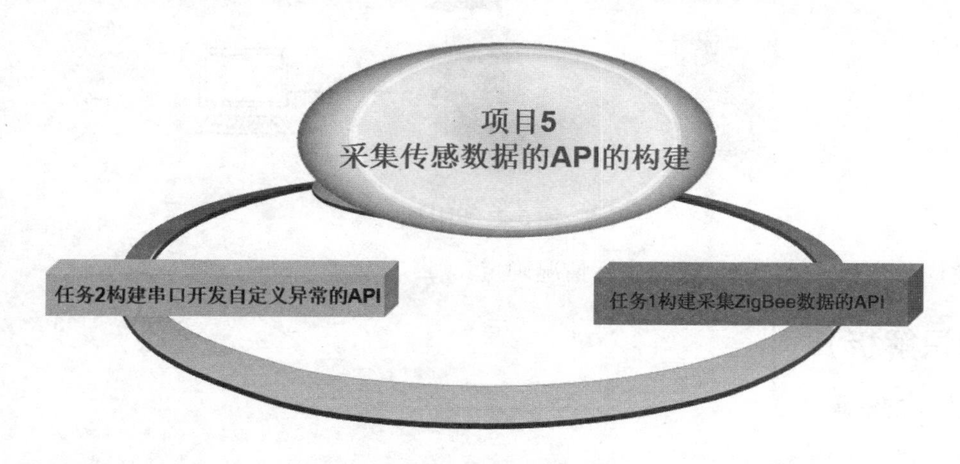

图5-1　项目5任务列表

【学习目标】

知识目标：理解继承的概念与好处；理解什么是多态；理解接口的定义与使用；理解异常的概念。

技能目标：能利用继承简化项目代码；能通过继承与接口封装自己的 API；能定义和使用自定义异常。

任务 1　构建采集 ZigBee 数据的 API

扫码观看视频讲解

【任务描述】

不同的传感器，不论是有线传感器还是无线传感器都有其自身的特点。例如，ZigBee 的传感器获取的数据格式都是类似的，可以将数据的解析过程，模拟量的转换过程进行整合。本任务将通过 Java 继承、接口等机制对 ZigBee 数据采集的代码进行模块化封装。构建采集 ZigBee 数据的 API。任务清单如表 5-1 所示。

表 5-1　任务清单

任务课时	4 课时	任务组员数量	建议 3 人
任务组采用设备	1 个 ZigBee 光照传感器 1 个 ZigBee 协调器 1 台 PC 相关电源，导线，工具		

【拓扑图】

本任务拓扑图如图 5-2 所示。

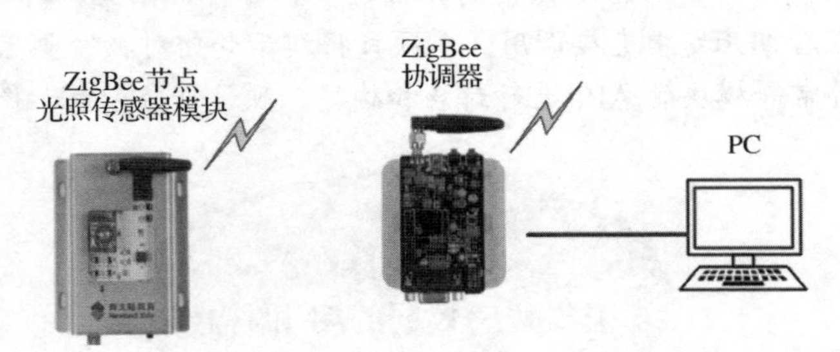

图 5-2　拓扑图

备注：本任务 COM 口根据实际情况确定。

【知识解析】

1. 类的继承

(1)　继承的概念。

继承是面向对象程序设计的一个基本特征，通过继承可以实现代码的复用。通过继承得到的类称为子类，被继承的类称为父类(超类)，父类包括所有直接或者间接被继承的类。子类可继承父类中的变量及方法，也可定义其自身特有的变量及方法。Java 语言不支持多重继承，只支持单一继承。

在 Java 中，可以通过在类的声明中加入 extends 子句来创建一个类的子类。具体格式如下所示。

```
1.  class SubClass extends SuperClass{
2.      …
3.  }
```

上面格式中 extends 关键字的后面是要继承的父类，如果缺省 extends 子句，则该类默认为 java.lang.Object 的子类，Object 类是所有类的父类。

接下来通过一个例子来学习如何使用继承，具体代码如下所示。

父类 Animal 的代码如下所示。

```
1.  public class Animal {
2.      public String name = "动物";//定义动物的名称
3.
4.      public Animal() {
5.      }
6.
7.      public Animal(String name) {
8.          this.name = name;
9.      }
10.
11.     //定义动物的叫法
12.     public void bark() {
13.         System.out.println("动物发出叫声...");
14.     }
15. }
```

子类 Dog 的代码如下所示。

```
1.  public class Dog extends Animal {
2.      // Dog 类中新增的成员变量
3.      int age = 2;
4.      String name = "旺财";
5.
6.      // Dog 类中新增的方法
7.      public void lookDoor() {
8.          System.out.println("狗狗还能看家！");
9.      }
10.
11.     public void testFun() {
12.         System.out.println("this.name=" + this.name);
13.         System.out.println("super.name=" + super.name);
14.     }
15.
16.     public static void main(String[] args) {
17.         Dog dog = new Dog();
18.         dog.bark();// bark 方法来自继承的父类 Animal
19.         dog.testFun();
20.     }
21. }
```

运行结果如图 5-3 所示。

图 5-3　运行结果

从结果中不难看出，子类 Dog 继承了父类定义的 name 属性和 bark()方法，同时子类还新增了 age 属性和 lookDoor()方法。

(2) 方法的重写。

方法重写指的是在继承关系中，子类定义了与父类同名的方法，且该方法的参数列表与返回值类型都与父类一致，子类重写该方法的内容。接下来修改上一个例子 Dog 类中的代码，在 Dog 类中重写 bark()方法，具体代码如下所示。

```
1.  public class Dog2 extends Animal {
2.      // Dog 类中新增的成员变量
3.      int age;
4.  
5.      // Dog 类中新增的方法
6.      public void lookDoor() {
7.          System.out.println("狗狗还能看家！");
8.      }
9.  
10.     // 重写父类的 bark 方法
11.     public void bark() {
12.         System.out.println("汪汪汪~~~");
13.     }
14. 
15.     public static void main(String[] args) {
16.         Dog2 dog = new Dog2();
17.         dog.name = "旺财";      // name 变量来自继承的父类 Animal
18.         dog.age = 2;            // age 变量来自 Dog 子类自身
19.         System.out.println(dog.name + "今年" + dog.age + "岁了！");
20.         dog.bark();             // bark 方法来自继承的父类 Animal
21.         dog.lookDoor();         // lookDoor 方法来自 Dog 子类自身
22.     }
23. }
```

运行结果如图 5-4 所示。

```
Problems Javadoc Declaration Console
<terminated> Dog2 [Java Application] D:\java\bin\javaw.exe
旺财今年2岁了！
汪汪汪~~~
狗狗还能看家！
```

图 5-4　运行结果

从运行结果中可以看出，Dog 类对象调用 bark()方法时，调用的是子类中重写的方法，而不是父类中定义的方法。

需要注意的是，子类重写父类的方法时，重写的方法不能比父类被重写的方法有更严格的访问权限、不能比被重写的方法抛出更多的异常。关于异常的知识会在后面相关内容中做详细介绍。

(3) super 关键字。

当子类重写父类的方法后，子类对象将无法访问父类被重写的方法。为了解决这个问题，Java 提供一个 super 关键字用于访问父类的成员。super 的用法和 this 类似，可以使用 super([参数])调用父类的构造方法(super 不写的情况，子类构造方法默认调用父类无参数的构造方法)，同样可以使用 super 加上符号“.”调用实例方法和实例变量。接下来通过一个例子学习 super 关键字的这三种用法，具体代码如下所示。

```
1.  public class Dog3 extends Animal {
2.      // Dog类中新增的成员变量
3.      int age = 2;
4.      String name = "旺财";
5.
6.      public Dog3() {
7.          //默认调用父类无参构造方法
8.          super();
9.      }
10.
11.     // Dog类中新增的方法
12.     void lookDoor() {
13.         System.out.println("狗狗还能看家！");
14.     }
15.
16.     // 重写父类的bark方法
17.     public void bark() {
18.         super.bark();// 调用父类定义的bark方法
19.         System.out.println("汪汪汪~~~");
20.         // 调用父类的变量
21.         System.out.println("super.name=" + super.name);
22.     }
23.
24.     public static void main(String[] args) {
25.         Dog3 dog = new Dog3();
26.         dog.bark();// bark方法来自继承的父类Animal
27.     }
28. }
```

运行结果如图 5-5 所示。

```
Problems  Javadoc  Declaration  Console
<terminated> Dog3 [Java Application] D:\java\bin\javaw.exe
Animal无参构造方法被调用了
动物发出叫声...
汪汪汪~~~
super.name=动物
```

图 5-5　运行结果

(4) final 关键字。

final 关键字可用于修饰类、变量和方法，它表示最终状态的意思。final 修饰类、变量和方法有如下几个特性。

- final 修饰的类不能被继承。
- final 修饰的方法不能被重写。
- final 修饰的变量是常量，只能赋值一次。

final 修饰类后，该类不能被继承，具体的代码如下所示。

```
1.   final class SuperClass {
2.
3.   }
4.
5.   public class TestFinal extends SuperClass{
6.
7.   }
```

编译错误结果如图 5-6 所示。

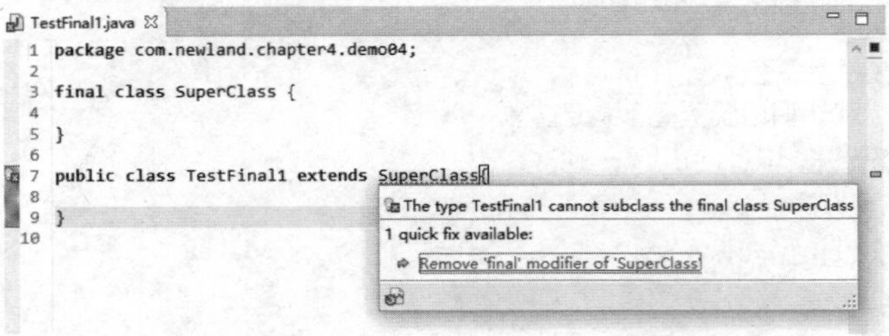

图 5-6　编译错误

从运行结果中可以看出，SuperClass 类为 final 修饰，TestFinal 是不能继承 final 修饰的 SuperClass 类的。

用 final 关键字修饰的方法不能被子类重写。接下来修改代码如下所示。

```
1.   class SuperClass2 {
2.      final void fun() {
3.      }
4.
5.   public class TestFinal extends SuperClass2 {
6.      void fun() {
7.      }
8.   }
```

编译错误结果如图 5-7 所示。

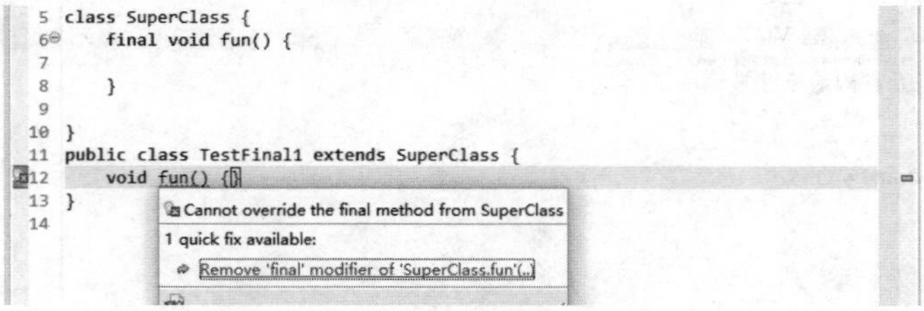

图 5-7　编译错误

从运行结果可以看出，final 修饰的方法不能被子类重写。

final 关键字用来修饰变量，变量在第一次赋值后就不能再改变值，相当于常量。

```
1.   public class TestFinal2 extends SuperClass {
2.   public static void main(String[] args) {
3.       final int num = 1;
4.       num = 2;
5.   }
```

编译错误结果如图 5-8 所示。

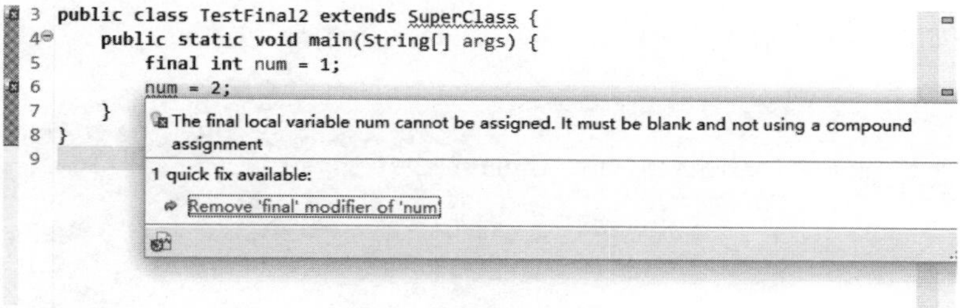

```
3 public class TestFinal2 extends SuperClass {
4     public static void main(String[] args) {
5         final int num = 1;
6         num = 2;
7     }
8 }
9
```

The final local variable num cannot be assigned. It must be blank and not using a compound assignment

1 quick fix available:

Remove 'final' modifier of 'num'

图 5-8　编译错误

从运行结果可以看出，final 修饰的局部变量在进行第二次赋值时，程序编译错误。同理，final 修饰的成员变量在第一次赋值后就不能再改变值了。

2. 抽象类和接口

(1) 抽象类。

在面向对象的概念中，所有的对象都是通过类来描绘的，但是反过来，并不是所有的类都是用来描绘对象的。如果一个类中没有包含足够的信息来描绘一个具体的对象，这样的类就是抽象类。例如，前面在定义 Animal 类时，bark()方法是用来表示动物的叫声的，不同的动物，叫声也不同，因此在 Animal 类中的 bark()方法无法准确表示具体动物的叫声。即 Animal 类没有包含足够的信息来描绘一个具体的对象，此时 Animal 类就可以定义成抽象类，而这个无法准确表示行为的方法则可以定义成抽象方法。

在 Java 中，定义抽象方法时不写方法体，并且抽象方法使用 abstract 关键字来修饰，具体示例如下。

```
1.       abstract void bark();
```

当一个类中包含了抽象方法时，该类必须定义成抽象类。抽象类也使用 abstract 关键字修饰。具体示例如下。

```
1.   abstract class Animal2{
2.       abstract void bark();
3.   }
```

需要注意的是，包含抽象方法的类必须定义成抽象类，但是抽象类中可以不包含抽象方法。另外，抽象类不能实例化，如果想调用抽象类中定义的方法，可以创建一个子类继承该抽象类，在子类中实现抽象类的所有抽象方法，即该子类包含足够的信息来描绘一个具体的对象，那么就可以通过子类的对象调用抽象类定义的方法。

接下来通过一个例子来学习抽象类与抽象方法的使用，具体代码如下所示。

```
1.  class Dog4 extends Animal2 {
2.      //实现抽象方法
3.      void bark() {
4.          System.out.println("汪汪汪～～～");
5.      }
6.      public static void main(String[] args) {
7.          Dog4 dog = new Dog4();
8.          dog.bark();
9.      }
10. }
```

运行结果如图 5-9 所示。

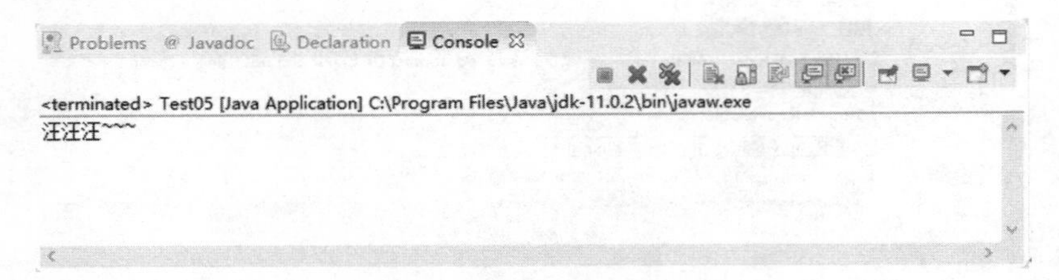

图 5-9　运行结果

(2) 接口。

在 Java 语言中，接口是常量和抽象方法的集合。接口并不是类，但接口与类很相似，只是它们的概念不同。类描述对象的属性和方法，接口则包含类必须实现的方法。

Java 中使用 interface 关键字来声明接口，其语法格式如下所示。

```
1.  [public] interface 接口名 [extends 接口 1,接口 2…]{
2.      [public] [static] [final] 数据类型 常量名 = 常量值;
3.      [public] [abstract] 返回值 抽象方法名(参数列表);
4.  }
```

在上面的语法中，一个接口可以有多个父接口，它们之间用逗号隔开。接口中的变量默认使用 public static final 来修饰，即全局常量。接口中定义的方法默认使用 public abstract 来修饰，即抽象方法。

接口不能实例化，但是可以被类实现。一个类要实现接口，必须实现接口中的所有方法，否则它就必须声明为抽象类。

接下来通过一个例子来学习接口的使用，具体代码如下所示。

```
1.  interface IFly {
2.      //定义飞翔方法
3.      void fly();
4.  }
5.  //鸟类实现飞翔接口
6.  class Bird implements IFly{
7.      //实现飞翔方法
8.      public void fly() {
9.          System.out.println("小鸟拍翅膀飞啊飞");
10.     }
11. }
12. //飞机类实现飞翔接口
13. class Airplane implements IFly{
```

```
14.     //实现飞翔方法
15.     public void fly() {
16.         System.out.println("飞机飞～～～～");
17.     }
18. }
19.
20. //测试类
21. public class TestInterface {
22.     public static void main(String[] args) {
23.         Bird bird = new Bird();
24.         bird.fly();
25.         Airplane airplane = new Airplane();
26.         airplane.fly();
27.     }
28. }
29.
```

运行结果如图5-10所示。

```
Problems  Javadoc  Declaration  Console
<terminated> TestInterface [Java Application] D:\java\bin\javaw.exe
小鸟拍翅膀飞啊飞
飞机飞~~~~
```

图5-10　运行结果

从上面的程序可以看出，类Bird与类Airplane分别实现了接口IFly，并实现了接口中的抽象方法。接口主要描述的是功能，如果一个类想要扩展功能，可以考虑实现接口的形式。当然一个类在继承另外一个类的同时也可以实现接口。继承关键字extends放在接口实现关键字implements之前。

(3) 内部类。

在Java语言定义类中，还有一种比较特殊的情况，即在一个类的内部再定义类，这样的类被称为内部类。内部类可以分为成员内部类、静态内部类、方法内部类。在这里仅针对成员内部类进行讲解。

在类体中定义的内部类，且该内部类没有static修饰，则该内部类为成员内部类。成员内部类可以访问外部类的所有成员。接下来通过一个例子来学习如何使用成员内部类，具体代码如下。

```
1.  public class Outer {
2.     private int count = 1;//定义类的成员变量
3.
4.     //该方法创建一个成员内部类对象，并调用对象的show方法
5.     void test() {
6.        Inner inner = new Inner();
7.        inner.show();
8.     }
9.
10.    class Inner {
11.       void show() {
```

```
12.             //在成员内部类中可以访问外部类的成员变量
13.             System.out.println("成员内部类访问外部类成员变量count=" + count);
14.         }
15.     }
16.
17.     public static void main(String[] args) {
18.         Outer outer = new Outer();
19.         outer.test();
20.     }
21. }
```

运行结果如图 5-11 所示。

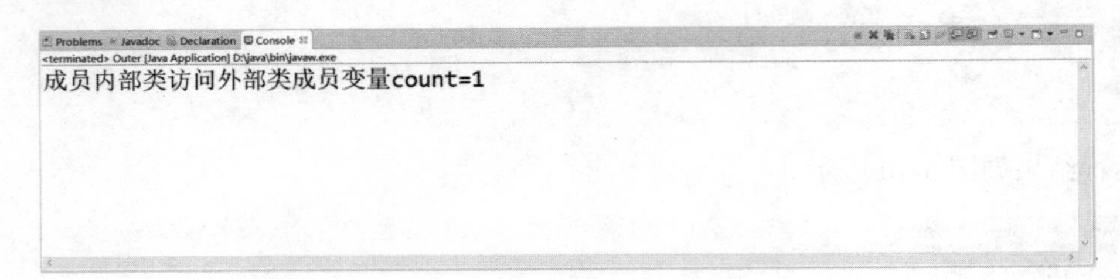

图 5-11　运行结果

上面的程序中，在外部类 Outer 内定义了一个成员内部类 Inner，并且 Inner 内部类可以访问外部类的成员变量。在外部类的成员方法中也可以通过内部类创建对象，如果想要在其他类中创建内部类对象，具体的语法格式如下。

```
1.   外部类名.内部类名 变量名 = new 外部类名().new 内部类名();
```

接下来通过一个例子来理解在其他类中如何创建内部类对象。

```
1.   public class TestInner {
2.       public static void main(String[] args) {
3.           Outer.Inner inner = new Outer().new Inner();
4.           inner.show();
5.       }
6.   }
```

如果不想让其他类访问成员内部类，可以对该成员内部类施加 private 权限。这样该成员内部类只能被它的外部类访问。

3. Lambda 表达式

Lambda 表达式，也可称为闭包，它是推动 Java 8 发布的最重要的新特性。Lambda 允许把函数作为一个方法的参数(函数作为参数传递进方法中)。使用 Lambda 表达式可以使代码变得更加简洁紧凑。

Lambda 表达式的语法格式如下所示。

```
(parameters)->expression
```

或

```
(parameters)->{statements;}
```

以下列出 Lambda 表达式的重要特性。

- 可选类型声明：不需要声明参数类型，编译器可以统一识别参数值。

- 可选的参数圆括号：一个参数无须定义圆括号，但多个参数需要定义圆括号。
- 可选的大括号：如果主体包含一个语句，就不需要使用大括号。
- 可选的返回关键字：如果主体只有一个表达式返回值，则编译器会自动返回值，大括号需要指明表达式返回了一个数值。

接下来列出 Lambda 表达式的一些简单的例子，如下所示。

```
1.  // 1. 不需要参数，返回值为 3
2.  () -> 3
3.
4.  // 2. 接收一个参数(数字类型)，返回其 3 倍的值
5.  x -> 3 * x
6.
7.  // 3. 接受 2 个参数(数字)，并返回它们的差值
8.  (x, y) -> x - y
9.
10. // 4. 接收 2 个 int 型整数，返回它们的和
11. (int x, int y) -> x + y
12.
13. // 5. 接收一个 string 对象，并在控制台打印，不返回任何值
14. (String s) -> System.out.print(s)
```

在 Java 8 以前的接口编程时，常常会用到匿名内部类，这会使代码变得冗长、不美观。而 Lambda 表达式主要用来定义行内执行的方法类型接口，可以替代以前匿名内部类的形式，使代码变得更简洁。

4. 多态

(1) 对象的类型转换。

在学习多态前，先学习对象的类型转换。对象能够进行类型转换的前提是：两种类型需要有继承关系。对象的类型转换分为“向上转型”和“向下转型”。对象的“向上转型”又称为“上溯”，即将子类的对象当作父类型使用。

上溯的对象不能通过父类变量去调用子类中特有的方法。接下来通过一个例子来演示对象的上溯情况，具体代码如下所示。

```
1.  //动物定义成抽象类
2.  abstract class Animal3 {
3.      // 叫声方法定义成抽象方法
4.      abstract void bark();
5.  }
6.  //狗继承动物
7.  class Dog5 extends Animal3 {
8.      // 实现动物 Animal 的抽象方法
9.      void bark() {
10.         System.out.println("汪汪~~");
11.     }
12.     void eat() {
13.         System.out.println("狗吃狗粮~~");
14.     }
15. }
16. public class TestPolymorphic {
17.
18.     public static void main(String[] args) {
```

```
19.         //上溯：将子类对象当作父类使用
20.         Animal3 animal = new Dog5();
21.         //可以调用父类中有定义的实例方法
22.         animal.bark();
23.         //不可以调用子类新增的方法
24.  //      animal.eat();
25.     }
26.
27.  }
```

上面代码中，子类 Dog 对象上溯为 Animal 类型时，对象可以访问父类中有定义的方法，但是不能访问子类新增的方法，如果想要重写访问中子类新增的方法，就要用到对象的“向下转型”，也称“下溯”。

对象的下溯需要进行类型的强制转换。接下来通过一个例子来学习对象的下溯，具体代码如下。

```
1.   //动物定义成抽象类
2.   abstract class Animal3 {
3.    // 叫声方法定义成抽象方法
4.    abstract void bark();
5.   }
6.   //狗继承动物
7.   class Dog6 extends Animal3 {
8.    // 实现动物 Animal 的抽象方法
9.    void bark() {
10.    System.out.println("汪汪~~");
11.   }
12.   void eat() {
13.    System.out.println("狗吃狗粮~~");
14.   }
15.  }
16.  public class TestPolymorphic2 {
17.   public static void main(String[] args) {
18.    Animal3 animal = new Dog6();
19.    Dog6 dog = (Dog6) animal;// 将 animal 对象下溯为 Dog 类型
20.    dog.bark();
21.    dog.eat();
22.   }
23.
```

运行结果如图 5-12 所示。

图 5-12　运行结果

从运行结果可以看出，下溯需要进行强制类型转换，对象成功下溯后，就可以访问子类新增的成员方法了。需要注意的是，必须是该子类的上溯对象，才能成功地下溯成该子类对象。下面通过一个例子演示下溯错误的情况，具体代码如下所示。

```
1.   public class TestPolymorphic3 {
2.    public static void main(String[] args) {
3.     Animal3 animal = new Dog6();
4.     Cat cat = (Cat) animal;// 将animal对象下溯为Cat类型
5.     cat.bark();
6.    }
7.   }
```

运行结果如图 5-13 所示。

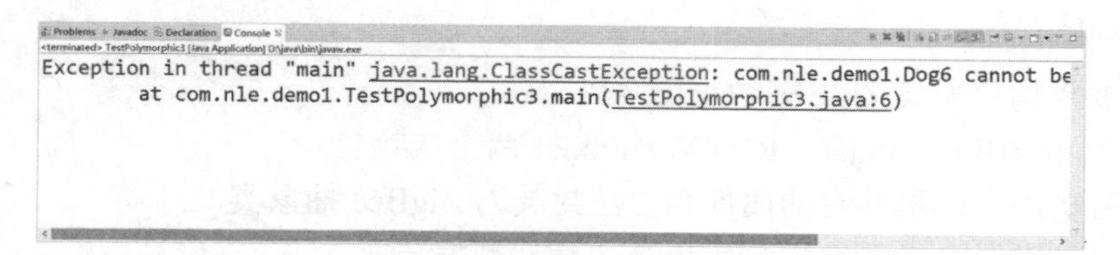
Exception in thread "main" java.lang.ClassCastException: com.nle.demo1.Dog6 cannot be
at com.nle.demo1.TestPolymorphic3.main(TestPolymorphic3.java:6)

图 5-13　运行结果

从运行结果可以看出，程序运行时出现了一个 java.lang.ClassCastException(类转换异常)。原因是 animal 对象由 Dog 类型对象上溯而来，此时下溯为 Cat 类型的对象会出现问题。为了避免下溯过程出现问题。Java 语言提供了 instanceof 关键字用于判断一个对象是否是某个类型的。语法格式如下。

```
1.   对象 instanceof 类(或接口)
```

修改上面例子的代码，如下所示。

```
1.   public static void main(String[] args) {
2.       Animal animal = new Dog();
3.       if (animal instanceof Cat) {
4.           Cat cat = (Cat) animal;// 将animal对象下溯为Cat类型
5.           cat.bark();
6.       }else {
7.           System.out.println("animal对象不是Cat类型的");
8.       }
9.   }
```

运行结果如图 5-14 所示。

animal对象不是Cat类型的

图 5-14　运行结果

从运行结果可以看出，使用 instanceof 判断 animal 对象是否是 Cat 类型的，如果是就转换成 Cat 类型，并调用 bark 方法，如果不是则打印结果“animal 对象不是 Cat 类型的”。因为 animal 是 Dog 类型，不是 Cat 类型对象，所以得到结果“animal 对象不是 Cat 类型的”。

(2)　多态的概念。

当一个类有很多子类，并且这些子类都重写了父类中的某个实例方法时，那么当子类创建的对象引用放到一个父类的对象时(即上转型对象)，如果这个上转型对象调用这个方法时就具有多种形态，因为子类在重写父类方法时可能产生不同行为。

多态是指父类的某个方法被子类重写时，可以产生自己的功能行为。多态性中父类一般抽象成一个接口，由多个类实现接口中的某个方法。

【任务实施】

1. 任务分析

本任务要求构建采集 ZigBee 数据的 API。

(1) 添加 SerialPortLib.jar、RXTXcomm.jar 两个依赖包。

(2) 将各类型传感器共有的属性和方法封装为 ZigBee 抽象类。

(3) 定义各类型传感器子类继承 ZigBee，重写数据转换方法。

(4) 添加数据操作接口向调用者提供获取数据实际值的方法。

2. 任务实施

(1) 新建 Java 工程 project5_task1，并在 src 下创建 com.nle.device 包。

(2) 在根目录下创建 libs 文件夹，将 RXTXcomm.jar、SerialPortLib.jar 包放入文件夹中，并添加到构建路径，如图 5-15 所示。

```
project5_task1
  > JRE System Library [JavaSE-1.8]
  v src
    > com.nle
  > Referenced Libraries
  v libs
      RXTXcomm.jar
      SerialPortLib.jar
```

图 5-15 创建工程

(3) 添加 com.nle.device.base 包，在目录下创建 ZigBee 抽象类，作为无线设备的父类。

(4) 定义传感器类型字符串常量，本任务只定义了三个，读者可根据实际设备自行添加。

```
1.   /** 光照传感器 */
2.   public static final String LIGHT = "21";
3.   /** CO 传感器(空气质量) */
4.   public static final String CO = "22";
5.   /** 温湿度传感器 */
6.   public static final String TEMP_HUMI = "01";
```

(5) 定义通用属性：传感器类型、真实数据、串口对象、数据字符串。

```
1.   /** ZigBee 类型  */
2.   protected String type;
3.   /**真实数据*/
4.   protected double value;
5.   /**串口对象*/
6.   protected SerialPort port;
7.   /**从串口获取的数据字符串*/
8.   protected String dataVal;
```

(6) 定义 ZigBee 串口数据解析方法 parseData()。

```
1.   /**
2.    * 解析从串口接收到的字节数据
3.    *
4.    * @param data 字节数据(只含类型、数据、校验码)
```

```
5.     */
6.    public void parseData(byte[] dataByte) {
7.       if(dataByte == null || dataByte.length == 0) return;
8.
9.       String dataStr =
      ByteUtils.byteArrayToHexString(dataByte).toUpperCase();
10.      if(!dataStr.contains("FE")) return;
11.
12.      String[] records = dataStr.split("FE");// ZigBee 协议数据头，固定为 0xfe
13.      try {
14.         for (int i = 0; i < records.length; i++) {
15.            if (records[i].length() == 0) { continue; }
16.
17.            String record = records[i];
18.            type = record.substring(32, 34);// 截取传感器的类型
19.            dataVal = record.substring(34); // 从数值开始截取
20.            System.out.println(dataVal);
21.         }
22.      } catch (Exception e) {
23.        e.printStackTrace();
24.      }
25.   }
```

(7) 定义模拟量数据转真实数据的抽象方法 convert()。由于不同的传感器解析数据的方法不同，所以将 convert 方法定义为抽象方法，由不同的传感器子类重写该方法来将模拟量转换为实际数值。

```
1.    /**
2.     * 将模拟量转换为实际数值
3.     */
4.    public abstract void convert();
```

(8) 新建数据控制接口 ValueController，添加 valueHandler 方法，供调用者实现数据访问。

```
1.    /**
2.     * 数据访问接口
3.     * @author admin
4.     *
5.     */
6.    public interface ValueController {
7.       /**访问数据*/
8.       public void valueHandler(String type,double value);
9.
10.   }
```

(9) 在 ZigBee 类中添加数据控制接口对应的方法 valuectrl()，当程序的某个功能需要使用该传感器数据时，通过实例化接口便可在重写的 valueHandler()方法中使用该传感器的 type、value 属性。

```
1.    /**数据控制接口*/
2.    public void valueCtrl(ValueController controller) {
3.       controller.valueHandler(type, value);
4.    }
```

(10) 新建一氧化碳传感器 ZigBeeCO 类，继承自 ZigBee 类，重写 convert 方法，添加根据一氧化碳传感器的 ZigBee 数据传输协议解析出实际数值的代码。

```
1.   /**
2.    * 一氧化碳传感器
3.    * @author admin
4.    *
5.    */
6.   public class ZigBeeCO extends ZigBee{
7.
8.      public ZigBeeCO(SerialPort port) {
9.         super(CO,port);
10.     }
11.
12.     @Override
13.     public void convert() {
14.        String hexVal = dataVal.substring(2,4) + dataVal.substring(0, 2);
15.        int value = Integer.parseInt(hexVal, 16);
16.        this.value = value / 100.0;
17.     }
18.
19.  }
```

(11) 新建光照传感器 ZigBeeLight 类，继承自 ZigBee 类，重写 convert 方法，添加根据光照传感器的 ZigBee 数据传输协议解析出实际数值的代码。

```
1.   /**
2.    * 光照传感器
3.    * @author admin
4.    *
5.    */
6.   public class ZigBeeLight extends ZigBee{
7.
8.      public ZigBeeLight(SerialPort port) {
9.         super(LIGHT, port);
10.     }
11.
12.     @Override
13.     public void convert() {
14.        String hexVal = dataVal.substring(2,4) + dataVal.substring(0, 2);
15.        int value = Integer.parseInt(hexVal, 16);
16.        double stemp = value / 100.0;
17.        double light = Math.pow(10, ((1.78 - Math.log10(33 / stemp - 10))
           / 0.6));
18.
19.       light = new BigDecimal(light).setScale(2, BigDecimal.ROUND_ HALF_UP).
           doubleValue();
20.        this.value = light;
21.
22.     }
23.
24.  }
```

(12) 新建温湿度传感器 ZigBeeTempHumi 类，继承自 ZigBee 类，重写 convert 方法，添加根据温湿度传感器的 ZigBee 数据传输协议解析出实际数值的代码。

```
1.   /**
2.    * 温湿度传感器
3.    * @author admin
4.    *
5.    */
6.   public class ZigBeeTempHumi extends ZigBee{
```

```
7.
8.      private double temperature;
9.      private double humidity;
10.
11.     public ZigBeeTempHumi(SerialPort port) {
12.         super(TEMP_HUMI, port);
13.     }
14.
15.     @Override
16.     public void convert() {
17.         //截取温度数据
18.         String hexVal = dataVal.substring(2,4) + dataVal.substring(0, 2);
19.         int value = Integer.parseInt(hexVal, 16);
20.         temperature = value / 10.0;
21.         //logger.info("temperature = " + temperature);
22.
23.         //截取湿度数据
24.         hexVal = dataVal.substring(6, 8) + dataVal.substring(4, 6);
25.         value = Integer.parseInt(hexVal, 16);
26.         humidity = value / 10.0;
27.
28.     }
29.
30.     public double getTemperature() {
31.         return temperature;
32.     }
33.
34.     public double getHumidity() {
35.         return humidity;
36.     }
37.
38. }
```

(13) 创建 com.nle.test 包，添加 Test 类，编写测试方法，读者可在 ZigBee 节点上更换不同的传感器模块对不同的类进行测试。

```
1.  public class Test {
2.      public static void main(String[] args) throws Exception {
3.          //打开串口
4.          SerialPort port = SerialPortManager.openPort("COM200", 38400);
5.          //初始化设备
6.          ZigBee zigbeeLight = new ZigBeeLight(port);
7.          //添加串口监听
8.          SerialPortManager.addListener(port, new SerialPortEventListener() {
9.
10.             @Override
11.             public void serialEvent(SerialPortEvent e) {
12.                 switch (e.getEventType()) {
13.                 case SerialPortEvent.DATA_AVAILABLE:
14.                     try {
15.                         //读取串口数据
16.                         byte[] res = SerialPortManager.readFromPort(port);
17.                         //解析串口数据
18.                         zigbeeLight.parseData(res);
19.                         //转换模拟量
20.                         zigbeeLight.convert();
21.                         //通过接口获取数据使用
22.                         zigbeeLight.valueCtrl(new ValueController() {
```

```
23.                         @Override
24.                         public void valueHandler(String type, double value) {
25.                             System.out.println("光照: "+value);
26.                         }
27.                     });
28.                 } catch (Exception e1) {
29.                     e1.printStackTrace();
30.                 }
31.                 break;
32.             }
33.         }
34.     });
35.   }
36. }
```

3. 运行结果

运行程序，结果如图 5-16 所示。

```
<terminated> MainApp (18) [Java Application] D:\java\bin\javaw.exe
光照:17251.60
光照:459.71
光照:860.69
光照:6704.28
光照:11191.43
光照:1312.40
光照:3209.33
```

图 5-16　运行结果

任务 2　构建串口开发自定义异常的 API

扫码观看视频讲解

【任务描述】

在物联网开发中，串口通信的过程要经过复杂的流程，因此导致串口打开失败的原因各式各样。程序中对串口管理工具进行封装后，异常直接在管理工具中进行捕获处理了，但是调用串口管理工具的地方并不知道程序异常的原因。为了既能对程序分层，又能在各层之间更加直观地体现导致程序出错的原因，可以利用 Java 的继承机制，自定义串口开发的异常，将错误信息进行描述后告知调用者，让程序结构更加完善。本任务要求读者自定义串口异常类，将不同的串口异常情况进行描述。任务清单如表 5-2 所示。

表 5-2　任务清单

任务课时	2 课时	任务组员数量	建议 1 人
任务组采用设备	无		

【知识解析】

1. 认识 Java 异常

在程序的运行过程中，可能会发生各种非正常的状况，例如数组索引超出边界、磁盘空间不足等。针对这种情况，Java 以异常类的形式对这些非正常状况进行封装。接下来通过一个例子认识一下什么是异常，具体代码如下。

```
1.  public class TestException {
2.      public static void main(String[] args) {
3.          int[] arr = new int[2];
4.          System.out.println(arr[3]);
5.      }
6.  }
```

运行结果如图 5-17 所示。

```
Exception in thread "main" java.lang.ArrayIndexOutOfBoundsException: 3
	at com.nle.demo2.TestException.main(TestException.java:5)
```

图 5-17　运行结果

从运行结果中可以看出，程序发生了一个 ArrayIndexOutOfBoundsException 异常，原因是程序第 4 行代码中的索引 3 超出了数组的边界。

接下来通过一张图展示异常 Throwable 类的继承关系，如图 5-18 所示。

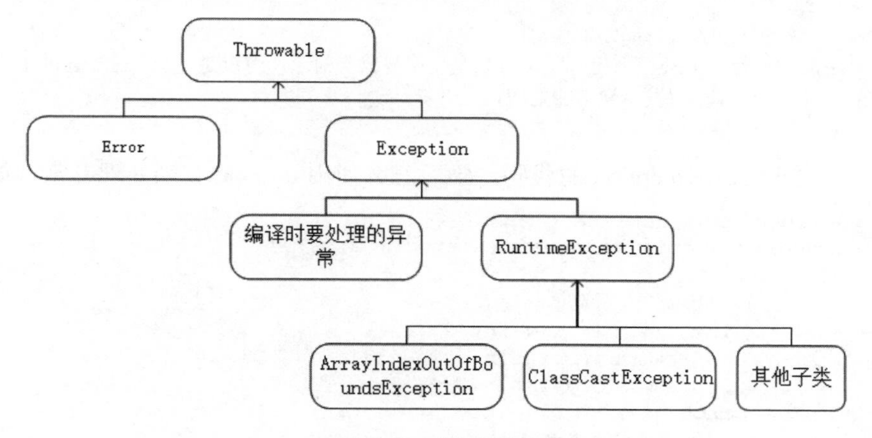

图 5-18　继承关系

从图 5-18 中可以看出，异常父类 Throwable 有两个子类 Error 和 Exception。其中，Error 是比较严重的错误，Exception 是程序中产生的异常。Exception 又分为编译期需要检查的异常与编译期不需要检查的异常(运行时异常)。

先模拟一下出现 Error 的情况，如下所示。

```
1.  public class TestError {
2.      // 认识Error
3.      // 递归算法：程序在执行过程中重复之前的情况
4.      static void print(int i) {
5.          System.out.println(i);
6.          i++;
7.          // StackOverflowError:当应用程序递归太深而发生堆栈溢出时，抛出该错误
8.          print(i);
9.      }
```

```
10.
11.     public static void main(String[] args) {
12.        print(1);
13.     }
14.  }
```

从上面的程序中可以看出，在入口方法 main()中调用 print()方法，而在 print()方法内部又调用了 print()方法，这样在方法内部调用自身方法，称为递归。在该程序中，没有结束递归的标记，print()方法被一直调用，且不会结束，这样程序将会因为内存溢出而出现 StackOverflowError 错误。程序中出现 Error 错误是比较严重的，这种错误无法使用 Java 的异常处理，只能找到出错的代码进行修改。

2. 处理 Java 异常

从图 5-18 中可以看出，Exception 分为两类：一种是编译时要处理的异常，这种异常在编译期必须处理，否则编译会出错；另一种是编译时不需要检查的异常 RuntimeException，这种异常代码在编译期不会检查。但不管 Exception 的哪种异常，都是在运行时出现异常。为了解决这种问题，Java 提供了一种处理异常的方式——捕获异常。捕获异常使用 try...catch 语句，具体语法如下所示。

```
1.   try {
2.         //代码块(可能发生异常)
3.   } catch (Exception e) {//参数是一个异常类对象，可以是 Exception 的子类对象
4.         //代码块(发生异常的处理)
5.   }
```

接下来，修改 TestException 的代码，学习怎么使用 try...catch 来处理异常，如下所示。

```
1.   public class TestException2 {
2.      public static void main(String[] args) {
3.         try {
4.            //可能产生异常的代码
5.            int[] arr = new int[2];
6.            System.out.println(arr[3]);
7.            System.out.println("---------");
8.         } catch (Exception e) {
9.            //处理异常的代码
10.           System.out.println("程序出现异常了！");
11.        }
12.     }
13.  }
```

程序运行结果如图 5-19 所示。

图 5-19 运行结果

在上面的代码中，在 try 语句块中使用数组对象 arr 访问索引为 3 的元素，程序会发生数组索引越界异常。发生异常时，try 代码块剩余的代码将不会执行，程序跳入 catch 语句块执行。所以在控制台上可以看到 catch 代码段中执行的打印效果。

当 try 语句块中可能发生多种异常时，可以使用多个 catch 代码块处理，接下来，修改 TestException2 的代码，学习多个 catch 代码块处理异常的情况，如下所示。

```
1.   public class TestException3 {
2.
3.       public static void main(String[] args) {
4.           try {
5.               //可能产生异常的代码
6.               int[] arr = new int[2];
7.               System.out.println(arr[1]);
8.
9.               Object o = null;
10.              System.out.println(o.toString());
11.              System.out.println("---------");
12.          } catch (ArrayIndexOutOfBoundsException e) {
13.              //处理异常的代码
14.              System.out.println("程序发生数组索引越界异常了！异常信息为："+
                     e.getMessage());
15.          }catch (NullPointerException e) {
16.              //处理异常的代码
17.              System.out.println("程序发生空指针异常了！");
18.              e.printStackTrace();
19.          }catch (Exception e) {
20.              System.out.println("程序发生异常了！");
21.          }
22.      }
23.  }
```

程序运行结果如图 5-20 所示。

```
Problems  Javadoc  Declaration  Console
<terminated> TestException3 [Java Application] D:\java\bin\javaw.exe
0
程序发生空指针异常了！
java.lang.NullPointerException
	at com.nle.demo2.TestException3.main(TestException3.java:12)
```

图 5-20 运行结果

从运行结果可以看出，数组的元素 arr[1]可以正常打印，打印对象 o 的 toString()结果时出现异常了。这是因为 o 对象的引用为空，所以发生了空指针异常，这时程序将跳入 NullPointException 异常对象的 catch 代码块中处理。需要注意的是，如果 try...catch 结构中 catch 出现多次，当有继承关系时，子类对象修饰的 catch 代码块需要放在父类对象修饰的 catch 代码块前面；如果没有继承关系，则不同 catch 代码块的先后位置不做要求。

在程序中，有时候会希望有些语句无论程序是否发生异常都要执行，这时就可以在 try...catch 语句后面加上一个 finally 代码块。接下来通过一个例子来学习 finally 代码块的使用，如下所示。

```
1.  public class TestFinally {
2.
3.      public static void main(String[] args) {
4.          try {
5.              // 可能产生异常的代码
6.              int[] arr = new int[2];
7.              System.out.println(arr[1]);
8.              return;
9.          } catch (ArrayIndexOutOfBoundsException e) {
10.             // 处理异常的代码
11.             System.out.println("程序发生数组索引越界异常了！异常信息为：" +
                  e.getMessage());
12.         } finally {
13.             System.out.println("finally 代码块被执行了！");
14.         }
15.     }
16. }
```

程序运行结果如图 5-21 所示。

```
0
finally代码块被执行了！
```

图 5-21 运行结果

从运行结果可以看出，try 代码块正常结束的情况，finally 代码块被执行了，且不受 try 代码块中 return 的影响。同理，可以测试当 try 代码块发生异常时，finally 代码块最后也会被执行。所以在 try...catch...finally 结构中，finally 总是会最后执行。因此 finally 代码块一般用来关闭网络连接、关闭流等系统资源的释放。

除了 try...catch 结构可以用来捕获异常外，还有另一种处理异常的方式——throws 关键字声明(抛出)异常。有时在写一个方法时，这个方法有可能发生异常，但是在该方法中又不想直接捕获处理异常，这时就可以在方法声明后面使用 throws 声明异常，具体异常的处理交给该方法的调用者完成。

throws 关键字声明(抛出)异常的语法格式如下。

```
1.  修饰符 返回值类型 方法名(参数列表) throws 异常类列表{
2.  }
```

接下来通过一个例子来学习 throws 关键字声明(抛出)异常，代码如下所示。

```
1.  public class TestThrows {
2.      //该方法是两个参数的除法并返回结果，使用 throws 关键字声明(抛出)异常
3.      public static int divide(int a, int b) throws Exception{
4.          return a/b;
5.      }
6.
7.      public static void main(String[] args) {
8.          try {
9.              int result = divide(1, 0);
```

```
10.             System.out.println(result);
11.         } catch (Exception e) {
12.             e.printStackTrace();
13.         }
14.     }
15. }
```

程序运行结果如图 5-22 所示。

```
java.lang.ArithmeticException: / by zero
	at com.nle.demo2.TestThrows.divide(TestThrows.java:6)
	at com.nle.demo2.TestThrows.main(TestThrows.java:11)
```

图 5-22 运行结果

从上面的代码可以看出，写了一个 divide()方法用来计算两个参数的除法结果，在方法的后面使用了 throws 声明异常，那么在调用 divide()方法的 main()方法中就需要处理异常，这时可以使用 try...catch 进行捕获处理，也可以在 main()方法后面继续声明异常，最后会由 Java 虚拟机把异常信息打印输出。

3. 自定义异常类

JDK 中定义了大量的异常类用来描述编程时可能出现的异常情况，但是还是可以自定义异常类用来描述编程过程中出现的业务异常。例如，可以定义一个判断工资是否满足要求的异常类，该异常继承 Exception 或其子类，如下所示。

```
1.  public class SalaryException extends Exception{
2.      //Java 程序员可以按照自己的业务逻辑来处理异常_自定义异常类能提供更复杂更详细的方法
3.      //这样能极大地提高软件的健壮性
4.      //自定义异常：1、写类继承自 Exception 或者 Throwable
5.      //2、业务代码中引发异常.....
6.      public SalaryException(String msg) {
7.          super(msg);
8.      }
9.  }
```

那么这个自定义的异常类怎么使用呢？这时就需要用到 throw 关键字抛出该异常类的对象，其语法格式如下。

```
1.  throw 异常类 异常对象
```

接下来通过一个例子，学习自定义异常类的使用，如下所示。

```
1.  public class TestThrow {
2.      static void testSal (int sal) throws SalaryException{
3.          if(sal<1500){
4.              throw new SalaryException("工资不满足要求! ...");
5.          }
6.          System.out.println("工资满足要求...");
7.      }
8.      public static void main(String[] args) {
9.          //测试自定义异常类
```

```
10.         try {
11.             testSal (1000);
12.         } catch (SalaryException e) {
13.             e.printStackTrace();
14.         }
15.     }
16. }
```

程序运行结果如图 5-23 所示。

```
Problems  Javadoc  Declaration  Console
<terminated> TestThrow [Java Application] D:\java\bin\javaw.exe
com.nle.demo2.SalaryException: 工资不满足要求！...
	at com.nle.demo2.TestThrow.testSal(TestThrow.java:6)
	at com.nle.demo2.TestThrow.main(TestThrow.java:13)
```

图 5-23　运行结果

上面的代码中，在 testSal()方法中判断，当传入的工资小于 1500 时，抛出一个自定义的异常对象 new SalaryException()。测试工资值 1000，小于 1500，所以可以在控制台中打印出异常信息。

4. Java 中的类加载和反射技术

(1) 类加载和反射技术的概念。

一个对象在运行时有两种类型，一种是编译时类型，另一种是运行时类型。如果在编译和运行时都知道类型的具体信息，则可以将一个对象转换为运行时的类型。但是当编译时无法预知对象的类型到底是属于哪些类时，那么程序就只有依靠运行时的信息来发现对象和类的真实信息了，这时就必须要用到反射技术。

一个类如果可以被执行，那么对于 JVM 来说，它的执行流程为：类的加载、连接、初始化。通过这种方式，才可以获取一个类的类对象，即 java.lang.Class 对象，并在此基础上获取该类的成员变量、方法和构造器等。

那么什么是类加载呢？类的加载就是将类的 class 文件读入内存，并为之创建一个 java.lang.Class 对象，也就是说当程序使用任何类时，系统都会为之建立一个 java.lang.Class 对象。接下来列出 java.lang.Class 的特点，如下所示。

- Class 是一个类，一个描述类的类(也就是描述类本身)，封装了描述方法 Method，描述字段 Field，描述构造器 Constructor 等属性。
- 对象反射后可以得到的信息：某个类的数据成员名、方法和构造器、某个类到底实现了哪些接口。
- 对于每个类而言，JRE 都为其保留一个不变的 Class 类型的对象。Class 对象包含某个特定类的有关信息。
- Class 对象只能由系统建立。
- 一个类在 JVM 中只会有一个 Class 实例。

接着看一下类的连接，类的连接主要负责把类的二进制数据合并到 JRE 中，可以分为三个步骤：第一，验证，检验被加载的类是否有正确的内部结构；第二，准备，负责为类的类变量分配内存，并设置默认初始值；第三，解析，将类的二进制数据中的符号引用替换成直接引用。

最后完成类的初始化，类的初始化主要对类变量进行初始化。

(2) 获取类的 Class 对象。

使用 Java 的类反射技术，首先要获取一个类的 Class 对象，主要有以下三种方式。

- 使用 Class 类的 forName(String clazzName)静态方法。
- 调用某个类的 class 属性获取该类对应的 Class 对象。
- 调用某个类的 getClass()方法。

接下来，通过一个例子来学习如何获取 Class 对象。首先创建一个测试实体类，如下所示。

```
1.  public class Person {
2.
3.      public String name;
4.      private int age;
5.
6.      public Person() {
7.      }
8.
9.      protected Person(String name, int age) {
10.         this.name = name;
11.         this.age = age;
12.     }
13.
14.     public String getName() {
15.         return name;
16.     }
17.
18.     public void setName(String name) {
19.         this.name = name;
20.     }
21.
22.     public int getAge() {
23.         return age;
24.     }
25.
26.     public void setAge(int age) {
27.         this.age = age;
28.     }
29.
30.     private void privateMethod() {
31.         System.out.println("这是一个私有方法");
32.     }
33.
34. }
```

创建 GetClassTest 类，测试获取 Class 对象的方式，代码如下所示。

```
1.  public class GetClassTest {
2.      public static void main(String[] args) {
3.          //Class 是一个描述类的类，对每个类而言只有一个不变的 Class 与其对应
4.          Class cls = null;
5.
6.          //1.使用 Class 类的 forName(String clazzName)静态方法
7.          try {
8.              cls = Class.forName("com.newland.chapter4.demo18.Person");
```

```
9.          System.out.println("使用Class类的forName()方法获取Class对象："+cls);
10.       } catch (ClassNotFoundException e) {
11.          e.printStackTrace();
12.       }
13.
14.       //2.调用某个类的class属性获取该类对应的Class对象
15.       cls = Person.class;
16.       System.out.println("直接通过类名获取Class对象："+cls);
17.
18.       //3.调用某个类的getClass()方法
19.       Person person = new Person();
20.       cls = person.getClass();
21.       System.out.println("调用某个类的getClass()方法获取Class对象："+cls);
22.    }
23. }
```

(3) 获取类的成员。

获取类的 Class 对象后，就可以通过 Class 对象获取类的构造方法、成员方法与成员变量。接下来，通过一个例子学习如何通过 Class 对象获取类的成员，代码如下所示。

```
1.  public class GetMemberTest {
2.
3.     // 构造方法
4.     public static void constructorTest(Class cls) throws
         NoSuchMethodException {
5.        System.out.println("获取指定的public构造方法");
6.        System.out.println(cls.getConstructor());
7.        System.out.println("获取全部的public构造方法");
8.        Constructor[] constructors1 = cls.getConstructors();
9.        for (Constructor constructor : constructors1) {
10.          System.out.println(constructor);
11.       }
12.       System.out.println("获取指定的构造方法，不受访问修饰符的限制");
13.       System.out.println(cls.getDeclaredConstructor(String.class,
          int.class));
14.       System.out.println("获取全部的构造方法，不受访问修饰符的限制");
15.       Constructor[] constructors2 = cls.getDeclaredConstructors();
16.       for (Constructor constructor : constructors2) {
17.          System.out.println(constructor);
18.       }
19.    }
20.
21.    // 成员方法
22.    public static void mothodTest(Class cls) throws NoSuchMethodException {
23.       System.out.println("获取指定的public方法");
24.       System.out.println(cls.getMethod("getName"));
25.       System.out.println("获取全部的public方法");
26.       Method[] methods1 = cls.getMethods();
27.       for (Method method : methods1) {
28.          System.out.println(method);
29.       }
30.       System.out.println("获取指定的方法，不受访问修饰符的限制");
31.       System.out.println(cls.getDeclaredMethod("getAge"));
32.       System.out.println("获取全部的方法，不受访问修饰符的限制");
33.       Method[] methods2 = cls.getDeclaredMethods();
```

```
34.          for (Method method : methods2) {
35.             System.out.println(method);
36.          }
37.       }
38.
39.       // 属性
40.       public static void filedTest(Class cls) throws NoSuchFieldException {
41.          System.out.println("获取指定的public属性");
42.          System.out.println(cls.getField("name"));
43.          System.out.println("获取全部的public属性");
44.          Field[] fields1 = cls.getFields();
45.          for (Field field : fields1) {
46.             System.out.println(field);
47.          }
48.          System.out.println("获取指定的属性，不受访问修饰符的限制");
49.          System.out.println(cls.getDeclaredField("name"));
50.          System.out.println("获取全部的属性，不受访问修饰符的限制");
51.          Field[] fields2 = cls.getDeclaredFields();
52.          for (Field field : fields2) {
53.             System.out.println(field);
54.          }
55.       }
56.
57.       public static void main(String[] args) throws NoSuchMethodException,
        NoSuchFieldException, IllegalAccessException,
58.             IllegalArgumentException, InvocationTargetException,
              InstantiationException {
59.          Class cls = Person.class;
60.
61.          constructorTest(cls);
62.          System.out.println("==========================================");
63.          mothodTest(cls);
64.          System.out.println("==========================================");
65.          filedTest(cls);
66.       }
67.    }
```

(4) 使用类反射技术执行指定方法。

通过上面的例子，已经可以用Class对象获取类的指定方法，接下来介绍如何获取类的实例，并且调用该实例方法，代码如下所示。

```
1.    public class Test {
2.       public static void main(String[] args) throws InstantiationException,
        IllegalAccessException, NoSuchMethodException,
3.             SecurityException, IllegalArgumentException,
              InvocationTargetException {
4.          Class cls = Person.class;
5.          // 通过Class对象获取类的实例
6.          Object obj = cls.newInstance();
7.          // 执行实例的setName方法
8.          Method setName = cls.getMethod("setName", String.class);
9.          // 调用invoke，执行setName方法，第一个参数为对象实例，后面的参数为指定方法
              的参数
10.         setName.invoke(obj, "小新");
11.
12.         // 获取实例的getName方法
13.         Method getName = cls.getMethod("getName");
```

```
14.         // 执行 getName 方法，得到 String 类型返回值
15.         String name = (String) getName.invoke(obj);
16.         System.out.println("通过类反射执行 getName 方法，name = " + name);
17.     }
18. }
```

运行结果如图 5-24 所示。

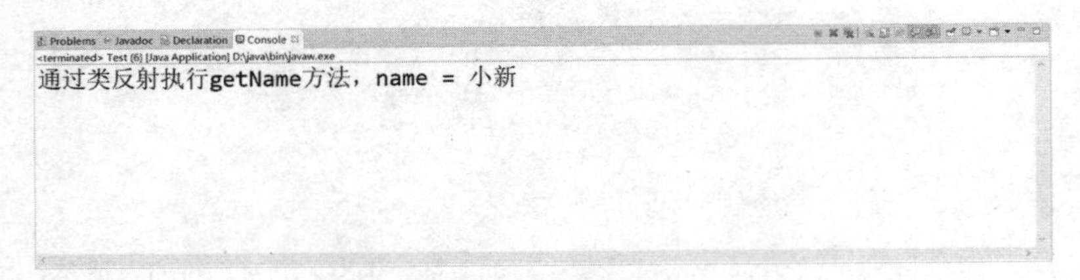

图 5-24　运行结果

在上面的代码中，通过 Class 对象的 getMethod()方法可以获得指定的公有方法，通过 Class 对象的 newInstance()可以获得 Class 对象对应类的实例，最后通过获取的 Method 对象调用 invoke()方法可以执行类对象的指定方法。

(5) 类加载和反射技术总结。

类的反射可以使代码更具灵活性，但是反射也会消耗更多的系统资源，所以如果不需要动态创建一个对象，那么就不需要用到反射。

类的反射技术的优点和缺点如下所示。

优点：

- 提高程序的灵活性和拓展性，能够在运行时动态地获取类的实例；
- 和 Java 的动态编译相结合，可以提供更强大的功能；
- 无须提前硬编码，便可以通过类名获取对应类的实例，进而操作该实例。

缺点：

- 性能较低，反射是一种接入式的操作，需要找到对应的字段和方法，比起直接的代码赋值，要慢得多；
- 使用反射应该在一个相对安全的环境下进行；
- 使用时会破坏类的封装性，破坏 OOP 的整体设计思想。

【任务实施】

1. 任务分析

本任务的目的是构建串口开发自定义异常的 API。

(1) 自定义异常类继承 Exception 或 RuntimeException。

(2) 为自定义异常类添加带有 String 类型参数的构造函数。

(3) 在需要处理异常的方法中捕获异常，并在不同的异常 catch 块中抛出自定义异常。

(4) 通过 throws 为抛出自定义异常的方法声明异常。

2. 任务实施

(1) 新建 Java 工程 project5_task2，并在 src 下创建 com.nle.task2 包。

(2) 在根目录下创建 libs 文件夹，将 RXTXcomm.jar 包放入文件夹中，并构建路径，如图 5-25 所示。

```
project5_task2
  JRE System Library [JavaSE-1.8]
  src
  Referenced Libraries
  libs
    RXTXcomm.jar
```

图 5-25 创建工程

(3) 创建 com.nle.serialport.exception 包，并创建 SerialPortException 类继承 Exception。然后为 SerialPortException 类添加带有 String 类型参数的构造函数。

```
1.  /**
2.   * 自定义串口异常类
3.   * @author admin
4.   *
5.   */
6.  public class SerialPortException extends Exception {
7.      private static final long serialVersionUID = 1876389493898520901L;
8.      /**
9.       * 带有字符串参数构造函数
10.      * @param message
11.      */
12.     public SerialPortException(String message) {
13.         super(message);
14.     }
15. }
```

(4) 添加 com.nle.serialport.manager 包，在该目录下添加串口管理工具类 SerialPortManager，添加打开串口的方法 openPort，参数为串口名称和波特率。

```
1.  public class SerialPortManager {
2.      /**
3.       * 打开串口
4.       *
5.       * @param portName
6.       *            端口名称,如:COM1
7.       * @param baudrate
8.       *            波特率,如:9600
9.       */
10.     public  SerialPort openPort(String portName, int baudrate){
11.
12.         return null;
13.     }
14.
15. }
```

(5) 完成打开串口的代码，并处理异常。根据不同的异常类型给表示异常信息的变量 errorMessage 赋值。

```
1.  public  SerialPort openPort(String portName, int baudrate){
2.      String errorMessage = null;
3.      try {
4.          // 通过端口名识别端口
```

```
5.          CommPortIdentifier portIdentifier = CommPortIdentifier.
              getPortIdentifier(portName);
6.
7.          // 打开端口，并给出端口名字和一个 timeout(打开操作的超时时间)
8.          CommPort commPort = portIdentifier.open(portName, 2000);
9.
10.         // 判断是否为串口
11.         if (commPort instanceof SerialPort) {
12.
13.             // 设置串口参数"波特率、数据位:8、停止位:1、校验位:NONE"
14.             SerialPort  serialPort = (SerialPort) commPort;
15.
16.             serialPort.setSerialPortParams(
17.                     baudrate,
18.                     SerialPort.DATABITS_8,
19.                     SerialPort.STOPBITS_1,
20.                     SerialPort.PARITY_NONE);
21.
22.             System.out.println(portName + "已打开");
23.             return  serialPort;
24.         }
25.         errorMessage = "端口指向设备不是串口类型！打开串口操作失败！";
26.     } catch (PortInUseException e) {
27.         errorMessage = "当前打开的串口已被使用";
28.     } catch (NoSuchPortException e) {
29.         errorMessage = "当前打开的串口不存在";
30.     } catch (UnsupportedCommOperationException e) {
31.         errorMessage = "不支持的操作";
32.     }
33. }
```

(6) 利用 errorMessage 实例化自定义异常，通过 throw 关键字在方法的最后抛出自定义异常，在方法参数后通过 throws 声明抛出的异常类型。

```
1.  public class SerialPortManager {
2.      /**
3.       * 打开串口
4.       *
5.       * @param portName
6.       *            端口名称,如:COM1
7.       * @param baudrate
8.       *            波特率,如:9600
9.       *
10.      * @throws SerialPortException
11.      *             打开串口失败
12.      */
13.     public  SerialPort openPort(String portName, int baudrate) throws
          SerialPortException{
14.       String errorMessage = null;
15.       try {
16.           // 通过端口名识别端口
17.           CommPortIdentifier portIdentifier = CommPortIdentifier.
                getPortIdentifier(portName);
18.
19.           // 打开端口，并给出端口名字和一个 timeout(打开操作的超时时间)
20.           CommPort commPort = portIdentifier.open(portName, 2000);
21.
```

```
22.             // 判断是否为串口
23.             if (commPort instanceof SerialPort) {
24.
25.                 // 设置串口参数"波特率、数据位:8、停止位:1、校验位:NONE"
26.                 SerialPort serialPort = (SerialPort) commPort;
27.
28.                 serialPort.setSerialPortParams(
29.                         baudrate,
30.                         SerialPort.DATABITS_8,
31.                         SerialPort.STOPBITS_1,
32.                         SerialPort.PARITY_NONE);
33.
34.                 System.out.println(portName + "已打开");
35.                 return serialPort;
36.             }
37.             errorMessage = "端口指向设备不是串口类型！打开串口操作失败！";
38.         } catch (PortInUseException e) {
39.             errorMessage = "当前打开的串口已被使用";
40.         } catch (NoSuchPortException e) {
41.             errorMessage = "当前打开的串口不存在";
42.         } catch (UnsupportedCommOperationException e) {
43.             errorMessage = "不支持的操作";
44.         }
45.         throw new SerialPortException(errorMessage);
46.     }
47.
48. }
```

(7) 创建测试类MainApp，测试打开串口的方法。

```
1.  /**
2.   * 测试类
3.   * @author admin
4.   *
5.   */
6.  public class MainApp {
7.      public static void main(String[] args) {
8.          SerialPortManager manager = new SerialPortManager();
9.          try {
10.             SerialPort port = manager.openPort("COM200", 9600);
11.         } catch (SerialPortException e) {
12.             e.printStackTrace();
13.         }
14.     }
15. }
```

3. 运行结果

程序运行结果如图5-26所示。

```
com.nle.serialport.exception.SerialPortException: 当前打开的串口不存在
        at com.nle.serialport.manage.SerialPortManager.ope[...]ager.
        at com.nle.serialport.test.MainApp.main(MainApp.ja[...])
```

当打开的串口不存在时，可以直观的看出问题

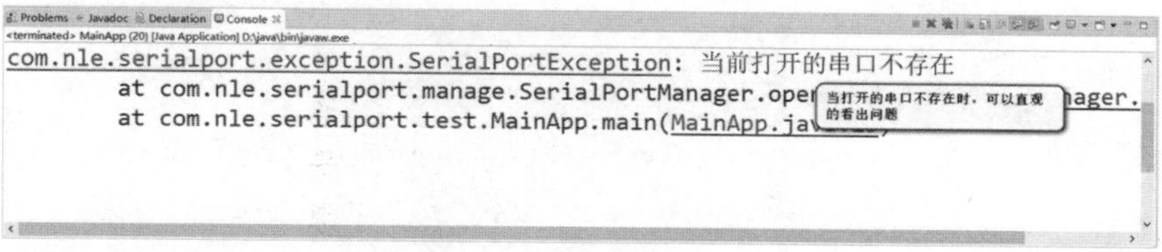

图5-26 运行结果

思考与练习

1. 简述什么是抽象。
2. 简述异常与错误的区别。
3. 思考如何通过反射机制修改对象的私有化属性。

项目 6 认识系统常用类

【项目描述】

Java API 为开发者提供了丰富的类进行常用的操作，在程序设计中，合理和充分利用类库提供的类和接口，不仅可以完成字符串处理、绘图、网络应用、数学计算等多方面的工作，而且可以大大提高编程效率，使程序简练、易懂。这不仅要求开发者熟练掌握 Java 的语法，还必须对系统常用的类所包含的功能有一定的了解。本项目将向读者演示如何利用字符串、数据类型包装类等系统常用类实现字符串操作、数据类型转换、字符串验证等功能。具体任务列表如图 6-1 所示。

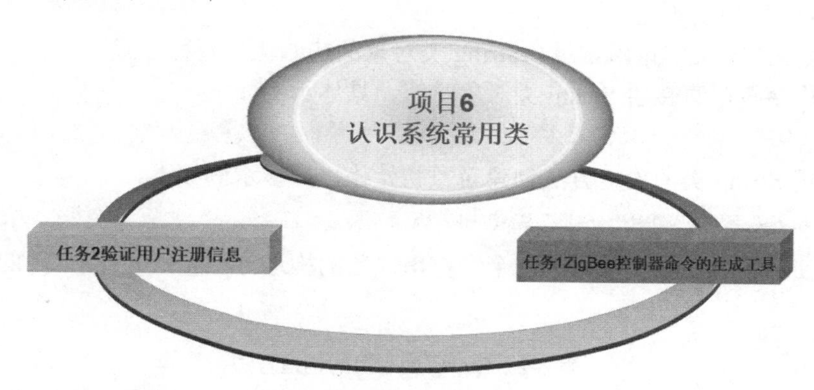

图 6-1　项目 6 任务列表

【学习目标】

知识目标：掌握 Java 常用 API；掌握字符串工具类；掌握 Math 常用方法；掌握日期工具类。

技能目标：能查阅 Java API 帮助文档；能使用常用工具类解决程序设计问题。

任务 1　ZigBee 控制器命令的生成工具

扫码观看视频讲解

【任务描述】

前面的任务中 4150 采集器的 DO 口指令和 ZigBee 指令的生成是利用数组来完成的。本任务将通过 String 类提供的切割、截取、替换、连接等方法对字符串进行操作，生成 ZigBee 继电器的操作指令，并将它封装成一个工具类。任务清单如表 6-1 所示。

表 6-1　任务清单

任务课时	4 课时	任务组员数量	建议 1 人
任务组采用设备	无		

【知识解析】

1. 字符串概述

在前面的知识中已经简单接触了 Java 的字符串。所谓字符串就是指一连串的字符，即由多个字符连接而成。字符串使用英文双引号("")表示，如"Hello Java"。在 Java 中定义了 String、StringBuffer 与 StringBuilder 三个类封装字符串，并提供了一系列操作字符串的方法，它们都在 java.lang 包下，所以在使用时不需要导入包。

2. String 类

(1) String 类的初始化。

String 类用于定义字符串常量。String 类对象的初始化有两种方式。

- 使用字符串直接给 String 类对象赋值，具体示例如下。

```
1.   String str = "abc";
```

- 使用 String 类的构造方法创建对象初始化，具体示例如下。

```
1.   String str = new String("abc");
```

该示例使用了 String 类提供的一个字符串参数的构造方法。String 类常用的构造方法如表 6-2 所示。

表 6-2　String 类常用的构造方法

构造方法声明	描　述
String()	创建一个字符串内容为空的 String 对象
String(String value)	创建一个指定字符串内容的 String 对象
String(char[] value)	创建一个指定字符数组的 String 对象

接下来通过一个例子来学习 String 类的使用，具体代码如下。

```
1.   public class Test01 {
2.       public static void main(String[] args) {
3.           // 使用""初始化 String 字符串对象
4.           String str1 = "abc";
5.           // 创建一个空字符串 String 对象
6.           String str2 = new String();
7.           // 创建一个内容为"abc"的 String 对象
8.           String str3 = new String("abc");
9.           // 创建一个字符串内容为数组{'a', 'b', 'c'}的 String 对象
10.          char[] arr = new char[] { 'a', 'b', 'c' };
11.          String str4 = new String(arr);
12.          System.out.println("str1=" + str1);
13.          System.out.println("str2=" + str2);
14.          System.out.println("str3=" + str3);
15.          System.out.println("str4=" + str4);
16.      }
17.  }
```

运行结果如图 6-2 所示。

```
str1=abc
str2=
str3=abc
str4=abc
```

图 6-2　运行结果

从运行结果中可以看出，第 4 行、第 8 行、第 11 行代码创建的 String 对象的字符串内容都为"abc"，但是 str1、str3、str4 指向的内存地址不一样。str1 指向的是“常量池”中的字符串"abc"，str3 与 str4 指向的是堆空间中的 String 对象。关于字符串的内存状态，这里不再详解。

(2)　String 类的常见操作。

String 类提供了丰富的方法用来操作字符串，String 类的常用方法如表 6-3 所示。

表 6-3　String 类常用的方法

方法声明	描　述
int indexOf(int ch)	返回指定字符在此字符串中第一次出现处的索引
int lastIndexOf(int ch)	返回指定字符在此字符串中最后一次出现处的索引
int indexOf(String str)	返回指定字符串在此字符串中第一次出现处的索引
int lastIndexOf(String str)	返回指定字符串在此字符串中最后一次出现处的索引
char charAt(int index)	返回字符串中 index 位置上的字符，其中，index 的取值范围是 0～(字符串长度-1)
boolean endWith(String suffix)	判断字符串是否以自定的字符串结尾
int length()	返回字符串的长度
boolean equals(Object anObject)	将字符串与指定的字符串比较

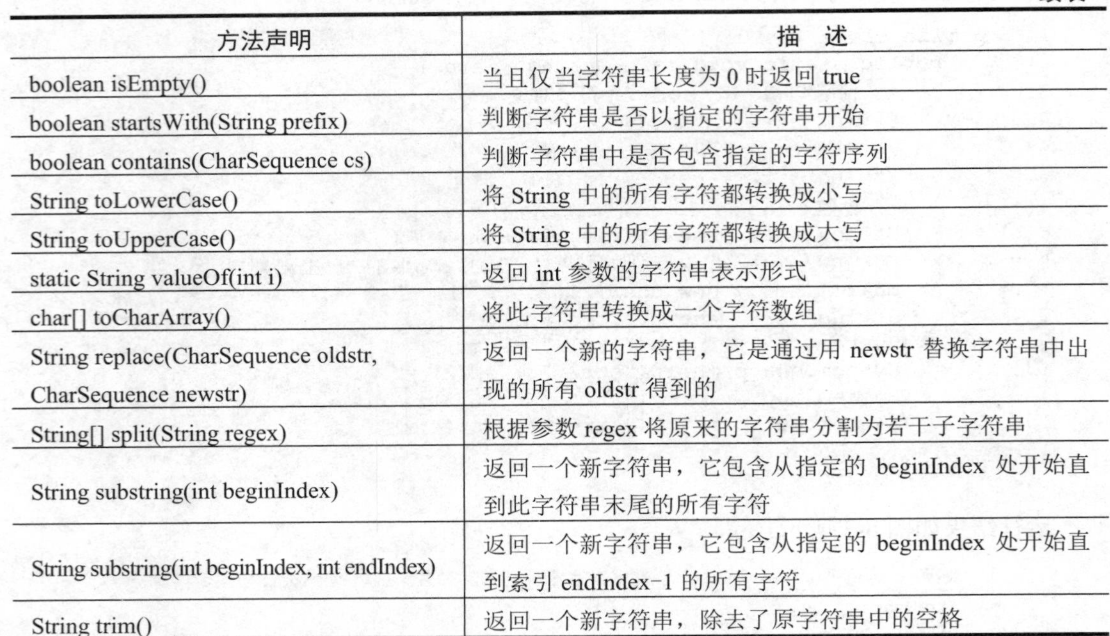

续表

方法声明	描　述
boolean isEmpty()	当且仅当字符串长度为 0 时返回 true
boolean startsWith(String prefix)	判断字符串是否以指定的字符串开始
boolean contains(CharSequence cs)	判断字符串中是否包含指定的字符序列
String toLowerCase()	将 String 中的所有字符都转换成小写
String toUpperCase()	将 String 中的所有字符都转换成大写
static String valueOf(int i)	返回 int 参数的字符串表示形式
char[] toCharArray()	将此字符串转换成一个字符数组
String replace(CharSequence oldstr, CharSequence newstr)	返回一个新的字符串，它是通过用 newstr 替换字符串中出现的所有 oldstr 得到的
String[] split(String regex)	根据参数 regex 将原来的字符串分割为若干子字符串
String substring(int beginIndex)	返回一个新字符串，它包含从指定的 beginIndex 处开始直到此字符串末尾的所有字符
String substring(int beginIndex, int endIndex)	返回一个新字符串，它包含从指定的 beginIndex 处开始直到索引 endIndex-1 的所有字符
String trim()	返回一个新字符串，除去了原字符串中的空格

接下来通过一个例子学习字符串常用方法的使用，具体代码如下所示。

```
1.   public class Test02 {
2.       public static void main(String[] args) {
3.           String s = "hello world, hello java";
4.           System.out.println("字符串的长度为;"+s.length());
5.           System.out.println("字符串中第一个字符："+s.charAt(0));
6.           System.out.println("字符 l 第一次出现的位置："+s.indexOf('l'));
7.           System.out.println("字符 l 最后一次出现的位置："+s.lastIndexOf('l'));
8.           System.out.println("子字符第一次出现的位置："+s.lastIndexOf("lo"));
9.           System.out.println("子字符最后一次出现的位置："+s.lastIndexOf("lo"));
10.          System.out.println("-------字符串的转换工作-----------");
11.          String str1 = "java";
12.          System.out.print("将字符串"+str1+"转换为字符数组后的结果：");
13.          char[] arr = str1.toCharArray();
14.          for (int i = 0; i < arr.length; i++) {
15.              //如果 i 是最后一个不要加逗号，其他需要加逗号
16.              if(i!=arr.length-1){
17.                  System.out.print(arr[i]+", ");
18.              }else {
19.                  System.out.println(arr[i]);
20.              }
21.          }
22.          System.out.println("将 int 值转换为 String 类型之后的结果："+
               String.valueOf(123));
23.          System.out.println("将字符串"+str1+"转换成大写之后的结果："+str1.
               toUpperCase());
24.          System.out.println("---------字符串替换操作------------");
25.          String str2 = "helloJava";
26.          System.out.println("将字符串 helloJava 替换成 nihaoJava 的结果："+
               str2.replace("hello", "nihao"));
```

```
27.         System.out.println("---------字符串去空格操作(只取出前后空格，字符串中间
              空格保留)------------");
28.         String str3 = "  hello java   ";
29.         System.out.println(str3+"去除空格后的结果："+str3.trim());
30.         System.out.println("---------字符串的判断操作------------");
31.         String s1 = "nihao";
32.         String s2 = "ni";
33.         System.out.println("判断是否以字符串 ni 开头："+s1.startsWith("ni"));
34.         System.out.println("判断是否以字符串 hao 结尾："+s1.endsWith("hao"));
35.         System.out.println("判断字符串是否包含 ih："+s1.contains("ih"));
36.         System.out.println("判断字符串是否为空："+s1.isEmpty());
37.         System.out.println("判断字符串是否相等："+s1.equals(s2));
38.         System.out.println("---------字符串的截取和分割------------");
39.         String ss = "192.168.1.1";
40.         //字符串的截取
41.         System.out.println("截取字符串第 2 个字符(包含)到第 4 个字符(不包含)的
              结果："+ss.substring(2, 4));
42.         System.out.println("截取字符串第 2 个字符到最后的结果："+ss.substring(2));
43.         //字符串的分割
44.         System.out.print("按点分割字符串后的元素为：");
45.         String[] strArr = ss.split("\\.");//字符.为特殊字符，如果要表示.需要
              进行转义\\.
46.         for (int i = 0; i < strArr.length; i++) {
47.             if(i!=strArr.length-1) {
48.                 System.out.print(strArr[i]+", ");
49.             }else {
50.                 System.out.print(strArr[i]);
51.             }
52.         }
53.
54.     }
55. }
```

运行结果如图 6-3 所示。

```
Problems  Javadoc  Declaration  Console
<terminated> Test02 [Java Application] D:\java\bin\javaw.exe
字符串的长度为;22
字符串中第一个字符：h
字符1第一次出现的位置：2
字符1最后一次出现的位置：15
子字符第一次出现的位置：15
子字符最后一次出现的位置：15
-------字符串的转换工作-----------
将字符串java转换为字符数组后的结果：j,a,v,a
将int值转换为String类型之后的结果：123
将字符串java转换成大写之后的结果：JAVA
---------字符串替换操作------------
将字符串helloJava替换成nihaoJava的结果：nihaoJava
---------字符串去空格操作（只取出前后空格，字符串中间空格保留）-------
  hello java   去除空格后的结果：hello java
---------字符串的判断操作------------
判断是否以字符串ni开头：true
判断是否以字符串hao结尾：true
判断字符串是否包含ih：true
判断字符串是否为空：false
判断字符串是否相等：false
---------字符串的截取和分割------------
截取字符串第2个字符（包含）到第4个字符（不包含）的结果：2.
截取字符串第2个字符到最后的结果：2.168.1.1
按点分割字符串后的元素为：192,168,1,1
```

图 6-3 运行结果

需要注意的是，equals 比较的是两个对象的“值”是否相等，而对象使用“==”比较的是内存地址。

3. StringBuffer 类与 StringBuilder 类

由于字符串是常量，一旦创建后，其内容和长度就固定了，因此要想对一个字符串进行修改，就只能创建新的字符串。在 JDK 中提供了 StringBuffer 类、StringBuilder 类(也称字符串缓存区)，这两个类可以方便地对字符串进行修改。StringBuffer 与 StringBuilder 的区别在于，StringBuffer 是线程安全的，StringBuilder 不是线程安全的。但在执行速率上，StringBuilder 比 StringBuffer 快。StringBuffer/StringBuilder 与 String 最大的区别就是，前者的内容和长度是可以改变的。

StringBuffer 类与 StringBuilder 类提供的方法类似，这里只介绍 StringBuffer 的常用方法，具体如表 6-4 所示。

表 6-4　StringBuffer 的常用方法

方法声明	描　述
StringBuffer append(char c)	添加参数到 StringBuffer 对象中
StringBuffer insert(int offset,String str)	在字符串中的 offset 位置插入字符串 str
StringBuffer deleteCharAt(int index)	移除此序列指定位置的字符
StringBuffer delete(int start,int end)	删除 StringBuffer 对象中指定范围的字符或者字符串序列
StringBuffer replace(int start,int end,String s)	在 StringBuffer 对象中替换指定的字符或字符串序列
void setCharAt(int index,char ch)	修改指定位置 index 处的字符序列
String toString()	返回 StringBuffer 缓冲区的字符串
StringBuffer reverse()	将字符序列用其反转形式取代

接下来通过一个例子学习 StringBuffer 的常用方法，具体代码如下。

```
1.   public class Test03StringBuffer {
2.      public static void main(String[] args) {
3.          StringBuffer sb = new StringBuffer("abc");
4.          System.out.println("字符串内容: " + sb);
5.          System.out.println("-----增加-----");
6.          // 在末尾添加一个字符串
7.          sb.append("def");
8.          System.out.println("append 添加后的结果: " + sb);
9.          // 在指定位置插入字符串
10.         sb.insert(3, "--");
11.         System.out.println("insert 插入后的结果: " + sb);
12.         System.out.println("-----删除-----");
13.         // 删除指定范围
14.         sb.delete(3, 5);
15.         System.out.println("删除指定范围的字符串结果: " + sb);
16.         // 删除指定位置
17.         sb.deleteCharAt(3);
18.         System.out.println("删除指定位置的字符结果: " + sb);
19.         System.out.println("-----修改-----");
20.         // 修改指定位置的字符
```

```
21.         sb.setCharAt(2, 'V');
22.         System.out.println("修改指定位置字符结果: " + sb);
23.         // 替换指定范围的字符串
24.         sb.replace(0, sb.length(), "nihao");
25.         System.out.println("替换指定范围的字符串结果: " + sb);
26.         System.out.println("-----反转字符串-----");
27.         System.out.println("反转字符串的结果: " + sb.reverse());
28.         System.out.println("-----StringBuffer 中获取 String 字符串-----");
29.         System.out.println("字符串内容为: " + sb.toString());
30.
31.     }
32. }
```

运行结果如图 6-4 所示。

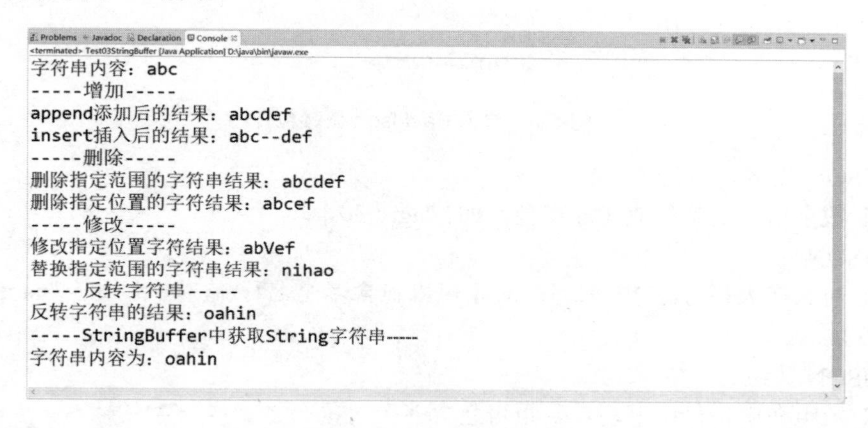

图 6-4 运行结果

上面的代码测试了 StringBuffer 对字符串的增加、删除、修改、反转等操作。StringBuffer 类(或 StringBuilder 类)与 String 类有许多相似的地方，初学者在使用时比较容易混淆。下面简单归纳两者的不同。

(1) String 类用于表示字符串常量，一旦创建，内容和长度都被固定。而 StringBuffer 类(或 StringBuilder 类)是用于存储字符串的，其内容和长度可以改变，如果需要经常修改字符串内容，请使用 StringBuffer 类(或 StringBuilder 类)创建字符串变量对象。

(2) String 类对象可以使用操作符“+”进行拼接，而 StringBuffer 类(或 StringBuilder 类)对象之间不能拼接。

(3) String 类重写了 Object 类的 equals 方法，可以使用 String 类的 equals 方法判断字符串内容是否相同。而 StringBuffer 类(或 StringBuilder 类)没有重写 Object 类的 equals 方法。

4. JSON 字符串解析

网络两端进行数据交互时，简单的数据可以直接发送，当数据较为复杂时，发送端与接收端必须有统一的格式。发送端以约定的格式组织好数据后发送出去，接收端接收到数据后按照约定格式解析数据，将数据转化为可用数据。

JSON(JavaScript Object Notation)是一种轻量级的数据交换格式。因为其易于阅读和编写，同时也易于机器解析和生成的特点，被广泛应用于互联网数据交换。Java 中并没有内置 JSON 的解析类库，因此使用 JSON 需要借助第三方类库。下面是几个常用的 JSON 解析类库。

- Gson：谷歌开发的 JSON 库，功能十分全面。
- FastJson：阿里巴巴开发的 JSON 库，性能十分优秀。
- Jackson：社区十分活跃且更新速度很快。

本案例基于 FastJson 讲解。

在工程中添加 fastjson-1.2.57.jar 包，并构建路径，如图 6-5 所示。

```
project6
  JRE System Library [JavaSE-1.8]
  src
    com.nle
  Referenced Libraries
    fastjson-1.2.57.jar
  libs
    fastjson-1.2.57.jar
```

图 6-5　导入 FastJson 依赖包

(1) JSON 数字。

JSON 数字可以是整型或者浮点型，如{ "age":30 }。

(2) JSON 对象。

JSON 对象在大括号({})中书写：对象可以包含多个名称/值对，例如{ "name":"温度", "value"; 30 }。

(3) JSON 数组。

JSON 数组在中括号中书写，数组可包含多个对象：

```
{
 "devices": [
              { "name":"温度" , "value":25.55 },
              { "name":"湿度" , "value":50.00},
              { "name":"光照" , "value":1234.56 }
   ]
}
```

在上面的例子中，对象 " devices " 是包含三个对象的数组。每个对象代表一条关于某个传感器(name、value)的数据。

(4) JSON 布尔值。

JSON 布尔值可以是 true 或者 false，如{ "flag":true }。

(5) JSON null。

JSON 可以设置 null 值，如{ "device":null }。

接下来通过一个例子学习 JSON 对象与字符串的相互转化，具体代码如下。

```
1.   public class TestJSON {
2.
3.      public static void main(String[] args) {
4.          // JSON 对象的组装
5.          JSONObject object1 = new JSONObject();
6.          // string
7.          object1.put("String", "string");
8.          // int
```

```
9.          object1.put("int", 2);
10.         // boolean
11.         object1.put("boolean", true);
12.         // array
13.         List<Integer> nums = Arrays.asList(1, 2, 3);
14.         object1.put("list", nums);
15.         // null
16.         object1.put("null", null);
17.
18.         System.out.println("对象转JSON: "+object1);
19.
20.         // JSON 字符串的解析
21.         JSONObject object2 = JSONObject
22.                 .parseObject("{\"boolean\":true, \"String\":\"string\",
                    \"list\":[1, 2, 3], \"int\":2}");
23.         // string
24.         String s = object2.getString("String");
25.         System.out.println(s);
26.         // int
27.         int i = object2.getIntValue("int");
28.         System.out.println(i);
29.         // boolean
30.         boolean b = object2.getBooleanValue("boolean");
31.         System.out.println(b);
32.         // list
33.         List<Integer> integers = JSON.parseArray(object2.getJSONArray
              ("list").toJSONString(), Integer.class);
34.         integers.forEach(System.out::println);
35.         // null
36.         System.out.println(object2.getString("null"));
37.     }
38. }
```

运行结果如图 6-6 所示。

```
对象转JSON: {"boolean":true,"string":"string","list":[1,2,3],"int":2}
string
2
true
1
2
3
null
```

图 6-6　运行结果

本书后面的内容会涉及通过网络将本地设备的数据与云平台进行交互，采用的数据交互格式便是 JSON 格式。下面以云平台用户登录为例。

(1)　访问 http://api.nlecloud.com/doc/api/，查看云平台开发账号 API，如图 6-7 所示。

(2)　API 说明的请求示例与响应示例分别如图 6-8、图 6-9 所示。

(3)　发送请求时，根据请求示例将账号、密码等信息组装为 JSONObject 对象，发送时便会自动转换为 JSON 格式，接收响应时，利用 FastJson 工具将应答的数据解析为 Java 对象，即可通过所需信息的关键字取出相应的值。

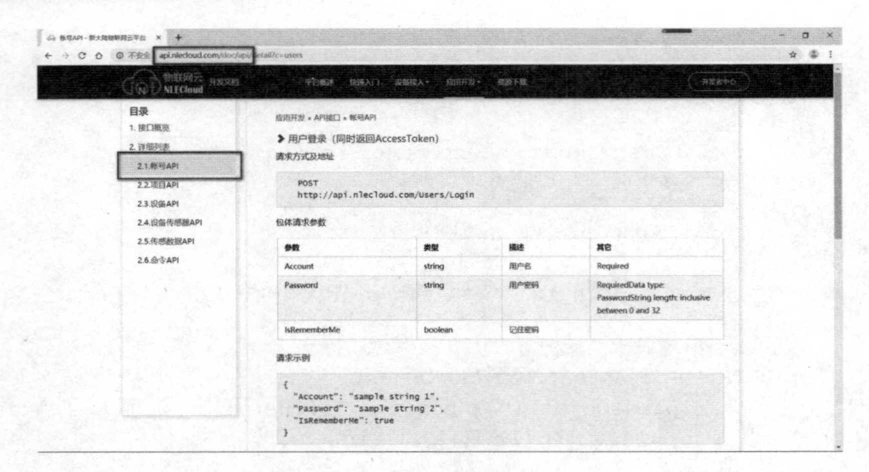

图 6-7　云平台 API

请求示例

```
{
  "Account": "sample string 1",
  "Password": "sample string 2",
  "IsRememberMe": true
}
```

图 6-8　请求示例

响应示例

```
{
  "ResultObj": {
    "UserID": 1,
    "UserName": "sample string 2",
    "Email": "sample string 3",
    "Telphone": "sample string 4",
    "Gender": true,
    "CollegeID": 6,
    "CollegeName": "sample string 7",
    "RoleName": "sample string 8",
    "RoleID": 9,
    "AccessToken": "sample string 10",
    "AccessTokenErrCode": 11,
    "ReturnUrl": "sample string 12",
    "DataToken": "sample string 13"
  },
  "Status": 0,
  "StatusCode": 1,
  "Msg": "sample string 2",
  "ErrorObj": {}
}
```

图 6-9　响应示例

```
1.  public class TestJSON2 {
2.
3.      public static void main(String[] args) {
4.          //发送请求
5.          JSONObject userInfo = new JSONObject();
6.          userInfo.put("Account", "188XXXXXXXX");
7.          userInfo.put("Password", "123456");
8.          userInfo.put("isRememberMe", false);
9.          System.out.println("请求信息: "+userInfo);
10.         //接收响应
11.         String resStr = "{" +
12.                 "  \"ResultObj\": {" +
```

```
13.                 "    \"UserID\": 1, " +
14.                 "    \"UserName\": \"sample string 2\", " +
15.                 "    \"Email\": \"sample string 3\", " +
16.                 "    \"Telephone\": \"sample string 4\", " +
17.                 "    \"Gender\": true, " +
18.                 "    \"CollegeID\": 6, " +
19.                 "    \"CollegeName\": \"sample string 7\", " +
20.                 "    \"RoleName\": \"sample string 8\", " +
21.                 "    \"RoleID\": 9, " +
22.                 "    \"AccessToken\": \"sample string 10\", " +
23.                 "    \"AccessTokenErrCode\": 11, " +
24.                 "    \"ReturnUrl\": \"sample string 12\", " +
25.                 "    \"DataToken\": \"sample string 13\"" +
26.                 "  }, " +
27.                 "  \"Status\": 0, " +
28.                 "  \"StatusCode\": 1, " +
29.                 "  \"Msg\": \"sample string 2\", " +
30.                 "  \"ErrorObj\": {}" +
31.                 "}";
32.         //提取对象中的普通属性
33.         JSONObject resObj = JSONObject.parseObject(resStr);
34.         System.out.println("响应状态: "+resObj.get("StatusCode"));
35.         //提取对象属性中的属性
36.         JSONObject resultObj = JSONObject.parseObject(resObj.get
              ("ResultObj").toString());
37.         System.out.println("应答用户信息: "+resultObj.get("UserName"));
38.     }
39. }
```

运行结果如图 6-10 所示。

```
Problems  Javadoc  Declaration  Console
<terminated> TestJSON2 [Java Application] D:\java\bin\javaw.exe
请求信息：{"Account":"188XXXXXXXX","isRememberMe":false,"Password":"123456"}
响应状态：1
应答用户信息：sample string 2
```

图 6-10 运行结果

【任务实施】

1. 任务分析

本任务要完成 ZigBee 控制器命令的生成工具。

(1) ZigBee 控制器命令格式。

```
命令：Head + len + type + data + lrc
Head：2byte，固定为 0xFF,0XF5
Len：1byte，数据包的字节数,如 0x05
Type：数据类型，0x02 为继电器控制命令
Data：数据域，2byte 设备编号(低位在前，如 34 12 表示 0x1234 )
          2byte 命令：00 01---打开继电器输出
```

```
                                00 02---关闭继电器输出
                                00 03---取反继电器输出
                                00 04---查询继电器状态
                                其他----保留
lrc: 1byte  校验位
```

示例如下：

```
FF F5 05 02 34 12 00 01 LRC
FF F5 05 02 34 12 00 01 be
FF F5 05 02 34 12 00 03 bc
```

(2) 定义十六进制字符串转 byte 数组的方法、byte 转十六进制的方法。

(3) 定义获取校验位的方法，将前面的各位按字节累加，然后取反加 1。

(4) 封装方法根据给定的 ZigBee 序列号和开关类型生成控制器命令。

2. 任务实施

(1) 新建 Java 工程 project6_task1，并在 src 下创建 com.nle.task1 包。

(2) 在 com.nle.task1 目录下创建 ZigBeeTools 类，如图 6-11 所示。

project6_task1
JRE System Library [JavaSE-1.8]
src
com.nle.task1
ZigbeeTools.java

图 6-11 创建 ZigbeeTools 类

(3) 定义十六进制字符串转 byte 数组的方法，hexStrToByte()参数为十六进制字符串。

```
1.   /**
2.    * 函数名称：hexStr2Byte</br>
3.    * 功能描述：String 转数组
4.    *
5.    * @param hex
6.    * @return
7.    *
8.    */
9.   public static byte[] hexStrToByte(String hex) {
10.      // 移除字符串中的空格
11.      hex = hex.replace(" ", "");
12.
13.      ByteBuffer bf = ByteBuffer.allocate(hex.length() / 2);
14.      for (int i = 0; i < hex.length(); i++) {
15.          String hexStr = hex.charAt(i) + "";
16.          i++;
17.          hexStr += hex.charAt(i);
18.          byte b = (byte) Integer.parseInt(hexStr, 16);
19.          bf.put(b);
20.      }
21.      return bf.array();
22.  }
```

(4) 定义 byte 转十六进制的方法 byteToHex()，参数为 byte 类型。

```
1.   /**
2.    * 函数名称：byteToHex</br>
3.    * 功能描述：byte 转十六进制
4.    *
5.    * @param b
```

```
6.    * @return
7.    */
8.   public static String byteToHex(byte b) {
9.      String hex = Integer.toHexString(b & 0xFF);
10.     if (hex.length() == 1) {
11.        hex = '0' + hex;
12.     }
13.     return hex.toUpperCase(Locale.getDefault());
14. }
```

(5) 定义获取校验位的方法 getLrc()，参数为前面各位字节数组和校验码前面各位总长度。

```
1.   /**
2.    * 计算控制继电器动作的控制指令的校验码
3.    *
4.    * @param pSendBuf
5.    * @param nEnd
6.    * @return
7.    */
8.   private static byte getLrc(byte[] pSendBuf, int nEnd) {
9.      byte byLrc = 0;
10.     for (int i = 0; i < nEnd; i++) {
11.        byLrc += pSendBuf[i];
12.     }
13.     byLrc = (byte) ~byLrc;
14.     byLrc = (byte) (byLrc + 1);
15.     return byLrc;
16. }
```

(6) 封装获取 ZigBee 控制器命令生成方法 getCom()，参数为 ZigBee 配置的序列号和表示设备开关的 boolean 类型变量。通过 String 类提供的方法按照格式规律生成控制指令。

```
1.   /**
2.    * 通过序列号获取 ZigBee 继电器的控制指令
3.    *
4.    * @param seq：烧写 ZigBee 继电器时的序列号
5.    * @param flag：true，开；flase，关
6.    * @return：组装好的控制指令
7.    */
8.   public static byte[] getCom(String seq, boolean flag) // 0001 的序列号要
     变成 01 00
9.   {
10.     String com = "FF F5 05 02 00 00 00 00";
11.     String[] str = com.split(" ");
12.     if (seq.length() == 4) {
13.        str[4] = seq.substring(2, 4);
14.        str[5] = seq.substring(0, 2);
15.     }
16.     if (flag)// 为true 代表开
17.     {
18.        str[7] = "01";
19.
20.     } else {
21.        str[7] = "02";
22.     }
23.     com = Arrays.toString(str).replaceAll("[\\[\\], ]", "");
24.     byte lrc = getLrc(hexStrToByte(com), 8);
```

```
25.     com = com+byteToHex(lrc);
26.     return hexStrToByte(com);
27. }
```

(7) 创建测试类 MainApp，测试控制命令生成方法。

```
1.  public class MainApp {
2.
3.      // 测试命令生成是否正确
4.      public static void main(String[] args) {
5.          //获取 ZigBee 控制器命令
6.          byte[] com = ZigBeeTools.getCom("0003", true);
7.          System.out.print("序列号 0003 的开灯指令为：");
8.          for (int i = 0; i < com.length; i++) {
9.              System.out.print(ZigBeeTools.byteToHex(com[i])+" ");
10.         }
11.         System.out.println();
12.         try {
13.             //打开串口
14.             SerialPort port = SerialPortManager.openPort("COM200", 38400);
15.             //发送指令
16.             SerialPortManager.sendToPort(port, com);
17.             System.out.println("照明灯打开");
18.         } catch (Exception e) {
19.             e.printStackTrace();
20.         }
21.     }
22.
23. }
```

3. 运行结果

运行程序，结果如图 6-12 所示。

```
Problems  Javadoc  Declaration  Console
<terminated> MainApp (21) [Java Application] D:\java\bin\javaw.exe
序列号0003的开灯指令为：FF F5 05 02 03 00 00 01 01
照明灯打开
```

图 6-12　运行结果

任务 2　验证用户注册信息

扫码观看视频讲解

【任务描述】

用户在填写注册信息时，常常会由于各种原因产生输入错误。虽然无法完全避免，但是保存用户信息前对输入的内容进行验证是非常必要的。这样不仅能保证用户信息的准确性，也能够提高用户注册的效率。例如，限定用户名的长度、密码的字符组成、用户确认密码、电话号码是否正确、邮箱格式是否规范、身份证是否规范等。本任务将利用数据类型的封装类实现用户注册信息的验证。任务清单如表 6-5 所示。

表 6-5　任务清单

任务课时	4 课时	任务组员数量	建议 1 人
任务组采用设备	无		

【知识解析】

1. Date 类

Date 类位于 java.util 包下，用于封装当前的日期和时间。接下来通过表 6-6 学习 Date 类的常用方法。

表 6-6　Date 类的常用方法

方法声明	描　述
Date()	构造方法，使用当前日期和时间初始化对象
Date(long millisec)	构造方法，接收一个参数，该参数是从 1970 年 1 月 1 日起的毫秒数
boolean after(Date date)	若调用此方法的 Date 对象在指定日期之后则返回 true，否则返回 false
boolean before(Date date)	若调用此方法的 Date 对象在指定日期之前则返回 true，否则返回 false
long getTime( )	返回自 1970 年 1 月 1 日 00:00:00 GMT 以来此 Date 对象表示的毫秒数
void setTime(long time)	用自 1970 年 1 月 1 日 00:00:00 GMT 以后的 time 毫秒数设置时间和日期
String toString( )	将此 Date 对象转换为 String 形式

接下来通过一个例子学习 Date 类的常用方法，具体代码如下。

```
1.  public class Test04Date {
2.      public static void main(String[] args) {
3.          //Date 表示一个日期的瞬时，精确到毫秒
4.          Date date = new Date();
5.          System.out.println(date);//Tue Sep 26 10:50:52 CST 2017
6.          Date date2 = new Date(123718273);//1970 年 1 月 1 日 0 点 + 123718273 毫秒
7.          System.out.println("date2="+date2);
8.          boolean after=date2.after(date);
9.          boolean before=date2.before(date);
10.         System.out.println("after="+after);
11.         System.out.println("before="+before);
12.         //获取当前日期的时间戳
13.         long time = date.getTime();
14.         System.out.println("当前时间距离 1970 年 1 月 1 日零点的毫秒值为: "+time);
15.
16.     }
17. }
```

运行结果如图 6-13 所示。

```
Wed Jun 19 09:53:25 CST 2019
date2=Fri Jan 02 18:21:58 CST 1970
after=false
before=true
当前时间距离1970年1月1日零点的毫秒值为：1560909205262
```

图 6-13　运行结果

上面的程序创建了两个 Date 类对象，一个使用默认构造方法创建一个当前日期时间的 Date 类对象，另一个使用传入一个时间毫秒值的构造方法创建一个指定日期时间的 Date 类对象，并且测试了 Date 类的常用方法 after 方法、before 方法与 getTime 方法。自 1970 年 1 月 1 日 00:00:00 GMT 以来的时间毫秒值还可以使用 System 类的 currentTimeMillis 方法获取。接下来通过一个例子学习如何通过 System 类的 currentTimeMillis()方法判断一段程序执行的时间，具体代码如下。

```
1.  public class Test05CurrentTimeMillis {
2.      public static void main(String[] args) {
3.
4.          long time = System.currentTimeMillis();
5.          System.out.println("time=" + time);
6.
7.          // 计算一段程序执行的时间
8.          long start = System.currentTimeMillis();
9.          int sum = 0;
10.         for (int i = 0; i < 10000000; i++) {
11.             sum += i;
12.             System.out.print("");
13.         }
14.         long end = System.currentTimeMillis();
15.         System.out.println("循环执行的时间=" + (end - start) + "毫秒");
16.     }
17. }
```

运行结果如图 6-14 所示。

图 6-14　运行结果

2. SimpleDateFormat 类

前面学习了使用 Date 来表示日期时间，如果开发人员想按自定义的日期时间显示，可以使用日期格式化类完成这个效果。Java 封装了一个简单的日期格式化类 SimpleDateFormat，用于以区域设置敏感的方式格式化和解析日期。SimpleDateFormat 允许选择任何用户自定义日期时间格式来运行。SimpleDateFormat 使用日期和时间模式字符指定日期格式。常用的日期和时间模式如表 6-7 所示。

表 6-7　日期和时间模式

字　母	描　述	表　示	示　例
y	年	Year	1996; 96
M	年中的月份	Month	July; Jul; 07
d	月份中的天数	Number	10

续表

字母	描述	表示	示例
E	星期中的天数	Text	Tuesday; Tue
H	一天中的小时数(0～23)	Number	0
h	am/pm 中的小时数(1～12)	Number	12
m	小时中的分钟数	Number	30
s	分钟数中的秒数	Number	55
S	毫秒数	Number	957

SimpleDateFormat 类的常用方法如表 6-8 所示。

表 6-8 SimpleDateFormat 类的常用方法

方法声明	描述
String format(Date date)	将日期格式化成日期/时间字符串
Date parse(String source)	从给定字符串的开始解析文本以生成日期

接下来通过一个例子学习 SimpleDateFormat 的使用，具体代码如下。

```
1.  public class Test06SimpleDateFormat {
2.
3.      public static void main(String[] args) {
4.          // 格式化日期
5.          SimpleDateFormat simpleDateFormat = new SimpleDateFormat("yyyy年
              MM月dd日 HH时mm分ss秒SSS毫秒");
6.          // 格式化当前日期
7.          String dateString = simpleDateFormat.format(new Date());
8.          System.out.println("yyyy年MM月dd日 HH时mm分ss秒SSS毫秒-->" +
              dateString);
9.          SimpleDateFormat simpleDateFormat2 = new SimpleDateFormat
              ("yyyy/MM/dd HH:mm:ss");
10.         String dateString2 = simpleDateFormat2.format(new Date());
11.         System.out.println("yyyy/MM/dd HH:mm:ss-->" + dateString2);
12.         // 解析字符串日期： String--->Date
13.         try {
14.             // 字符串的格式必须和格式化对象的格式相同才能正确解析
15.             Date date3 = simpleDateFormat2.parse("1999/2/3 10:10:10");
16.             System.out.println(date3);
17.         } catch (ParseException e) {
18.             // TODO Auto-generated catch block
19.             e.printStackTrace();
20.         }
21.         // 方式二：解析字符串日期，当无法正确解析的时候，返回null值
22.         Date date4 = simpleDateFormat2.parse("1999/aasdf2/3 10:10:10", new
              ParsePosition(0));
23.         System.out.println(date4);
24.     }
25. }
```

运行结果如图 6-15 所示。

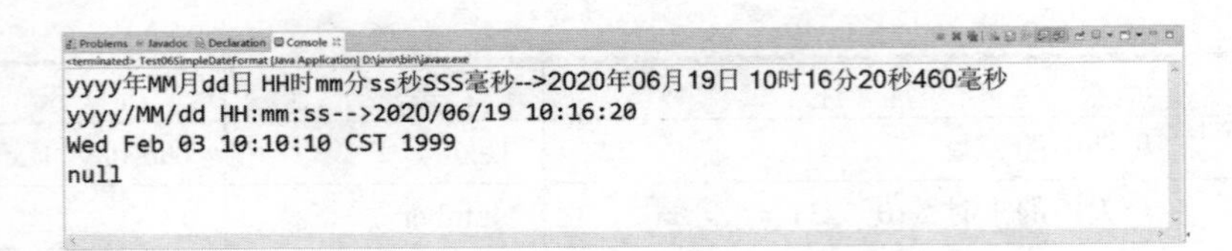

图 6-15　运行结果

上面的程序中，在创建 SimpleDateFormat 对象时，构造方法内传入模式字符。然后使用 format()方法可以格式化日期，返回字符串日期值。还可以使用 parse()方法把一个日期字符串值解析成 Date 类日期对象。当然，前提是这个日期字符串是符合解析模式的。

3. Calendar 类

前面学习了通过 Date 类可以获取日期时间，但是无法通过 Date 类设置和获取日期数据的特定部分。低版本的 JDK 中，Date 提供了相应的方法设置和获取日期数据的特定部分，但是目前已经不建议这样使用了，可以使用 Calendar 类操作日历字段。

Calendar 类是一个抽象类，不能直接实例化，可以使用它的一个静态方法 getInstance()来获取一个 Calendar 子类对象。Calendar 类的常用方法如表 6-9 所示。

表 6-9　Calendar 类的常用方法

方法声明	描　述
static Calendar getInstance()	返回一个日历子类对象
int get(int field)	获取指定字段的时间值
void set(int field，int value)	用给定的值设置时间字段
Date getTime()	获取日历当前时间
long getTimeInMillis()	获取用长整型表示的日历的当前时间

其中，set()方法有多个重载的方法，详细内容请参考 JDK 的 API(应用程序接口)。Calendar 提供了一些常量字段用来表示日期的特定部分，具体如表 6-10 所示。

表 6-10　Calendar 的常量字段

常　量	描　述
Calendar.YEAR	年份
Calendar.MONTH	月份
Calendar.DATE	日期
Calendar.DAY_OF_MONTH	日期，和上面的字段意义完全相同
Calendar.HOUR	12 小时制的小时
Calendar.HOUR_OF_DAY	24 小时制的小时
Calendar.MINUTE	分钟
Calendar.SECOND	秒
Calendar.DAY_OF_WEEK	星期几

接下来通过一个例子学习 Calendar 的使用，具体代码如下。

```
1.  public class Test07Canlendar {
2.      public static void main(String[] args) {
3.          //日历类Calendar抽象类，不能直接实例化
4.          Calendar calendar = Calendar.getInstance();//获取一个Calendar子类对象
5.          int year=calendar.get(Calendar.YEAR);//年份
6.          System.out.println("年year="+year);
7.          int month=calendar.get(Calendar.MONTH);//月份(0～11)
8.          System.out.println("月month="+(month+1));//真实月份=获取的月份+1
9.          int dayOfMonth=calendar.get(Calendar.DAY_OF_MONTH);//日期
10.         int dayOfWeek=calendar.get(Calendar.DAY_OF_WEEK);//星期几()
11.         System.out.println("日dayOfMonth="+dayOfMonth);
12.         System.out.println("星期dayOfWeek="+dayOfWeek);
13.         int hour=calendar.get(Calendar.HOUR);//12小时制
14.         int hourOfDay=calendar.get(Calendar.HOUR_OF_DAY);//24小时制
15.         System.out.println("时HOUR="+hour);
16.         System.out.println("HOUR_OF_DAY="+hourOfDay);
17.         System.out.println("分MINUTE="+calendar.get(Calendar.MINUTE));//分
18.         System.out.println("秒SECOND="+calendar.get(Calendar.SECOND));//秒
19.         //设置时间
20.         calendar.set(2008, 7, 8, 18, 8, 8);
21.         System.out.println("获取Date类对象："+ calendar.getTime());
22.         System.out.println("获取时间毫秒值："+calendar.getTimeInMillis());
23.     }
24. }
```

运行结果如图 6-16 所示。

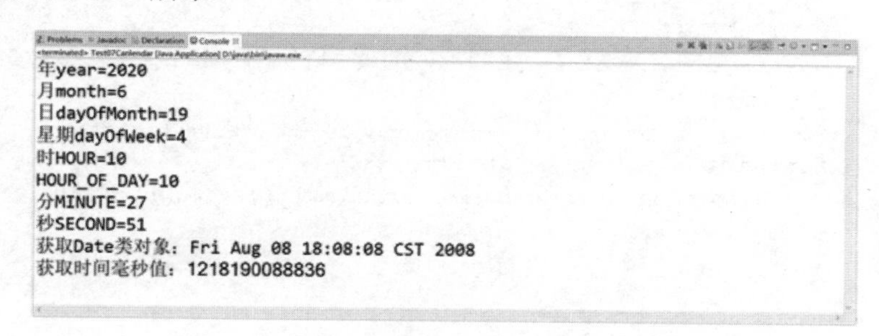

图 6-16　运行结果

4. Math 类

Java 的 Math 类包含用于执行基本数据运算的属性和方法，如初等指数、对数、平方根和三角函数。Math 的方法被定义为 static 形式，可以通过 Math 类直接调用。接下来通过一张表列出 Math 类常用的一些方法，如表 6-11 所示。

表 6-11　Math 类的常用方法

方法声明	描　述
static double abs(double a)	返回值为 double 的绝对值(有重载的方法)
static double ceil(double a)	返回大于或等于参数的最小(最接近负无穷大) double 值，等于一个数学整数

续表

方法声明	描 述
static double floor(double a)	返回小于或等于参数的最大(最接近正无穷大) double 值，等于一个数学整数
static double round(double a)	返回参数中最接近的 long，其中 long 四舍五入为正无穷大(有重载的方法)
static double max(double a,double b)	返回两个 double 值中的较大值(有重载的方法)
static double min(double a,double b)	返回两个 double 值中的较小值(有重载的方法)
static double pow(double a,double b)	将第一个参数的值返回到第二个参数的幂
static double random()	返回值为 double，值为正号，大于等于 0.0，小于 1.0

接下来通过一个例子学习 Math 的使用，如下所示。

```
1.   public class Test08Math {
2.       public static void main(String[] args) {
3.           System.out.println("圆周率 PI 的值：" + Math.PI); // 圆周率 PI
4.           System.out.println("-3.1 的绝对值：" + Math.abs(-3.1));// 绝对值
5.           System.out.println("3.4 的 ceil 值：" + Math.ceil(3.4));// 大于或等于参数的最大 double 值的整数
6.           System.out.println("3.4 的 floor 值:" + Math.floor(3.4));// 小于或等于参数的最大 double 值的整数
7.           System.out.println("3.4 的四舍五入值:" + Math.round(3.4));// 四舍五入值
8.           System.out.println("2 的 3 次方值:" + Math.pow(2, 3));// 幂值
9.           System.out.println("1 和 100 的最大值:" + Math.max(1, 100));// 两个参数的最大值
10.          System.out.println("1 和 100 的最小值:" + Math.min(1, 100));// 两个参数的最小值
11.          System.out.println("-------0.0 到 1.0 的随机数--------");
12.          for (int i = 0; i < 5; i++) {
13.              System.out.println(Math.random());// 输出 5 个随机数
14.          }
15.      }
16.  }
```

运行程序，结果如图 6-17 所示。

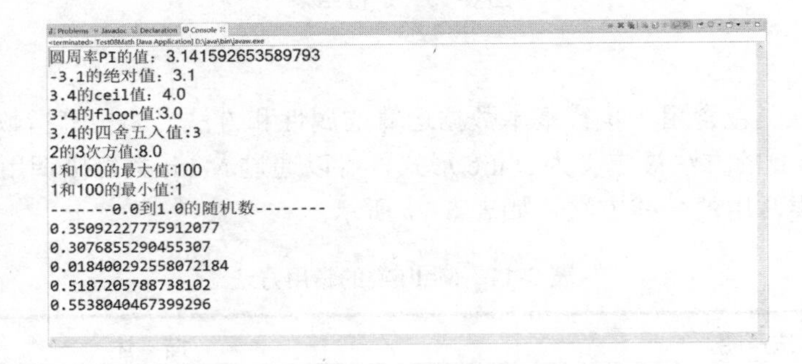

图 6-17　运行结果

思考：从运行结果中可以看出，使用 Math 类的 random()方法，可以获得 0.0～1.0 的随机数。能否模拟掷骰子的实验，使模拟随机数在 1～6 之间呢？

5. Random 类

除了使用 Math 类的 random()方法可以获得一个随机数外，Java 还提供了一个随机数封装类 Random。接下来介绍如何使用 Random 获得随机数，代码如下所示。

```
1.  public class Test09Random {
2.      public static void main(String[] args) {
3.          Random random = new Random();
4.          for (int i = 0; i < 5; i++) {
5.              int num = random.nextInt(6) + 1;// 获取1～6之间的随机数
6.              System.out.println(num);
7.          }
8.      }
9.  }
```

运行结果如图 6-18 所示。

图 6-18　运行结果

在上面的代码中，首先通过 Random 类创建一个实例，接着使用实例的 nextInt()方法，传入一个整数 6，可以得到 0～6 范围的随机数(包含 0 不包含 6)，得到的随机数再加上 1，就可以获得 1～6(包含 1 和 6)的随机数了。

6. 基本数据类型的封装类

在实际开发中，经常会遇到需要使用对象的情况，而不是直接使用基本数据类型。为了解决这一问题，Java 语言为每一个基本数据类型提供了对应的封装类。下面列出 Java 中基本数据类型的封装类，如表 6-12 所示。

表 6-12　基本数据类型的封装类

基本数据类型	封装类型
boolean	Boolean
byte	Byte
short	Short
int	Integer
long	Long
float	Float
double	Double
char	Char

接下来，通过一个例子学习基本数据类型与封装类型的转换以及基本数据类型和字符串的转换，代码如下所示。

```
1.   public class Test10 {
2.       public static void main(String[] args) {
3.           //基本数据类型--->封装类型
4.           int i =10;
5.           Integer integer = new Integer(i);
6.           Integer integer2 = Integer.valueOf(i);
7.           //JDK1.5 以上可以直接转换
8.           Integer integer3 = i;
9.   
10.          //封装类型--->基本数据类型
11.          int j = integer.intValue();
12.          // JDK 1.5 以上可以直接转换
13.          int z = integer;
14.  
15.          //基本数据类型-->字符串类型
16.          int a = 123;
17.          String str = 123 + "";
18.          String str2 = String.valueOf(a);
19.  
20.          //字符串类型-->基本数据类型
21.          String str3 = "123";
22.          double ii=Double.parseDouble(str3);
23.      }
24.  }
```

从上面的代码中可以看出以下内容。

- 基本数据类型转换成封装类型，可以使用它的封装类的静态方法 valueOf()，也可以使用 new 关键字创建。在 JDK1.5 及以上编译环境，还可以直接使用赋值符号进行转换。
- 封装类型转换成基本数据类型，可以使用对象的 xxxValue()方法将它转换成对应的基本数据类型。同样地，在 JDK1.5 及以上编译环境，可以直接使用赋值符号进行转换。
- 基本数据类型转换成字符串类型，直接在值后面加上一个空字符串即可，也可以使用 String 类的 valueOf()方法进行转换。
- 字符串类型转换成基本数据类型，使用对应的基本数据类型的封装类的 parseXxx()方法进行转换。需要注意，字符串必须是基本数据类型要求的格式，否则无法成功转换，将会出现 NumberFormatException 异常。

【任务实施】

1. 任务分析

本任务要求对用户提交的注册信息进行验证。

(1) 封装保存用户信息的 javabean 类，符合 javabean 规范。

(2) 封装用户注册信息验证工具类，对不同的属性按规则验证。

(3) 接收用户信息并验证通过后，用户注册成功。

2. 任务实施

(1) 新建 Java 工程 project6_task2，并在 src 目录下创建 com.nle.task2 包。

(2) 封装用户 javabean 类 User，用户的基本信息有用户名、用户密码、年龄、电话、生日。

```
1.  /**
2.   * 用户 javabean
3.   * @author admin
4.   *
5.   */
6.  public class User {
7.
8.      private String username;//用户名
9.      private String pwd;//密码
10.     private int age;//年龄
11.     private Date birthday;//生日
12.     private String phone;//电话
13.
14.     public User() {
15.
16.     }
17.
18.     public User(String username, String pwd, int age, Date birthday, String
          phone) {
19.         super();
20.         this.username = username;
21.         this.pwd = pwd;
22.         this.age = age;
23.         this.birthday = birthday;
24.         this.phone = phone;
25.     }
26.
27.     public String getUsername() {
28.         return username;
29.     }
30.
31.     public void setUsername(String username) {
32.         this.username = username;
33.     }
34.
35.     public String getPwd() {
36.         return pwd;
37.     }
38.
39.     public void setPwd(String pwd) {
40.         this.pwd = pwd;
41.     }
42.
43.     public int getAge() {
44.         return age;
45.     }
46.
47.     public void setAge(int age) {
48.         this.age = age;
49.     }
50.
51.     public Date getBirthday() {
52.         return birthday;
53.     }
54.
55.     public void setBirthday(Date birthday) {
```

```
56.         this.birthday = birthday;
57.     }
58.
59.     public String getPhone() {
60.         return phone;
61.     }
62.
63.     public void setPhone(String phone) {
64.         this.phone = phone;
65.     }
66.
67. }
```

(3) 创建用户信息验证工具类 ValidateTools，按照约定好的规则，对用户提交的注册信息字符串进行验证。验证通过后，利用数据类型包装类将用户信息字符串转换为方便保存的数据类型。

```
1.  /**
2.   * 注册信息验证工具
3.   * @author admin
4.   *
5.   */
6.  public class ValidateTools {
7.      /**
8.       * 用户名验证：用户名只能由大小写字母组成，长度为 6-12
9.       * @param username
10.      * @return
11.      */
12.     public boolean valUserName(String username) {
13.         //将用户输入字符串转换为字符数组
14.         char[] chars = username.toCharArray();
15.         boolean flag = true;
16.         for (int i = 0; i < chars.length; i++) {
17.             //字符类型包装类 Character 验证字符是否是字母
18.             if(!Character.isAlphabetic(chars[i])) {
19.                 flag = false;
20.                 System.out.println("用户名只能由字母组成！");
21.                 break;
22.             }
23.         }
24.
25.         if(username.length()<6||username.length()>12) {
26.             flag = false;
27.             System.out.println("用户名长度 6-12！");
28.         }
29.         return flag;
30.     }
31.     /**
32.      * 密码验证：必须包含字母，长度大于 6
33.      * 再次输入密码比较
34.      * @param pwd
35.      * @return
36.      */
37.     public boolean valPwd(String pwd) {
38.         char[] chars = pwd.toCharArray();
39.         //密码是否符合要求
40.         boolean flag = true;
41.         //是否包含字母
```

```
42.         boolean alphaCon = false;
43.         //验证是否只有数字或字母
44.         for (int i = 0; i < chars.length; i++) {
45.             //字符包装类验证字符是否是数字或字母
46.             if(!Character.isLetterOrDigit(chars[i])) {
47.                 System.out.println("密码只能包含数字或字母！");
48.                 flag = false;
49.             }
50.         }
51.         //必须包含字母
52.         for (int i = 0; i < chars.length; i++) {
53.             //字符包装类验证字符是否是字母
54.             if(Character.isAlphabetic(chars[i])) {
55.                 alphaCon = true;
56.                 break;
57.             }
58.         }
59.         if(!alphaCon) {
60.             System.out.println("密码必须包含字母!");
61.             return false;
62.         }
63.         //密码长度大于 6
64.         if(pwd.length()<6) {
65.             System.out.println("密码长度必须大于 6!");
66.             return false;
67.         }
68.
69.         //再次输入密码
70.         if(flag) {
71.             System.out.println("请再次输入密码：");
72.             Scanner scan = new Scanner(System.in);
73.             String pwdSure = scan.next();
74.             if(!pwd.equals(pwdSure)) {
75.                 flag = false;
76.                 System.out.println("密码输入不一致！");
77.             }
78.         }
79.         return flag;
80.     }
81.     /**
82.      * 验证年龄
83.      * @param age
84.      * @return
85.      */
86.     public boolean valAge(String age) {
87.         char[] chars = age.toCharArray();
88.         boolean flag = true;
89.         for (int i = 0; i < chars.length; i++) {
90.             //字符包装类验证字符是否是数字
91.             if(!Character.isDigit(chars[i])) {
92.                 flag = false;
93.                 System.out.println("年龄格式不正确！");
94.                 break;
95.             }
96.         }
97.         return flag;
98.     }
99.     /**
```

```
100.      * 验证电话
101.      * @param phone
102.      * @return
103.      */
104.     public boolean valPhone(String phone) {
105.        char[] chars = phone.toCharArray();
106.        boolean flag = true;
107.        for (int i = 0; i < chars.length; i++) {
108.           //字符包装类验证字符是否是数字
109.           if(!Character.isDigit(chars[i])) {
110.              flag = false;
111.              System.out.println("电话号码只能包含数字！");
112.              break;
113.           }
114.        }
115.
116.        if(phone.length()!=11) {
117.           flag = false;
118.           System.out.println("电话号码不正确！");
119.        }
120.
121.        return flag;
122.     }
123.     /**
124.      * 验证生日
125.      * @param date
126.      * @return
127.      */
128.     public boolean valBirthday(String birthday) {
129.        boolean flag = true;
130.        //正则方式验证生日格式
131.        Pattern pattern = Pattern
132.              .compile("^[1, 2]\\d{3}-(0[1-9]||1[0-2])-(0[1-9]||[1, 2]
                  [0-9]||3[0, 1])$");
133.        Matcher matcher = pattern.matcher(birthday);
134.        if (!matcher.matches()) {
135.           System.out.println("请按照提示格式输入生日！");
136.           flag = false;
137.        }
138.
139.        return flag;
140.     }
141.
142. }
```

(4) 创建测试类 MainApp，通过 Scanner 类接收用户从控制台输入的注册信息，实例化 ValidateTools 对输入的每个注册信息进行验证，将获取的信息转为对应的数据类型，赋值给实例化的 User 对象，验证通过后进行下一项输入。所有信息输入成功后注册成功。

```
1.   public class MainApp {
2.
3.      public static void main(String[] args) {
4.         ValidateTools tools = new ValidateTools();
5.         Scanner scan = new Scanner(System.in);
6.         User user = new User();
7.         //验证用户名
8.         while(true) {
9.            System.out.println("请输入用户名：");
```

```
10.             String username = scan.next();
11.             if(tools.valUserName(username)) {
12.                 user.setUsername(username);
13.                 break;
14.             }
15.         }
16.         //验证密码
17.         while(true) {
18.             System.out.println("请输入密码：");
19.             String pwd = scan.next();
20.             if(tools.valPwd(pwd)) {
21.                 user.setPwd(pwd);
22.                 break;
23.             }
24.         }
25.
26.         //验证年龄
27.         while(true) {
28.             System.out.println("请输入年龄：");
29.             String age = scan.next();
30.             if(tools.valAge(age)) {
31.                 //利用包装类将字符串转换为整型
32.                 user.setAge(Integer.parseInt(age));
33.                 break;
34.             }
35.         }
36.
37.         //验证电话
38.         while(true) {
39.             System.out.println("请输入电话：");
40.             String phone = scan.next();
41.             if(tools.valPhone(phone)) {
42.                 user.setPhone(phone);
43.                 break;
44.             }
45.         }
46.
47.         //验证生日
48.         while(true) {
49.             System.out.println("请输入生日(yyyy-MM-dd):");
50.             String birthday = scan.next();
51.             if(tools.valBirthday(birthday)) {
52.                 SimpleDateFormat sdf = new SimpleDateFormat("yyyy-MM-dd");
53.                 try {
54.                     //利用 SimpleDateFormat 将字符串转换为 Date 类型
55.                     Date birth = sdf.parse(birthday);
56.                     user.setBirthday(birth);
57.                 } catch (ParseException e) {
58.                     e.printStackTrace();
59.                 }
60.                 break;
61.             }
62.         }
63.
64.         System.out.println("恭喜，注册成功!");
65.         scan.close();
66.     }
67. }
```

3. 运行结果

运行程序，结果如图 6-19 所示。

图 6-19　运行结果

思考与练习

1. String 是不可变的有什么好处？
2. 写一个方法来判断一个 String 是否是回文(顺读和倒读都一样的词)。
3. 总结基本数据类型与包装类之间的转换规律。

项目7 智慧园区系统界面开发和事件处理

【项目描述】

在前面几个项目中，已经对JavaFX界面有了基本的运用。这些界面是如何制作出来的？界面之间是如何跳转的？它们又是如何让用户通过操作来实现程序的功能的呢？通过本项目，将带领大家通过JavaFX开发智慧园区项目的界面，实现界面之间的跳转，并对界面上的控件添加监听。让读者初步了解图形化界面的制作过程和原理，能够借助JavaFX制作界面，实现物联网PC端应用的开发。具体任务列表如图7-1所示。

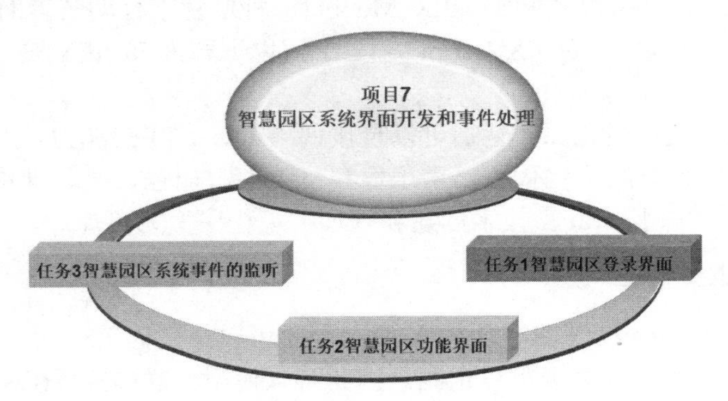

图7-1 项目7任务列表

【学习目标】

知识目标： 了解JavaFX的基本应用和JavaFX工程结构；会运用JavaFX的各种容器和

布局；会运用 JavaFX 的常用控件；会使用 SceneBuilder 制作界面；会为相应的 FXML 文件创建对应的界面控制类；会在界面控制类中为界面组件绑定监听方法；会运用 Stage 实现界面之间的跳转；会运用@FXML 注解获取界面控件。

技能目标：能使用 Eclipse 和 SceneBuilder 开发 JavaFX 应用界面；能为 JavaFX 界面控件添加监听实现设备控制和数据采集；能借助 JavaFX 开发物联网工程 PC 端应用。

任务 1　智慧园区登录界面

扫码观看视频讲解

【任务描述】

本任务要求读者掌握 JavaFX 设计制作的流程，能够独立搭建 JavaFX 的开发环境，熟悉常用的 JavaFX 控件。利用 Eclipse 和 JavaFX 的可视化工具 SceneBuilder 完成智慧园区登录界面的制作，其中包括用户名和密码的输入功能、登录按钮、注册按钮等。任务清单如表 7-1 所示。

表 7-1　任务清单

任务课时	4 课时	任务组员数量	建议 1 人
任务组采用设备	无		

【知识解析】

1. JavaFX 简介

JavaFX 是 Java 的下一代图形用户界面工具包。JavaFX 是一组图形和媒体 API，可以用它来创建和部署富客户端应用程序。JavaFX 允许开发人员快速构建丰富的跨平台应用程序。JavaFX 通过硬件加速图形支持现代 GPU。

JavaFX 允许开发人员在单个编程接口中组合图形、动画和 UI 控件。图表编程语言可用于开发互联网应用程序(RIA)。JavaFX 技术主要应用于创建 Rich Internet Applications(RIAs)。当前的 JavaFX 包括 JavaFX 脚本和 JavaFX Mobile(一种运营于行动装置的操作系统)，今后 JavaFX 将包括更多的产品。JavaFX Script 编程语言(以下称为 JavaFX)是一种声明性的、静态类型脚本语言。

JavaFX 技术有着良好的前景，包括可以直接调用 Java API 的能力。因为 JavaFX Script 是静态类型，它同样具有结构化代码、重用性和封装性，如包、类、继承和单独编译和发布单元，这些特性使得用 JavaFX 技术创建和管理大型程序变为可能。

2. JavaFX 的主要特征

(1)　JavaFX 是用 Java 编写的，JavaFX 应用程序可以从任何 Java 库引用 API。

(2)　JavaFX 应用程序的外观可以定制。因此可以使用级联样式表(CSS)来对 JavaFX 应用程序进行风格化。平面设计师可以通过 CSS 自定义外观和样式。

(3)　可以在 FXML 脚本语言中描述 UI 的表示方面，并使用 Java 对应用程序逻辑进行编码。

(4)　通过使用 JavaFX SceneBuilder，可以通过拖放来设计 UI。SceneBuilder 将创建可以

移植到集成开发环境(IDE)的 FXML 标记，以便开发人员可以添加业务逻辑。

(5) JavaFX 有一个称为 WebView 的控件，可以呈现复杂的网页。WebView 支持 JavaScript，可以通过 Java API 在 Web 页面中调用 JavaScript。WebView 还支持额外的 HTML5 功能，包括 Web 套接字、Web Workers 和 Web 字体，还可以从 WebView 打印网页。

(6) Swing 互操作性。现有的 Swing 应用程序可以使用 JavaFX 类，例如图表和 WebView。还可以使用 SwingNode 类将 Swing 内容嵌入应用程序中。

(7) 3D 图形功能。JavaFX 支持 Shape，如 Box、Cylinder、MeshView 和 Sphere 子类，SubScene、Material、PickResult、AmbientLight 和 PointLight。

(8) Canvas API。使用 Canvas API，可以在 JavaFX 场景上绘制。

(9) 打印 API。javafx.print 包提供了 JavaFX Printing API 的类。

(10) 富文本支持。JavaFX 支持增强的文本，包括双向文本和复杂的文本脚本，以及多行、多种风格的文本。

(11) 多点触控支持。JavaFX 提供对多点触摸操作的支持。

(12) JavaFX 支持 Hi-DPI 显示。

3. JavaFX 工程

在配置好开发环境后，要开发一个 JavaFX 应用程序，第一步就是在 Eclipse 中新建一个 JavaFX 工程。JavaFX 工程比 Java 工程多一个 JavaFX SDK 依赖库，如图 7-2 所示。

4. JavaFX 工程入口

新建的 JavaFX 工程会自动生成一个程序入口放在 application 包中，读者可以直接使用，也可以将其删除自行创建。本案例采用自行创建的方法，入口类必须继承 Application 类并重写 start 方法。start 方法会传入一个 Stage 参数，primaryStage 指向本应用的顶级容器，开发者可以将初始化好的场景通过 setScene 方法设置给 Stage。然后调用 Stage 的 show 方法。最后，在 main 函数中调用 Application 类的 launch 方法，要注意必须将 main 函数放在入口类中，如图 7-3 所示。

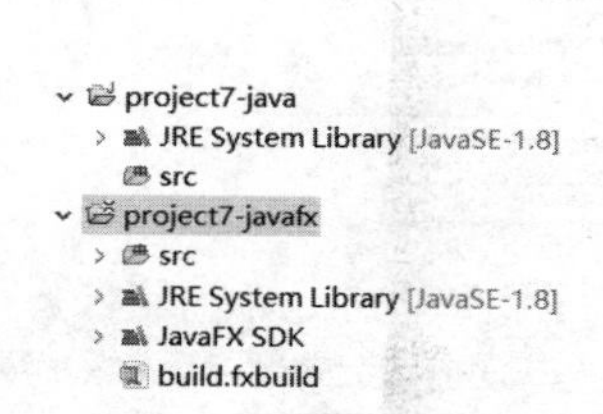

图 7-2　工程结构

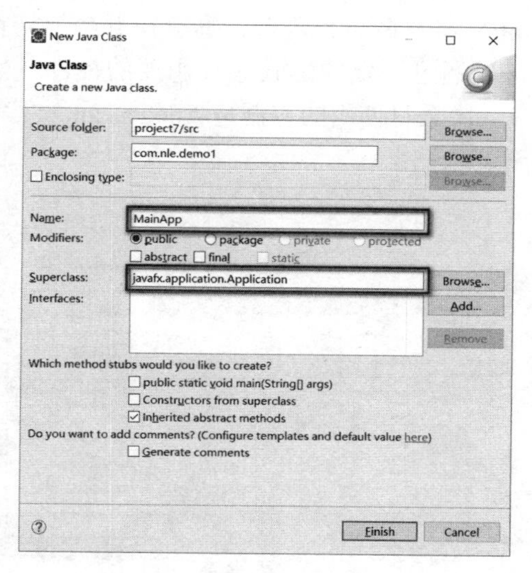

图 7-3　创建程序入口

```
1.  public class MainApp extends Application
2.  {
3.      @Override
4.      public void start(Stage primaryStage) {
5.          try {
6.              //初始化容器
7.              BorderPane root = new BorderPane();
8.              //初始化场景
9.              Scene scene = new Scene(root, 400, 400);
10.             //设置场景样式
                 scene.getStylesheets().add(getClass().getResource
                 ("application.css").toExternalForm());
11.             //设置显示场景
12.             primaryStage.setScene(scene);
13.             //显示舞台
14.             primaryStage.show();
15.         } catch(Exception e) {
16.             e.printStackTrace();
17.         }
18.     }
19.
20.     public static void main(String[] args) {
21.         launch(args);
22.     }
23. }
```

5. 创建 FXML 文件

JavaFX 界面的制作有两种方式，一种是通过 Java 代码的方式，另一种是通过 FXML 文件的方式，本文重点介绍后者。在入口类的同一目录下新建一个 FXML 文件。默认使用的布局为锚点布局。文件的命名一般以 View 结尾，如 XxxxxxView.fxml。文件建好后，通过 SceneBuilder 打开便可以开始添加控件，如图 7-4～图 7-7 所示。

6. SceneBuilder 添加控件

一个 FXML 文件表示应用的一个场景 Scene，场景可以设置给应用的顶级容器 Stage，JavaFX 将它命名为舞台，寓意在舞台上切换不同的场景。将 FXML 通过 SceneBuilder 打开后可以通过拖曳的形式添加想要的控件，然后对控件进行编辑和添加事件监听。为了方便拖曳，本项目使用的容器是 AnchorPane，读者可以根据自己的需求选用不同的容器，如图 7-8 所示。

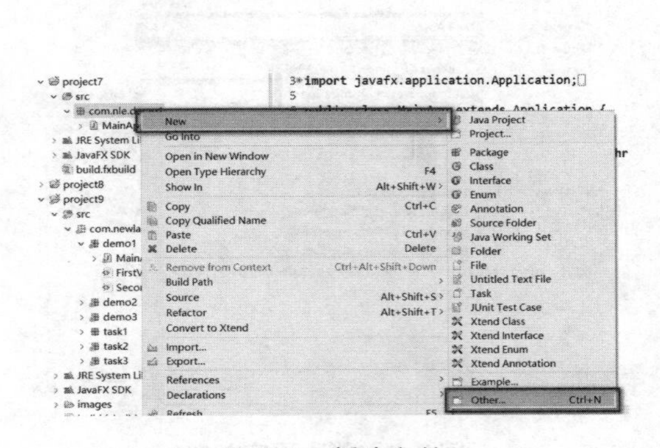

图 7-4　新建文件

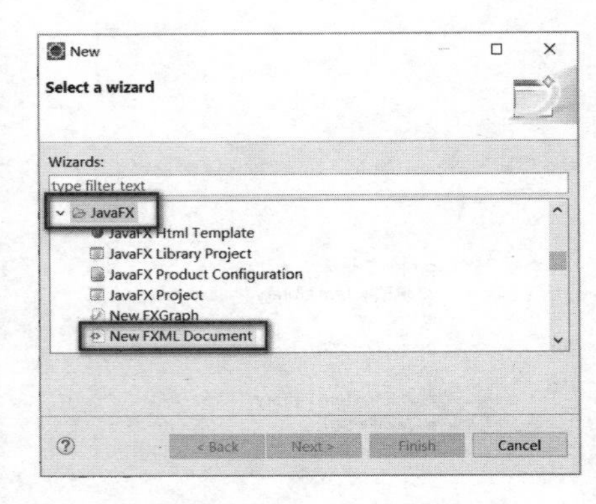

图 7-5　创建 FXML 文件

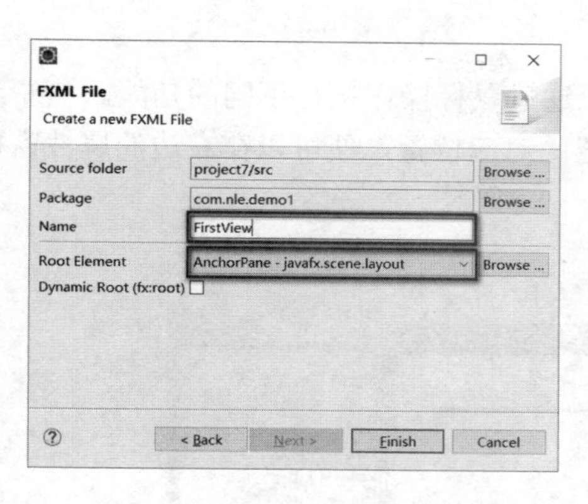

图 7-6　命名 FXML 文件

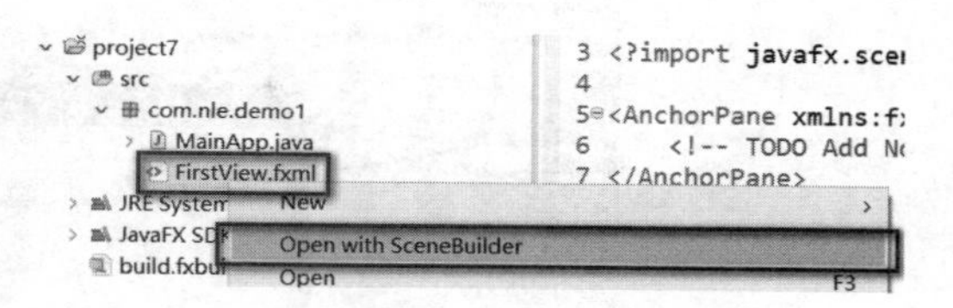

图 7-7　使用 SceneBuilder 打开 FXML 文件

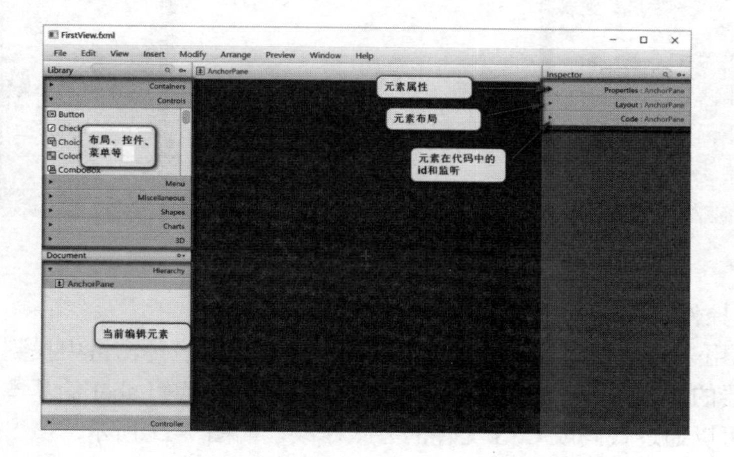

图 7-8　SceneBuilder 窗口布局

(1) AnchorPane 容器。

AnchorPane 容器类似于绝对布局，在 SceneBuilder 的左下方选中 AnchorPane 元素后，可以在右侧的 Layout 中自定义容器的尺寸，然后就可以将想要的控件拖放到容器的任何位置，方便快速地制作界面，如图 7-9 所示。

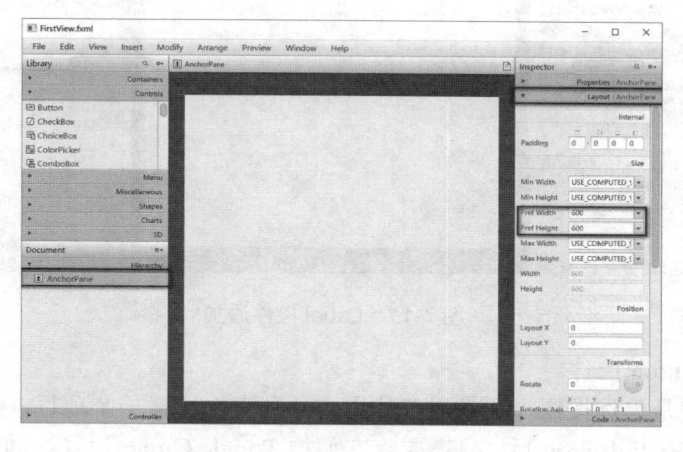

图 7-9　AnchorPane 容器属性

(2) Button 控件。

Button 是最常用的交互控件之一，用户可以通过鼠标操作执行不同的功能。从左边的控件窗格中找到 Button 控件，并将其拖放到容器的对应位置，便可以在右边的属性窗格中编辑 Button 的属性，如图 7-10 所示。

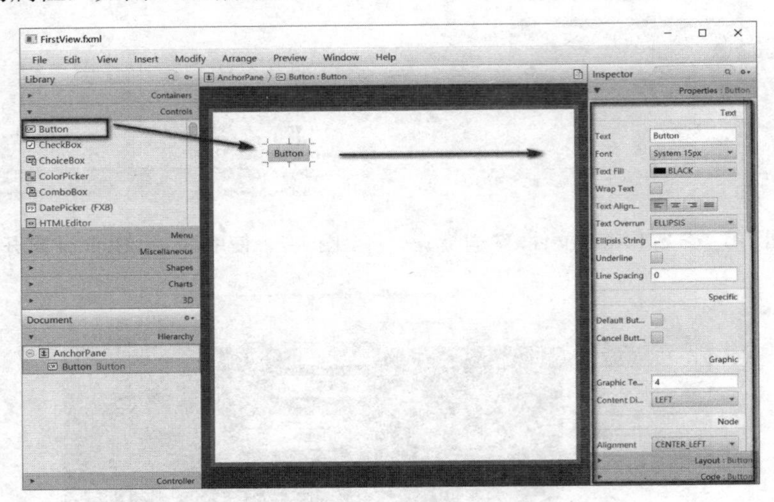

图 7-10　Button 控件添加

(3) Label 控件。

Label 可以用来显示一个文本或者图形图标。从左边的控件窗格中找到 Label 控件，并将其拖放到容器的对应位置，便可以在右边的属性窗格中编辑 Label 的基本属性。对于图形和图标属性，可以通过代码或 CSS 文件的方式修改，如图 7-11 所示。

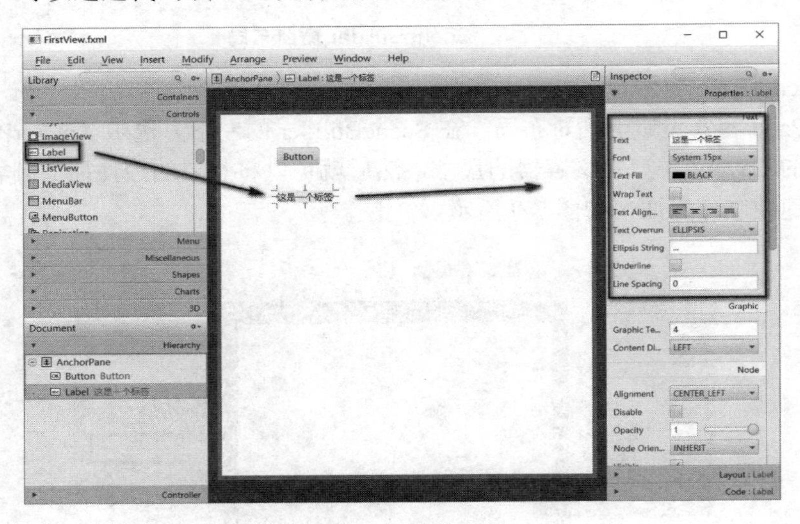

图 7-11　Label 控件添加

(4) RadioButton 控件。

RadioButton 是单选按钮，通常用来让用户从列表中选择一个项目，例如选择题中的单选题。添加多个 RadioButton 后，只需要将它们的 Toggle Group 设置为同一个值，用户就只能选中其中的一个项目，如图 7-12 所示。

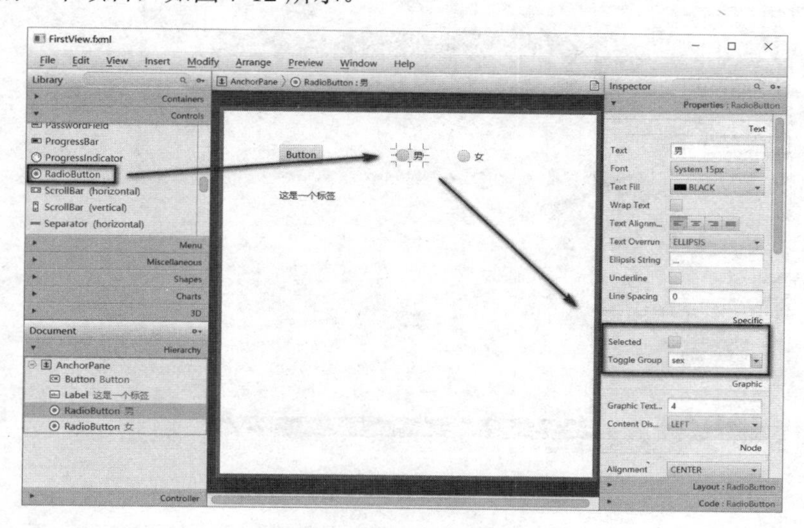

图 7-12　添加 RadioButton 控件

(5)　CheckBox 控件。

CheckBox 是复选框，允许用户同时选择多个项目。例如，让用户选择兴趣爱好，如图 7-13 所示。

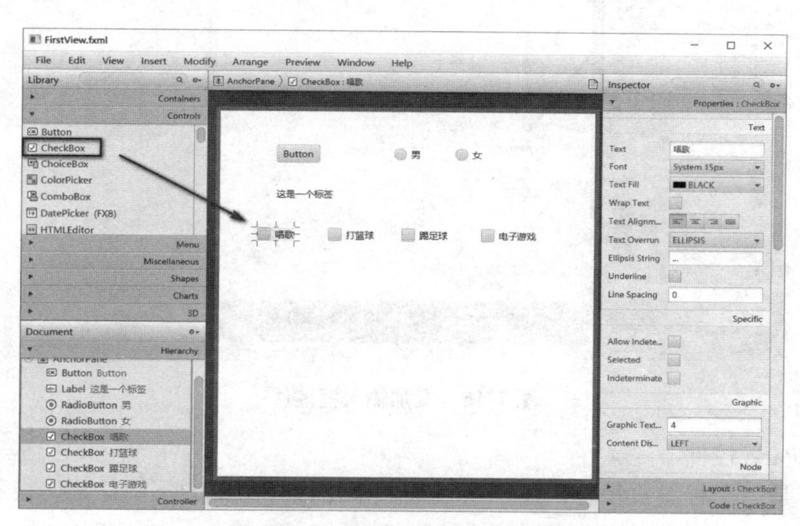

图 7-13　添加 CheckBox 控件

(6)　ComboBox 控件。

ComboBox 是一个组合框，用户可以通过下拉列表的形式来选择多个项目中的一个。组合框中的下拉列表内容必须在 Java 代码中进行添加，将在下一任务中介绍，如图 7-14 所示。

(7)　TextField 和 PasswordField 控件。

TextField 和 PasswordField 都可以用来接收用户输入的文本信息，不同的是，前者对用户来说输入内容是可见的，后者输入的内容对用户来说是不可见的，显示为黑色圆点。通常分别用来输入用户的用户名和密码，如图 7-15 所示。

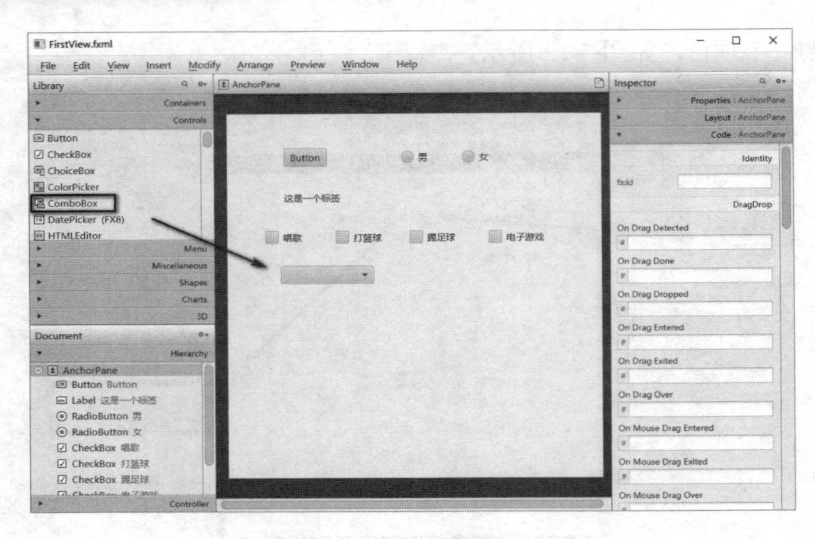

图 7-14 添加 ComboBox 控件

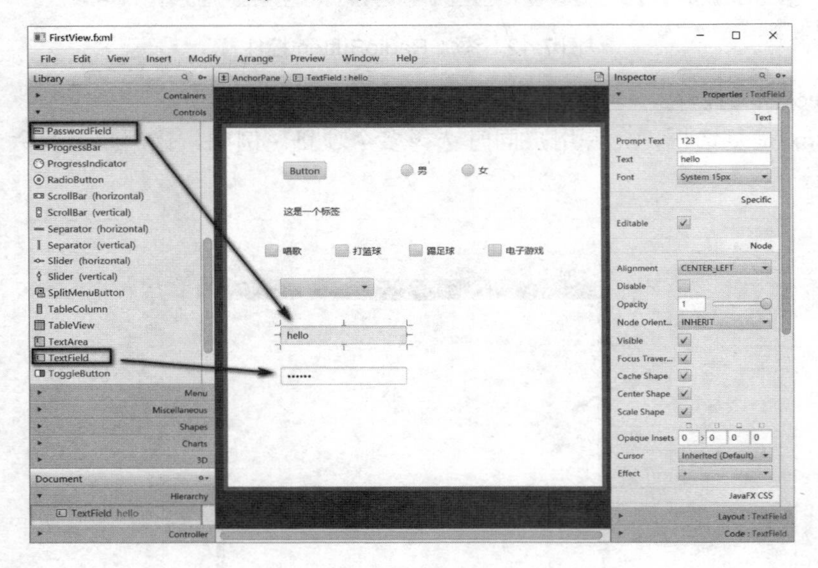

图 7-15 添加输入框控件

(8) TabPane 控件。

TabPane 是一个标签页控件，用户可以通过单击控件头部的不同标签切换不同的界面。当需要在一个窗口中为用户提供多个不同界面时可使用该控件，如图 7-16 所示。

7. 加载 FXML 文件

设计好界面后保存文件，在 Eclipse 中刷新工程以保证 FXML 文件的修改同步到工程中。在本项目知识解析中介绍了 JavaFX 工程的入口，可以将初始化好的场景设置到 Stage 中进行展示。现在就可以通过 fxml 文件初始化一个场景 Scene，修改入口类中 start 方法中的代码，先初始化一个 fxml 文件加载器 FXMLLoader，调用 setLocation 方法给加载器设置 FXML 文件路径。然后调用加载器的 load 方法获取场景最外层的容器对象。通过容器对象初始化一个场景，并将它设置给 Stage，最后调用 Stage 的 show 方法完成 FXML 文件到界面的渲染。

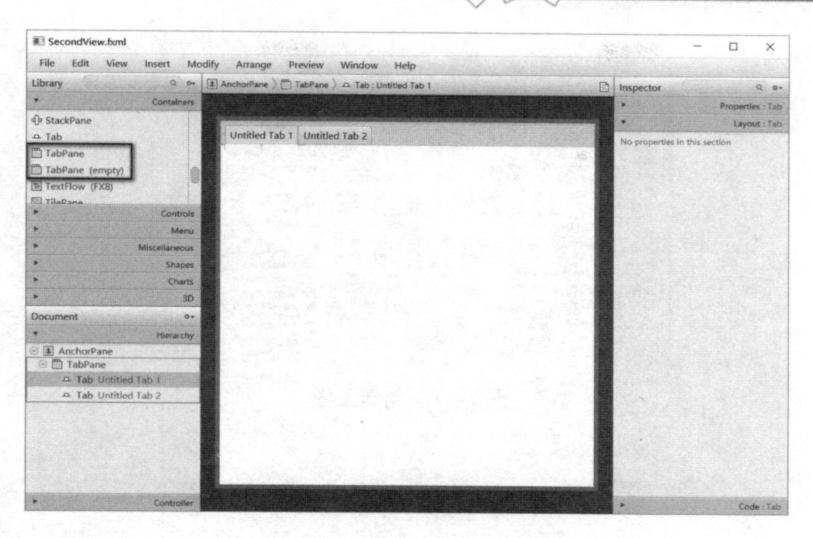

图 7-16 添加 TabPane 控件

```
1.  public void start(Stage primaryStage)
2.  {
3.      try {
4.          //初始化 fxml 文件加载器
5.          FXMLLoader loader = new FXMLLoader();
6.          //设置加载的 fxml 文件路径
7.          loader.setLocation(getClass().getResource("FirstView.fxml"));
8.          //获取 fxml 文件最外层容器
9.          AnchorPane view = loader.load();
10.         //初始化场景
11.         Scene scene = new Scene(view);
12.         //将场景设置给应用的顶级容器 Stage
13.         primaryStage.setScene(scene);
14.         //让 Stage 显示
15.         primaryStage.show();
16.
17.     } catch(Exception e) {
18.         e.printStackTrace();
19.     }
20. }
```

【任务实施】

1. 任务分析

本任务要完成智慧园区的登录界面。

(1) 程序界面元素：用户名输入框、密码输入框、登录按钮、注册按钮。

(2) 通过 SceneBuilder 编辑并保存登录界面的 FXML 文件。

(3) 在程序入口类的 start 方法中加载登录界面。

2. 任务实施

(1) 创建 project7_task1 工程，在 src 目录下新建 com.nle.task1 包，如图 7-17 和图 7-18 所示。

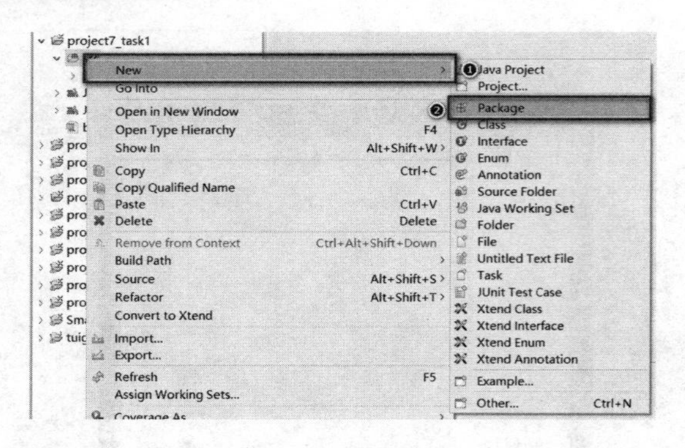

图 7-17　创建工程

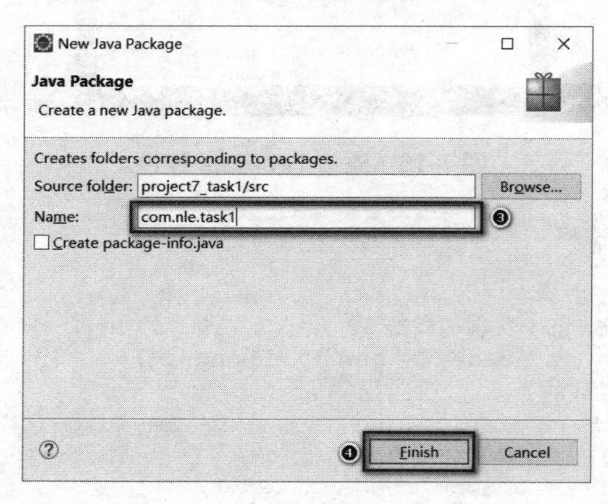

图 7-18　创建包

(2) 在 task1 包下新建入口类，命名为 MainApp，继承 Application，并重写 start 方法。在 MainApp 中新建 main 函数，并调用静态方法 launch，如图 7-19 所示。

New Java Class
Java Class
Create a new Java class.
Source folder: project7_task1/src Browse...
Package: com.nle.task1 Browse...
Enclosing type: Browse...
Name: MainApp
Modifiers: public package private protected
abstract final static
Superclass: javafx.application.Application Browse...
Interfaces: Add... Remove
Which method stubs would you like to create?
public static void main(String[] args)
Constructors from superclass
Inherited abstract methods
Do you want to add comments? (Configure templates and default value here)
Generate comments
Finish Cancel

图 7-19　创建程序入口类

```
1.  public class MainApp extends Application {
2.
3.      @Override
4.      public void start(Stage primaryStage) throws Exception {
5.      }
6.
7.      public static void main(String[] args) {
8.          launch(args);
9.      }
10.
11. }
```

(3) 在 task1 包下新建 FXML 文件，命名为 LoginView.fxml，使用 SceneBuilder 打开该文件进行界面的编辑，如图 7-20 和图 7-21 所示。

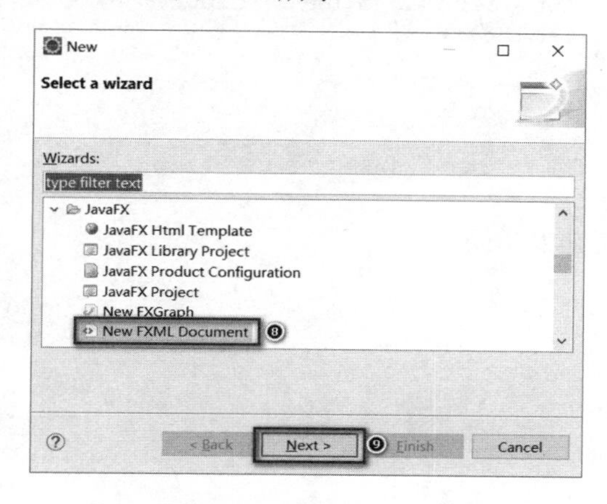

图 7-20　创建 FXML 文件

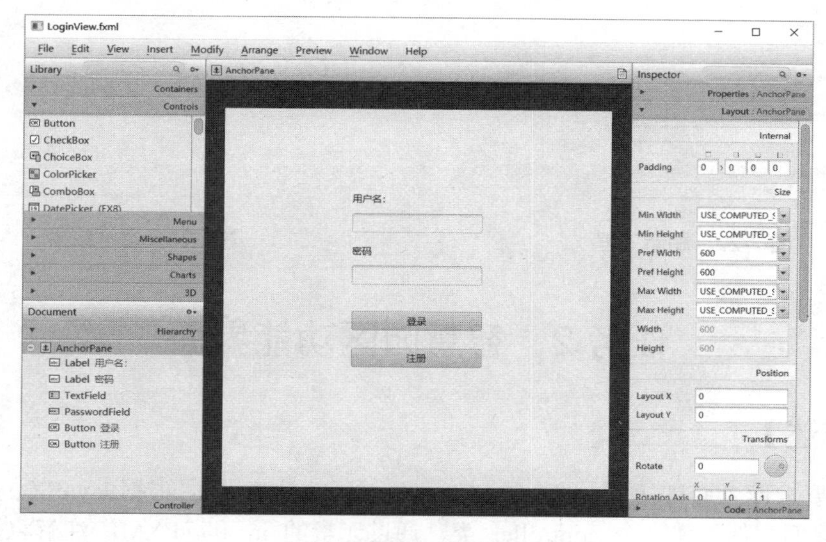

图 7-21　编辑登录界面

(4) 界面编辑完成后刷新工程，在 MainApp 的 start 方法中加载 FXML 文件，运行程序，查看界面效果，如图 7-22 所示。

```
1.  @Override
```

```
2.   public void start(Stage primaryStage) throws Exception {
3.
4.       try {
5.           //初始化 fxml 文件加载器
6.           FXMLLoader loader = new FXMLLoader();
7.           //加载 fxml 文件
8.           loader.setLocation(getClass().getResource("LoginView.fxml"));
9.           AnchorPane view = loader.load();
10.          Scene scene = new Scene(view);
11.          //设置窗口标题
12.          primaryStage.setTitle("智慧园区登录");
13.          primaryStage.setScene(scene);
14.          primaryStage.show();
15.      } catch (Exception e) {
16.          // TODO Auto-generated catch block
17.          e.printStackTrace();
18.      }
19.
20.  }
```

3. 运行结果

运行程序，结果如图 7-23 所示。

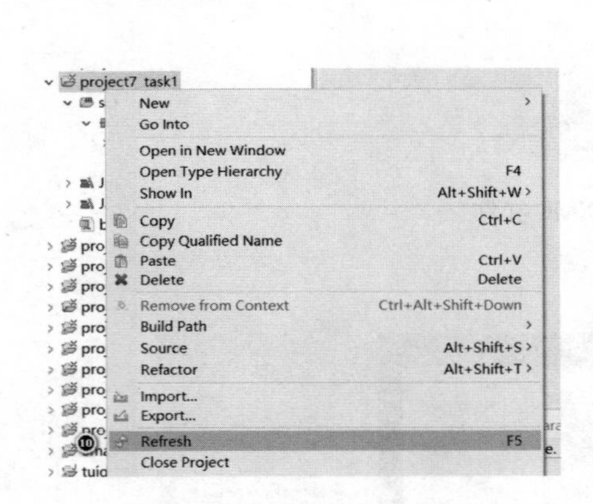

图 7-22　刷新工程

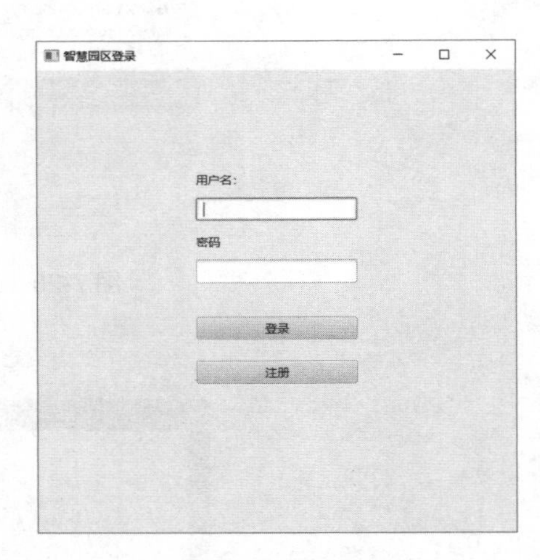

图 7-23　运行结果

任务 2　智慧园区功能界面

【任务描述】

本任务要求读者通过任务 1 的步骤制作智慧园区的功能界面。根据 JavaFX 的 mvc 设计，为 FXML 界面文件创建一个 controller 类，通过元素的 id 和@FXML 注解将控件映射到 controller 类中，便可以通过 API 对界面的控件进行编辑，包括控件的样式和各种属性。智慧园区的功能界面包括监控园区环境的人体检测、二氧化碳监测、噪声监测、温湿度监测，报警灯报警情况的显示，以及手动控制园区设备按钮等。控件添加好后在 controller 中美化界面控件。任务清单如表 7-2 所示。

表 7-2 任务清单

任务课时	4 课时	任务组员数量	建议 1 人
任务组采用设备	无		

【知识解析】

1. JavaFX 元素的 id

为了将界面上的各个控件映射到 Java 代码中的不同属性，需要为界面上的控件添加不同的 id。选中控件后，在 SceneBuilder 右侧的 Code 窗格中进行设置。映射到 Java 类中以后，元素的 id 作为类中的属性名，如图 7-24 和图 7-25 所示。

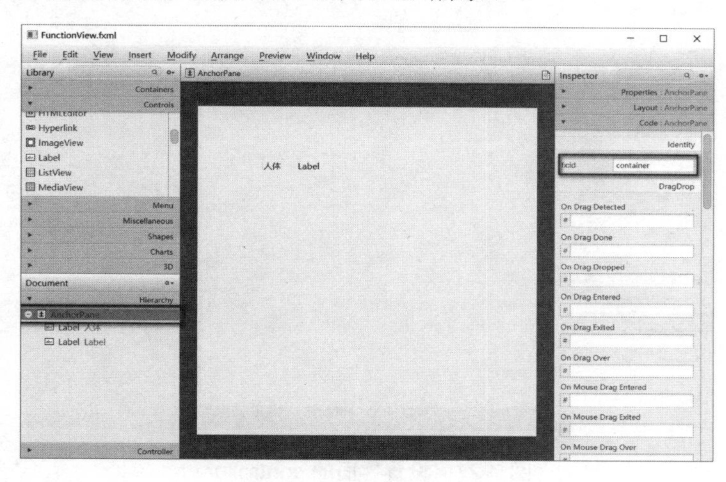

图 7-24 编辑控件 id

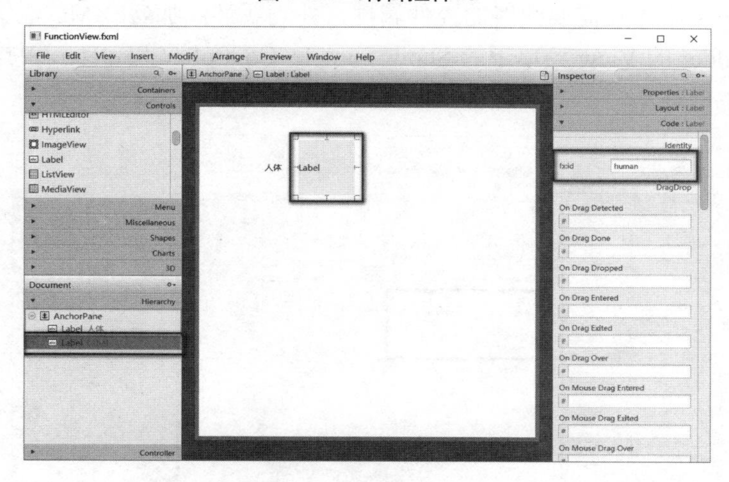

图 7-25 选择控件进行编辑

2. JavaFX 界面的 controller

按照 MVC 的设计模式将 JavaFX 应用程序划分为 3 个部分：模型层、视图层和控制层。按照任务 1 准备好程序的入口和 FXML 文件，JavaFX 应用程序需要为每个 FXML 文件创建一个控制类 controller，通常命名为 XxxxViewController。读者在实际开发中可以为每个部分分别建立自己的包。controller 类创建好后可以在 SceneBuilder 左下方的 Controller 窗格中将完整的类名填写到 Controller Class 中，如图 7-26 和图 7-27 所示。

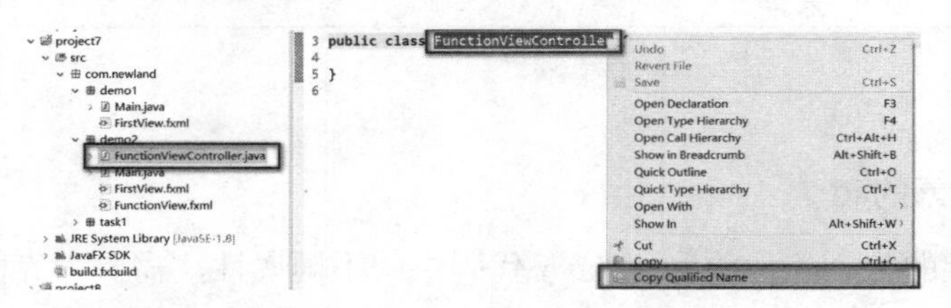

图 7-26　获取 controller 完整类名

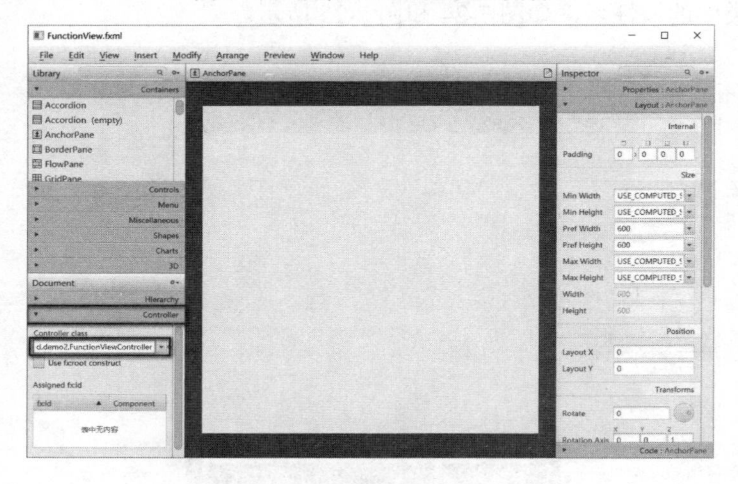

图 7-27　设置界面的 controller

在 controller 类中添加需要映射的控件属性，并为属性添加@FXML 注解。这些代码也可以在 SceneBuilder 的 View 菜单下的 Show Sample Controller Skeleton 中找到，如图 7-28 所示。

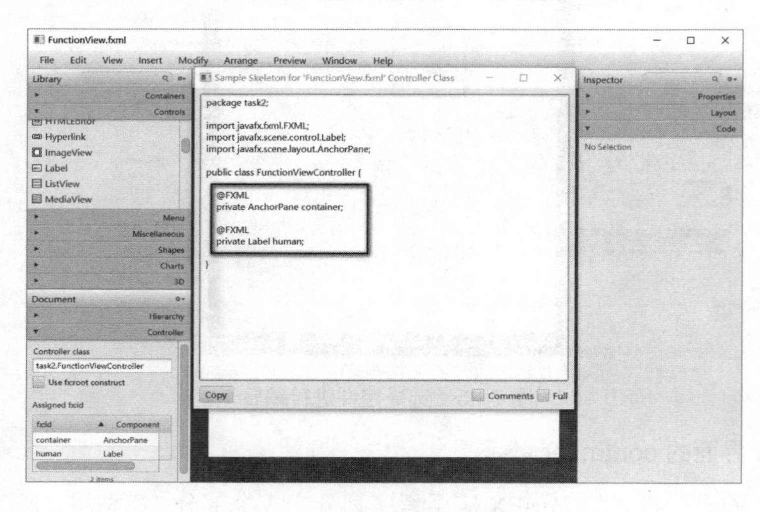

图 7-28　添加控件属性

当需要修改界面元素的样式或属性时，可以为 controller 添加 initialize 方法，并为方法添加@FXML 注解。然后在 initialize 方法中调用元素的方法来修改元素。

```
1.  public class FunctionViewController
2.  {
3.      @FXML
4.      private AnchorPane container;
5.      @FXML
6.      private Label bg;
7.      @FXML
8.      private Label human;
9.      @FXML
10.     private void initialize() {
11.         //设置元素属性
12.        human.setText("无人");
13.         //设置标签背景图片
14.         bg.setGraphic(new ImageView("file:images/bg.jpg"));
15.
16.     }
17.
18. }
```

【任务实施】

1. 任务分析

本任务要完成智慧园区的功能界面。

(1) 程序界面元素，包括温湿度、二氧化碳、噪音数值标签；报警灯报警情况显示标签；灯、风扇控制开关按钮。

(2) 将元素都拖放到适当位置后，添加不同的 id，创建 controller 类并在 initialize 方法中将各个元素样式设置好。

2. 任务实施

(1) 新建工程 project7_task2，新建包 com.nle.task2，并准备好程序入口类。

(2) 新建 FunctionView.fxml 文件，在 SceneBuilder 中设计好界面，如图 7-29 所示。

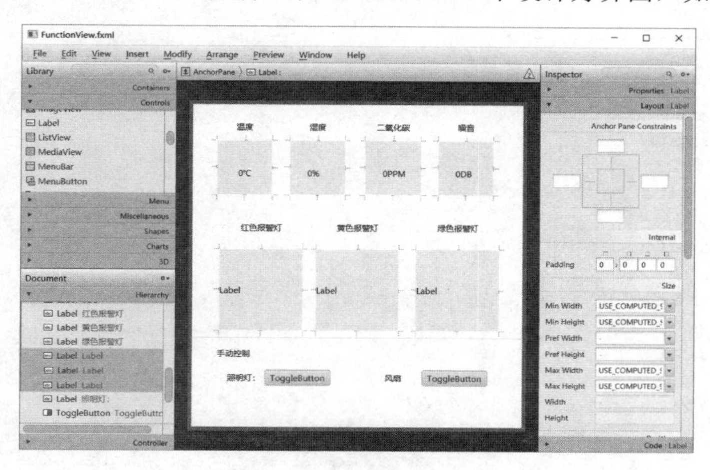

图 7-29 功能界面

(3) 为功能界面的各个控件添加 id，如图 7-30 所示。

(4) 在工程目录下创建 images 文件夹，将准备好的图片素材复制过来，如图 7-31 所示。

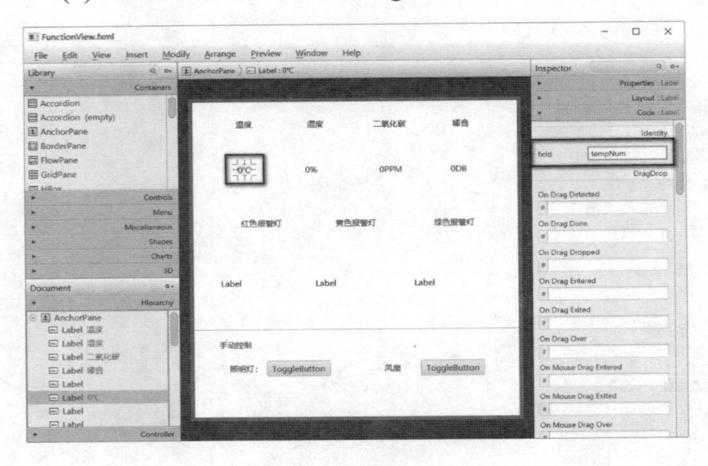

图 7-30 编辑控件 id

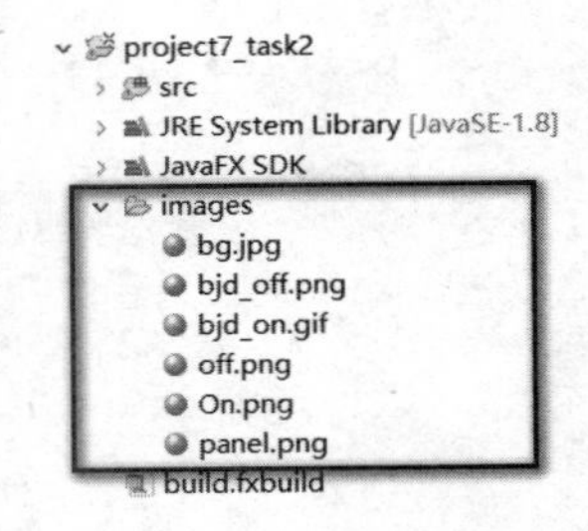

图 7-31 添加图片素材

(5) 创建功能界面的 controller 类，在 initialize 方法中初始化各个控件的样式和属性。

```
1.   public class FunctionViewController
2.   {
3.       @FXML
4.       private AnchorPane container;
5.       @FXML
6.       private Label yellowAlarmBg;
7.       @FXML
8.       private Label co2Num;
9.       @FXML
10.      private ToggleButton blowerBtn;
11.      @FXML
12.      private Label noiseNum;
13.      @FXML
14.      private Label noiseBg;
15.      @FXML
16.      private Label greenAlarmBg;
17.      @FXML
18.      private Label tempBg;
19.      @FXML
20.      private Label co2Bg;
21.      @FXML
22.      private Label tempNum;
23.      @FXML
24.      private Label redAlarmBg;
25.      @FXML
26.      private Label humiBg;
27.      @FXML
28.      private ToggleButton lightBtn;
29.      @FXML
30.      private Label humiNum;
31.      @FXML
32.      private void initialize() {
33.          //设置 Label 背景图片
34.          tempBg.setGraphic(new ImageView("file:images/panel.png"));
35.          humiBg.setGraphic(new ImageView("file:images/panel.png"));
```

```
36.         co2Bg.setGraphic(new ImageView("file:images/panel.png"));
37.         noiseBg.setGraphic(new ImageView("file:images/panel.png"));
38.         redAlarmBg.setGraphic(new ImageView("file:images/bjd_off.png"));
39.         yellowAlarmBg.setGraphic(new ImageView("file:images/bjd_off.png"));
40.         greenAlarmBg.setGraphic(new ImageView("file:images/bjd_off.png"));
41.         //设置开关按钮初始图片
42.         lightBtn.setBackground(null);
43.         lightBtn.setGraphic(new ImageView("file:images/on.png"));
44.         blowerBtn.setBackground(null);
45.         blowerBtn.setGraphic(new ImageView("file:images/on.png"));
46.     }
47. }
```

3. 运行结果

运行程序，结果如图 7-32 所示。

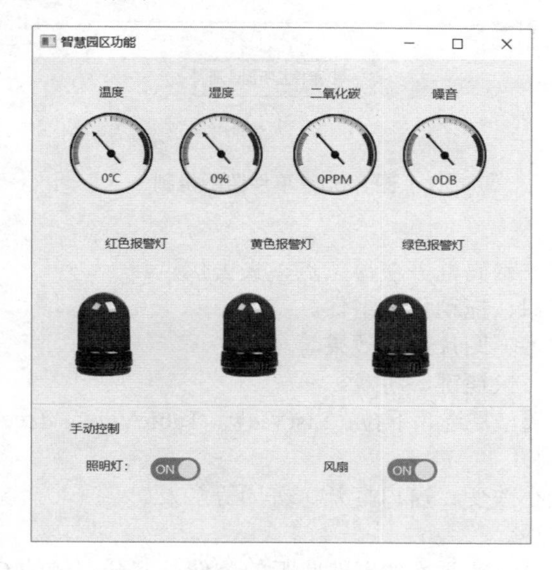

图 7-32 运行结果

任务 3 智慧园区系统事件的监听

扫码观看视频讲解

【任务描述】

对于应用程序的图形化界面来说，与用户的交互是通过事件监听来实现的。前两个任务已经设计完成智慧园区的登录界面和功能界面。本任务将为前两个任务设计好的界面控件添加监听，实现登录按钮的监听、获取输入框文本、界面之间的跳转、开关按钮的监听等交互功能。任务清单如表 7-3 所示。

表 7-3 任务清单

任务课时	4 课时	任务组员数量	建议 1 人
任务组采用设备	无		

【知识解析】

(1) 事件监听。

事件监听就是当用户通过鼠标或键盘等对程序控件进行操作时给控件添加对应的行为，使控件具备相应的功能。与 Swing、Android 不同，JavaFX 的控件事件是通过属性绑定的，更接近 HTML 的事件绑定方式。如图 7-33 所示为事件监听机制。

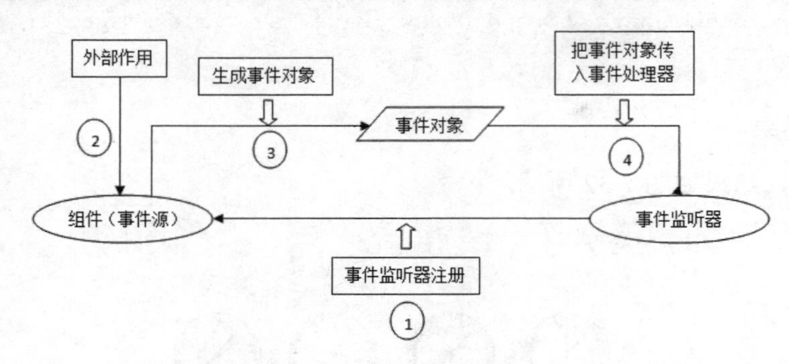

图 7-33 事件监听机制

(2) 常见事件。

焦点事件：文本框、密码框等获得焦点、失去焦点。

拖曳事件：拖动窗口、拖动目标组件。

鼠标事件：单击文本、图片进入效果等。

键盘事件：功能键、快捷键、游戏。

列表选项事件：单选、复选、下拉、ListView、TableView、TreeView 等选项发生改变，获得最新选取的值。

窗口事件：窗口大小改变、窗口打开、关闭等触发。

(3) 事件监听的实现。

在 SceneBuilder 中选中需要添加事件监听的控件，打开右侧的 Code 窗格。在 onAction 下输入事件响应的方法名，如图 7-34 所示。

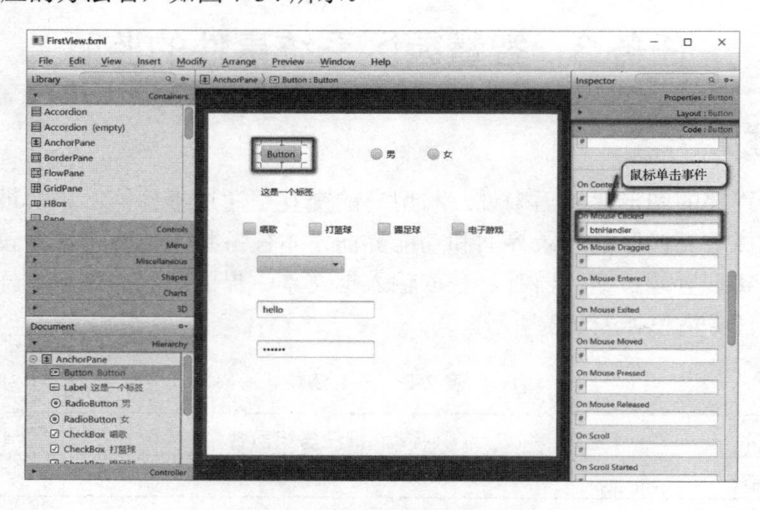

图 7-34 编辑控件事件监听

在 FXML 界面对应的 controller 中添加 btnHandler 方法，方法必须添加@FXML 注解，形式参数 XxxxxEvent 可选择性添加。事件对象中包含单击事件或键盘事件的属性参数，例如，单击的是鼠标左键还是鼠标右键可以通过添加 MouseEvent 参数，然后根据 event.getButton()来判断。

```
1.  @FXML
2.  void btnHandler(MouseEvent event)
3.  {
4.      if(event.getButton()==MouseButton.PRIMARY) {
5.          System.out.println("鼠标左键");
6.      }else if(event.getButton()==MouseButton.SECONDARY){
7.          System.out.println("鼠标右键");
8.      }
9.  }
```

(4) JavaFX 界面场景的跳转。

通常一个应用程序都包含一个以上的场景，用户在进行各种操作的过程中就必须实现场景的切换。JavaFX 的实现方式就是给 Stage 加载不同的场景 Scene。可以将这些代码进行封装，在需要跳转的地方调用即可。

```
1.  public static void showScene(Stage stage, String fxmlName)
2.  {
3.      try {
4.          //初始化 FXML 文件加载器
5.          FXMLLoader loader = new FXMLLoader();
6.          // 设置加载的 FXML 文件路径
7.          loader.setLocation(Main.class.getResource(fxmlName));
8.          // 获取 FXML 文件最外层容器
9.          AnchorPane view = loader.load();
10.         // 初始化场景
11.         Scene scene = new Scene(view);
12.         // 将场景设置给应用的顶级容器 Stage
13.         stage.setScene(scene);
14.         // 让 Stage 显示
15.         stage.show();
16.     } catch (IOException e) {
17.         // TODO Auto-generated catch block
18.         e.printStackTrace();
19.     }
20.
21. }
```

(5) JavaFX 的弹出框。

程序与用户交互过程中需要给用户提示信息时，弹出框是一个很好的选择。JavaFX 提供了 Alert 类用来显示信息、警告、错误、用户确认等内容的提示信息。

```
1.  //初始化弹框
2.  Alert alert = new Alert(AlertType.WARNING);
3.  //设置标题
4.  alert.setTitle("警告");
5.  //设置提示信息
6.  alert.setContentText("这是一个警告");
7.  //显示弹框
8.  alert.show();
```

弹框效果如图 7-35 所示。

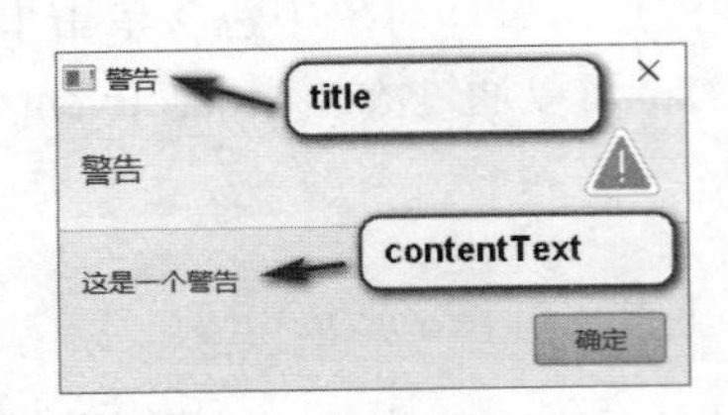

图 7-35　弹框

【任务实施】

1. 任务分析

界面：智慧园区登录界面、智慧园区功能界面鼠标。
功能：
(1)　单击“登录”按钮，获取文本框和密码框的文本。
(2)　单击“登录”按钮，判断用户名和密码，跳转到智慧园区功能界面或通过弹框提示错误信息。
(3)　为功能界面的开关按钮添加监听，单击切换背景图片。

2. 任务实施

(1)　新建工程 project7_task3，新建 com.nle.task3 包，准备好程序入口、登录界面、登录界面 controller 类、功能界面、功能界面 controller 类，如图 7-36 所示。

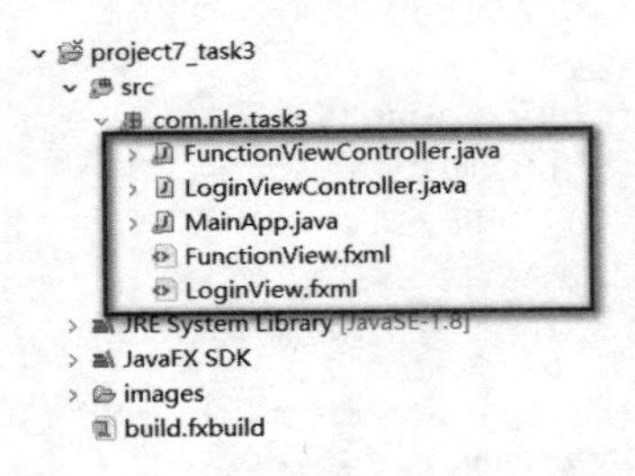

图 7-36　FXML 文件与 controller 类

(2)　为登录界面的登录按钮添加 onAction 监听 loginHandler 方法，在该方法中获取输入框的文本，并做简单的判断后通过 Alert 控件给用户发送提示。

```
1.   public class LoginViewController
2.   {
3.       @FXML
4.       private TextField userNameField;
5.       @FXML
6.       private Button loginBtn;
7.       @FXML
8.       private PasswordField pwdField;
9.       @FXML
10.      private Button regBtn;
11.      //提示信息弹窗
12.      private Alert alert;
```

```
13.     @FXML
14.     private void initialize() {
15.         alert = new Alert(AlertType.WARNING);
16.         alert.setTitle("温馨提示");
17.     }
18.
19.     @FXML
20.     void loginHandler(ActionEvent event) {
21.         String username = userNameField.getText();
22.         String pwd = pwdField.getText();
23.         if("admin".equals(username)&&"admin".equals(pwd)) {
24.
25.         }else {
26.             alert.setContentText("用户名或密码不正确，请重试！");
27.             alert.show();
28.         }
29.
30.     }
31. }
```

运行结果如图 7-37 所示。

图 7-37　运行结果

(3) 登录成功后通过自定义的 showScene 方法跳转至功能界面。

```
1.  @FXML
2.  void loginHandler(ActionEvent event) {
3.      String username = userNameField.getText();
4.      String pwd = pwdField.getText();
5.      if("admin".equals(username)&&"admin".equals(pwd)) {
6.              MainApp.showScene(MainApp.primaryStage, "FunctionView.fxml");
7.      }else {
8.          alert.setContentText("用户名或密码不正确，请重试！");
9.          alert.show();
10.     }
11.
12. }
```

(4) 为功能界面开关按钮添加监听，判断按钮的 select 状态切换背景图片。

```
1.  @FXML
```

```
2.    void lightHandler(ActionEvent event) {
3.        if(lightBtn.isSelected()) {
4.            lightBtn.setGraphic(new ImageView("file:images/off.png"));
5.        }else {
6.            lightBtn.setGraphic(new ImageView("file:images/on.png"));
7.        }
8.    }
9.
10.   @FXML
11.   void blowerHandler(ActionEvent event) {
12.       if(blowerBtn.isSelected()) {
13.           blowerBtn.setGraphic(new ImageView("file:images/off.png"));
14.       }else {
15.           blowerBtn.setGraphic(new ImageView("file:images/on.png"));
16.       }
17.   }
```

3. 运行结果

运行程序，结果如图 7-38 所示。

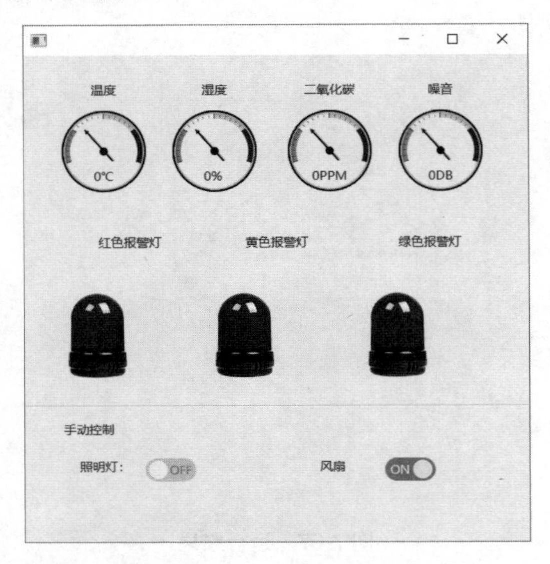

图 7-38　运行结果

思考与练习

1. 简述 JavaFX 环境的安装过程。
2. 思考观察者设计模式及其与监听设计模式的区别。
3. 通过 JavaFX 设计一个自己的应用界面。

【项目描述】

数据的存储是程序运行的基础，因此选用合适的存储方式对于编码过程至关重要。本项目着重介绍 Java 中各种集合的使用。通过智慧园区项目中的传感器数据存储功能、登录与注册功能、Map 存储采集器数据功能，讲解各个集合在开发过程中的应用，让读者了解不同集合的存储结构和特性。学会在不同的应用开发场景下，正确地选用集合来完成需求。具体任务列表如图 8-1 所示。

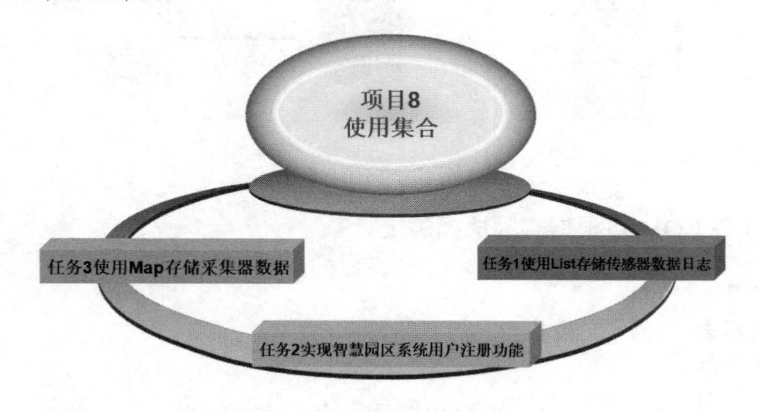

图 8-1　项目 8 任务列表

【学习目标】

知识目标： 会描述 Java 集合框架体系；会描述 List 接口的特点和使用；会描述 Set 接口的特点和使用；会描述迭代器与集合之间的关系；会描述 Map 接口的特点和使用。

技能目标：能使用 List 存储传感器数据；能使用 Set 实现用户的注册功能；能使用迭代器遍历各类集合；能使用 Map 接口实现键值对数据的存储。

任务 1　使用 List 存储传感器数据日志

扫码观看视频讲解

【任务描述】

本任务要求读者实现对智慧园区中传感器数据日志的存储。为了记录设备历史数据和运行情况，需要不断地通过采集器采集数据，并保存历史数据，定时将数据日志存储到日志文件。在此之前，存储相同类型的数据可以用数组来实现，但是传感器的数据日志是不断增加的，也就是说无法提前知道数据的长度。要实现动态地分配存储空间，就可以通过 Java 集合框架中 List 接口下的子类实现对传感器数据的存储。任务清单如表 8-1 所示。

表 8-1　任务清单

任务课时	4 课时	任务组员数量	建议 3 人
任务组采用设备	1 个 ZigBee 光照传感器 1 个 ZigBee 协调器 相关电源、导线、工具		

【拓扑图】

本任务拓扑图如图 8-2 所示。

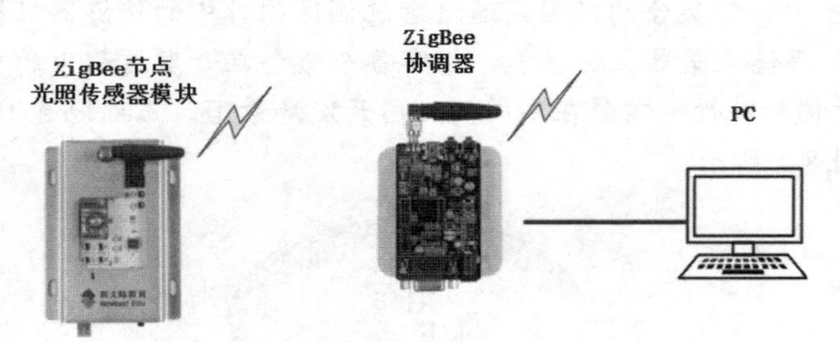

图 8-2　拓扑图

备注：本任务 COM 口根据实际情况确定。

【知识解析】

1. Java 集合概述

在实际应用开发过程中，如何组织数据是一个非常重要的问题，在之前的程序中使用数组是一个很好的选择，但需要事先知道将要保存的对象的数量，因为数组长度是不可变的，如果需要保存一个动态变化的数据集合，数组就显得力不从心了。为了在程序中可以保存数目不确定的对象，可以使用 Java JDK 中提供的集合类，这些类可以存储任意类型的对象并且长度可变。集合类位于 java.util 包中，在使用时要注意导包。

Java 集合类是指 Java 中设计用于容纳对象的各种数据结构，通常也称为容器类。Java 集合框架则是指由 Java 集合类以及相关操作构成的体系结构，包含接口、接口的实现类和对集合运算的算法，其层次结构如图 8-3 所示。

集合按照其存储结构可以分为两个不同的概念。

(1) Collection：用于存储一系列符合某种规则的元素。它有两个重要的子接口，分别是 List 和 Set。

(2) Map：用于存储具有键(key)、值(val)对关系的元素，可以通过指定的 key 找到对应的 val。Map 接口的主要实现类是 HashMap 和 TreeMap。

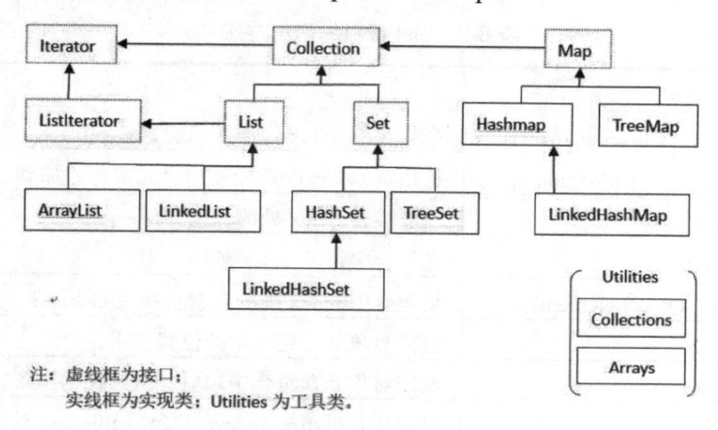

图 8-3　集合框架

2. Collection 接口

Collection 接口是 List 和 Set 的父接口，在 Collection 中定义了通用的一些方法用于操作 List 和 Set 集合的增删改查操作，如表 8-2 所示。

表 8-2　Collection 接口的常用方法

方　法	描　述
boolean add(Object o)	向集合中添加一个元素
boolean addAll(Collection c)	将指定 Collection 中的所有元素都添加到集合中
void clear()	移除集合中的所有元素
boolean remove(Object o)	在集合中移除指定的元素
boolean isEmpty()	判断集合是否为空
Iterator iterator()	迭代器，用于遍历集合中的元素
int size()	集合中成员的个数
boolean contains(Object o)	判断集合中是否包含元素 o

表 8-2 只是列出了 Collection 中的部分方法，选自 Java API 文档，后面所有的方法说明都来自 Java API 文档，未列出的方法请参考 Java API 文档。

3. List 接口

List 接口继承自 Collection 接口，通常称为 List 集合。在 List 集合中允许出现重复的元素，所有的元素都以一种线性方式进行存储，其特点如下。

- List(序列)，元素有序，并且可重复。
- List 可以精确地控制元素的插入位置，或删除指定位置的元素。
- List 作为 Collection 集合的子接口，除了继承 Collection 接口中的全部方法外，还增加了特有的可以操作下标的方法，如表 8-3 所示。

表 8-3　List 接口的常用方法

方　法	描　述
boolean add(int index，Object element)	在集合的 index 处插入元素 element
boolean addAll(int index，Collection c)	将指定 Collection 中的所有元素都添加到集合的 index 处
Object get(int index)	获取集合 index 处的元素
Object remove(int index)	在集合中移除 index 处的元素
Object set(int index，Object element)	将集合中 index 处的元素替换成 element
Int indexOf(Object o)	返回对象 o 在集合中的位置索引
Int lastIndexOf(Object o)	返回对象 o 在集合中最后一次出现的位置索引
List subList(int fromindex，int toindex)	返回从索引 fromindex(包括)到 toindex(不包括)处所有元素组成的子集合

4. ArrayList 集合

ArrayList 是 List 接口的实现类。每个 ArrayList 实例都有一个容量，该容量是指用来存储列表元素的数组的大小，它总是至少等于列表的大小，随着向 ArrayList 中不断添加元素，其容量也自动增长。接下来以一个例子来说明 ArrayList 的基本使用。

```
1.   public class ArrayListTest {
2.      public static void main(String[] args) {
3.          //声明一个 ArrayList 实现类对象
4.          ArrayList list=new ArrayList();
5.          //插入字符串 abc 到集合中
6.          list.add("abc");
7.          //插入整数 2 到集合中
8.          list.add(2);
9.          //在集合下标为 0 的位置插入整数 5
10.         list.add(0,5);
11.         //可以插入空元素
12.         list.add(null);
13.         //插入重复的整数 2
14.         list.add(2);
15.         System.out.println("集合中成员的个数="+list.size());
16.         //输入集合中的元素，因为 List 接口可以精确地操作下标，所以用下标控制
17.         for(int i=0;i<list.size();i++)
18.         {
19.            System.out.print(list.get(i)+" ");
20.         }
```

```
21.         //移除元素 abc
22.         list.remove("abc");
23.         System.out.println("\n*************");
24.         //再次输出 list 中的成员
25.         System.out.println(list);
26.     }
27. }
```

运行结果如图 8-4 所示。

```
集合中成员的个数=5
5 abc 2 null 2
*************
[5, 2, null, 2]
```

图 8-4　运行结果

在编译上述代码时会出现图 8-5 所示的警告，这是因为在使用 List 时没有显式指定集合中存储什么类型的元素，会产生安全隐患，后面将用泛型来解决这个安全隐患。

图 8-5　泛型声明

如果编译时出现红叉错误，这是因为没有导包，可以按提示进行导包，也可以用 Ctrl+Shift+O 组合键进行自动导包。图 8-6 中应该选择 Import 'List' (java.util)。

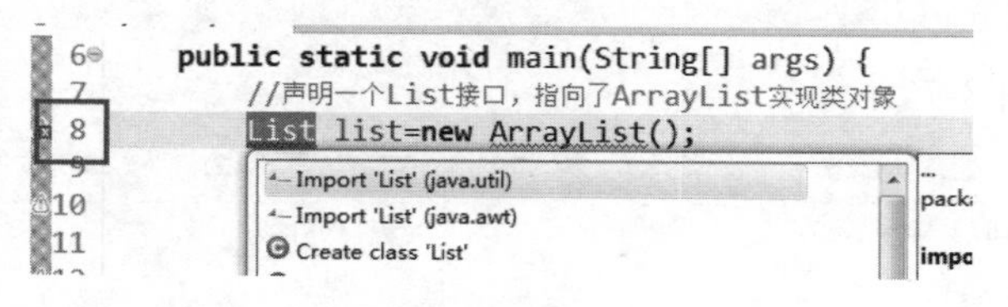

图 8-6　导包

5. LinkedList 集合

ArrayList 允许快速访问元素，但是在插入和删除元素时，元素会大量移动。List 接口的另一个实现类是 LinkedList，该集合内部维护了一个双向循环的链表，链表中的每一个元素都使用引用的方式来记住它的前一个元素和后一个元素，从而可以将所有的元素连接起来，添加和删除都只需要改变引用关系即可。LinkedList 中常用的方法如表 8-4 所示。

表 8-4　LinkedList 的常用方法

方　法	描　述
void addFirst(Object o)	在集合的开头插入元素 o
void addLast(Object o)	在集合的结尾插入元素 o

续表

方　法	描　述
Object getFirst()	获取集合中的第一个元素
Object getLast()	获取集合中的最后一个元素
Object removeFirst()	移除集合中的第一个元素
Object removeLast()	移除集合中的最后一个元素

接下来以一个例子来说明 LinkedList 的使用。

```
1.   public class LinkedListTest {
2.       public static void main(String[] args) {
3.           // 声明一个 LinkedList 实现类对象
4.           LinkedList list = new LinkedList();
5.           // 插入字符串 abc 到集合中
6.           list.add("abc");
7.           // 插入整数 2 到集合的最前面
8.           list.addFirst(2);
9.           // 在集合尾部位置插入整数 5
10.          list.addLast(5);
11.          System.out.println("集合中成员的个数=" + list.size());
12.          // 输入集合中的元素，因为 List 接口可以精确地操作下标，所以用下标控制
13.          for (int i = 0; i < list.size(); i++) {
14.              System.out.print(list.get(i) + " ");
15.          }
16.          System.out.println("\n*************");
17.          // 移除第一个元素
18.          list.removeFirst();
19.          // 再次输出 list 中的成员
20.          System.out.println(list);
21.      }
22.  }
```

运行结果，如图 8-7 所示。

图 8-7　运行结果

6. Iterator 接口

Iterator 接口主要用于迭代访问(即遍历)Collection 中的元素，又称为迭代器。它仅提供三个方法：hashNext()、next()、remove()。当调用集合的 iterator()时就会获得一个迭代器对象，该对象指向集合第一个元素的前面，通过 hasNext()判断集合中是否有下一个元素，如果有就用 next()取出该元素，如果没有则说明遍历到了集合的尾部；remove()则是用来通过迭代器删除集合中的成员的。需要特别说明的是，当通过迭代器获取 ArrayList 集合中的元素时，这些元素都被当成 Object 类型，如果想得到特定类型的元素，则需要进行强制类型转换。接下来以一个例子来说明 Iterator 的使用。

```
1.  public class IteratorTest {
2.      public static void main(String[] args) {
3.          ArrayList list=new ArrayList();
4.          list.add("a1");
5.          list.add("a2");
6.          list.add("a3");
7.          //获取集合 list 的迭代器对象
8.          Iterator it=list.iterator();
9.          //判断是否有下一个成员，如果有，返回真，循环继续，否则循环结束
10.         while(it.hasNext())
11.         {
12.             //next()可以让迭代器指向下一个成员并取出当前元素
13.             Object obj=it.next();
14.             System.out.print(obj+" ");
15.         }
16.     }
17. }
```

运行结果如图 8-8 所示。

```
<terminated> IteratorTest [Java Application] D:\java\bin\javaw.exe
a1 a2 a3
```

图 8-8　运行结果

在上述代码的 while 循环中，it.next()取出来的是 Object 类型，如果明知集合中是 String 类型，要用 String 类型输出，则必须强制类型转换成 String：

```
1.  while(it.hasNext())
2.  {
3.          //next()可以让迭代器指向下一个成员并取出
4.          //Object obj=it.next();
5.          String str=(String)it.next();
6.          System.out.print(str+" ");
7.  }
```

在使用迭代器 Iterator 对集合中的元素进行迭代时，如果同时调用了集合对象的 add 或 remove 方法，会出现异常。当用迭代器遍历到 a2 时，想在 a2 的后面添加 a99，程序运行就报错了。

```
1.  public class IteratorTest2 {
2.      public static void main(String[] args) {
3.          ArrayList list=new ArrayList();
4.          list.add("a1");
5.          list.add("a2");
6.          list.add("a3");
7.          Iterator it=list.iterator();
8.          while(it.hasNext())
9.          {
10.             String str=(String)it.next();
11.             //判断该元素是不是字符串 a2
12.             if(str.equals("a2"))
13.             {
14.                 //往集合中添加字符串 a99
```

```
15.                 list.add("a99");
16.             }
17.             System.out.print(str+" ");
18.         }
19.     }
20. }
```

运行结果如图 8-9 所示。

```
a1 a2 Exception in thread "main" java.util.ConcurrentModificationException
        at java.util.ArrayList$Itr.checkForComodification(ArrayList.java:909)
        at java.util.ArrayList$Itr.next(ArrayList.java:859)
        at com.nle.demo1.IteratorTest2.main(IteratorTest2.java:16)
```

图 8-9　并发访问异常

运行过程中出现了并发访问异常 ConcurrentModificationException，这个异常出现的原因是，集合中删除了元素会导致迭代器预期的迭代次数发生改变，导致迭代器的结果不准确。解决方式如下：迭代器可以改成专门针对 List 集合特有的 ListIterator，该迭代器本身多了一个 add 方法，这样就可以不用集合的 add 方法了，具体代码如下：

```
1.  public class IteratorTest2 {
2.      public static void main(String[] args) {
3.          ArrayList list=new ArrayList();
4.          list.add("a1");
5.          list.add("a2");
6.          list.add("a3");
7.          ListIterator  it=list.listIterator();
8.          while(it.hasNext())
9.          {
10.             String str=(String)it.next();
11.             //判断该元素是不是字符串 a2
12.             if(str.equals("a2"))
13.             {   //往集合中添加字符串 a99
14.                 it.add("a99");
15.             }
16.         }
17.         System.out.println(list);
18.     }
19. }
```

运行结果如图 8-10 所示。

图 8-10　运行结果

可以看到，确实已经实现在 a2 后面添加 a99。

7. forEach 遍历

除了可以用 Iterator 来遍历元素外，还可以用简化版的 foreach 循环来遍历数组或集合中的元素，语法如下：

```
1.   for(容器中元素类型 变量名：容器名){
2.       执行语句;
3.   }
```

foreach 循环会自动遍历容器中的每个元素，接下来用一个例子来说明 foreach 的使用。

```
1.   public class ForeachTest {
2.       public static void main(String[] args) {
3.           ArrayList list=new ArrayList();
4.           list.add("a1");
5.           list.add("a2");
6.           list.add("a3");
7.           for(Object obj:list)
8.           {
9.               System.out.print(obj+" ");
10.          }
11.      }
12.  }
```

运行结果如图 8-11 所示。

图 8-11 运行结果

【任务实施】

1. 任务分析

本任务要完成智慧园区项目的传感器数据日志输出。

(1) 界面包含元素：温度、湿度显示数值和标签，报警灯图片标签，实时日志显示文本框，采集开关按钮。

(2) 封装数据日志对象，包含采集时间、设备名称和数据信息。

(3) 导入所需 jar 包，采集设备数据。

(4) 将采集到的数据对象添加到 ArrayList 集合。本任务数据执行插入、删除操作较少，遍历较多，所以选择 ArrayList 作为数据容器。

2. 任务实施

(1) 新建 project8_task1 工程，在 src 目录下新建 com.nle.task1 包。导入教材配套资源提供的 jar 包，如图 8-12 所示。

(2) 在 task1 包下新建入口类，创建设备监控界面 FXML 文件 DeviceView.fxml，并创建对应的 controller 类，源码在配套资源\project8\com\nle\task1 下，如图 8-13 所示。

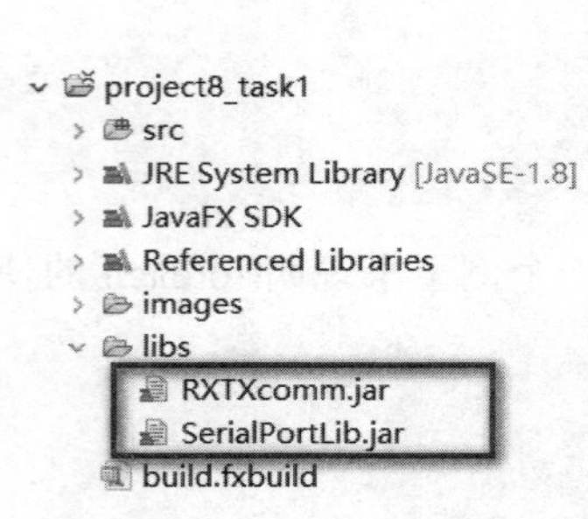

图 8-12 导入依赖包

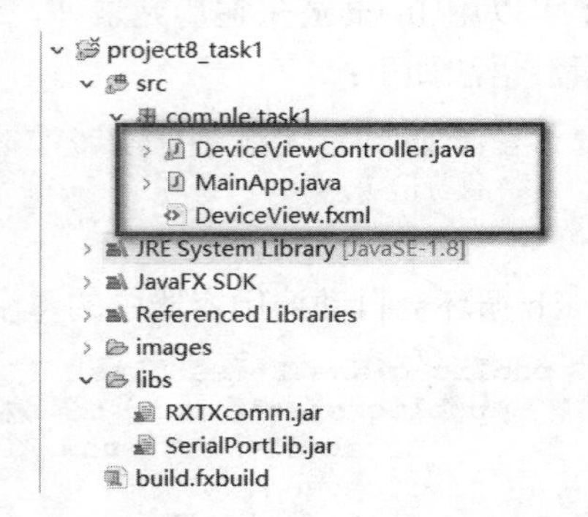

图 8-13 创建 FXML 文件和 controller 类

(3) 查看协调器连接串口名称。

(4) 在 task1 包下新建日志类 Log。

```
1.  public class Log {
2.      private String time;
3.      private String deviceName;
4.      private String msg;
5.      public Log(String time, String deviceName, String msg) {
6.          super();
7.          this.time = time;
8.          this.deviceName = deviceName;
9.          this.msg = msg;
10.     }
11.     public String getTime() {
12.         return time;
13.     }
14.     public void setTime(String time) {
15.         this.time = time;
16.     }
17.     public String getDeviceName() {
18.         return deviceName;
19.     }
20.     public void setDeviceName(String deviceName) {
21.         this.deviceName = deviceName;
22.     }
23.     public String getMsg() {
24.         return msg;
25.     }
26.     public void setMsg(String msg) {
27.         this.msg = msg;
28.     }
29.     @Override
30.     public String toString() {
31.         return time + ", 设备: " + deviceName + ", 信息: " + msg + "\n";
32.     }
33. }
```

(5) 在 task1 下创建串口监听类 ZigBeeListener，在重写的 serialEvent 方法中对读取到的串口数据进行解析，并保存历史数据。

```
1.  public class ZigBeeListener implements SerialPortEventListener {
2.
3.      private SerialPort portZigBee;
4.      private DeviceViewController dVcontroller;
5.      private List<Log> logList;
6.
7.      public ZigBeeListener(SerialPort portZigBee, DeviceViewController
          dVcontroller) {
8.          super();
9.          this.portZigBee = portZigBee;
10.         this.dVcontroller = dVcontroller;
11.         logList = new ArrayList<>();
12.     }
13.
14.     @Override
15.     public void serialEvent(SerialPortEvent arg0) {
16.
17.         switch (arg0.getEventType()) {
18.         case SerialPortEvent.DATA_AVAILABLE:
19.             byte[] res = SerialPortManager.readFromPort(portZigBee);
20.             String hexStr = ByteUtils.byteArrayToHexString(res);
21.             byte[] resData = ByteUtils.hexStrToByte(hexStr);
22.             if (resData[17] == 1) {
23.                 String tempHexStr = ByteUtils.byteToHex(resData[19]) +
                      ByteUtils.byteToHex(resData[18]);
24.                 int temp = Integer.parseInt(tempHexStr, 16);
25.                 // 获取报警温度
26.                 String tempStr = dVcontroller.getTempLine().getText();
27.                 int tempLine = 0;
28.                 if (tempStr != null && !"".equals(tempStr)) {
29.                     tempLine = Integer.parseInt(tempStr);
30.                 }
31.                 refreshAlarm((temp/10)>tempLine);
32.                 String humiHexStr = ByteUtils.byteToHex(resData[21]) +
                      ByteUtils.byteToHex(resData[20]);
33.                 int humi = Integer.parseInt(humiHexStr, 16);
34.                 addLog(new Log(null, "温湿度", "采集到温度"+temp/10+"℃, "+"湿
                      度"+humi/10+"%"));
35.                 refreshNum(temp / 10, humi / 10);
36.             }
37.             break;
38.         }
39.
40.     }
41.
42.     /**
43.      * 界面更新
44.      *
45.      * @param temp
46.      * @param humi
47.      */
48.     private void refreshNum(int temp, int humi) {
49.         Platform.runLater(new Runnable() {
50.             @Override
51.             public void run() {
52.                 dVcontroller.getTempNum().setText(temp + "℃");
53.                 dVcontroller.getHumiNum().setText(humi + "%");
```

```
54.                }
55.            });
56.
57.        }
58.
59.        /**
60.         *
61.         * 刷新报警灯
62.         * @param alarm
63.         */
64.        private void refreshAlarm(boolean alarm) {
65.            if (alarm) {
66.                Platform.runLater(new Runnable() {
67.                    @Override
68.                    public void run() {
69.                        addLog(new Log(null, "报警灯", "温度超过报警线"));
70.                        dVcontroller.getAlarmPic().setGraphic(new ImageView
                             ("file:images/bjd_on.gif"));
71.                    }
72.                });
73.            } else {
74.                Platform.runLater(new Runnable() {
75.                    @Override
76.                    public void run() {
77.                        dVcontroller.getAlarmPic().setGraphic(new ImageView
                             ("file:images/bjd_off.png"));
78.                    }
79.                });
80.            }
81.        }
82.
83.        /**
84.         * 添加日志
85.         *
86.         * @param log
87.         */
88.        private void addLog(Log log) {
89.            DateFormat bf = new SimpleDateFormat("HH:mm:ss");
90.            Date date = new Date();
91.            String time = bf.format(date);
92.            log.setTime(time);
93.            logList.add(log);
94.            Platform.runLater(new Runnable() {
95.                @Override
96.                public void run() {
97.                    dVcontroller.getLogArea().appendText(log.toString());
98.                }
99.            });
100.
101.       }
102.
103. }
```

3. 运行结果

运行程序，结果如图 8-14 所示。

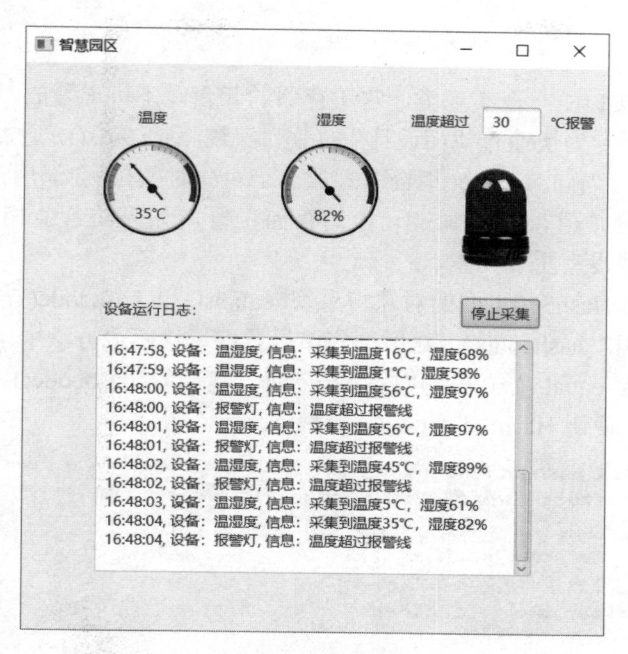

图 8-14　运行结果

任务 2　实现智慧园区系统用户注册功能

【任务描述】

扫码观看视频讲解

本任务要求读者通过集合模拟智慧园区系统用户的注册功能。用户注册时首先要获取用户的基本信息，包括昵称、电话、密码、确认密码。在保存用户信息之前需要对用户提交的信息做基本的验证，避免无效信息。保存用户时需要判断用户提交的信息是否重复。通常信息保存在服务器数据库中，本案例未用到数据库，数据保存在本地内存。Set 不可重复的特性正好可以满足判断注册用户是否重复这一需求。任务清单如表 8-5 所示。

表 8-5　任务清单

任务课时	4 课时	任务组员数量	建议 1 人
任务组采用设备	无		

【知识解析】

1. Set 接口

集合框架中的 Set 是指 Set 接口和所有 Set 接口的实现类，它们都继承自 Collection。其特点是元素无序，不可重复。

需要注意的是，加入 Set 的每个元素都必须是唯一的。Set 通过集合中的对象的 equals()方法来判断对象的唯一性，所以，加入 Set 的对象必须重写 Object 中的 equals()方法。Set 常用的实现类有 HashSet 和 TreeSet。

2. HashSet

HashSet 是 Set 接口的一个实现类，它所存储的元素是不可重复的，并且元素都是无序的。HashSet 利用了“专为快速查找而设计”的散列函数 hashcode()，通过对象的 hashcode() 决定存放的位置，因此当向 HashSet 集合中添加一个对象时，首先会调用该对象的 hashcode() 方法来计算对象的哈希值，从而确定元素的存储位置，如果哈希值相同，再调用对象的 equals()来确保该位置没有重复元素。

综上所述，放进 HashSet 里面的对象得重写 equals()和 hashcode()方法，同时在重写这个类的 equals()方法和 hashcode()方法时，应该尽量保证这两个方法中用相同的属性，也就是要求两个对象通过 equals()方法比较返回 true 时，它们的 hashcode()方法返回值也相等。接下来以一个例子来说明 HashSet 的基本使用。

```
1.   public class HashSetTest {
2.       public static void main(String[] args) {
3.           //创建 HashSet 对象，用接口引用指向这个对象
4.           Set set=new HashSet();
5.           //往集合中存放字符串"abc"
6.           boolean b=set.add("abc");
7.           System.out.println(b);
8.           set.add("def");
9.           //再次往集合中存放字符串"abc"
10.          b=set.add("abc");
11.          System.out.println(b);
12.          System.out.println(set);
13.      }
14.  }
```

运行结果如图 8-15 所示。

图 8-15　运行结果

从代码运行过程来看，当往 HashSet 中存放重复的 def 元素时，先用 HashCode()得到对象的地址，若发现这个地址有元素时，再用 equals()判断两个对象值是否相等，如果相等则认为是重复元素，则存放失败，add 方法的返回值为 false。

通过接收 add()方法的返回值来知道是否存入成功，true 为成功，false 为失败。上述代码是把 String 类对象存入 HashSet，String 类有重写过 equals()和 hashcode()，所以不能把重复的 String 对象存入 HashSet。如果将案例中的 Person 对象存入 HashSet 时，结果又如何呢？

```
1.   public class HashSetTest2 {
2.       public static void main(String[] args) {
3.           Set set = new HashSet();
4.           set.add(new Person("王五", "13375678982"));
```

```
5.         set.add(new Person("郭亮", "13809876543"));
6.         set.add(new Person("王五", "13375678982"));
7.         System.out.println(set);
8.     }
9. }
```

运行结果如图 8-16 所示。

```
[王五 13375678982, 王五 13375678982, 郭亮 13809876543]
```

图 8-16　运行结果

程序运行结果中出现了两个相同的 Person 信息“王五”，这样的信息应该被视为重复元素，不允许同时出现在 HashSet 集合中，之所以没有去掉重复的 Person，是因为 Person 类没有重写过 equals()和 hashCode()。现在来改写 Person 类，假设名字相同并且号码相同的就是同一个学生，Person 类添加以下方法：

```
1.  @Override
2.  public int hashCode() {
3.      final int prime = 31;
4.      int result = 1;
5.      result = prime * result + ((name == null) ? 0 : name.hashCode());
6.      result = prime * result + ((phone == null) ? 0 : phone.hashCode());
7.      return result;
8.  }
9.
10. @Override
11. public int compareTo(Object o) {
12.     if(!(o instanceof Person))
13.         return -1;//如果不是 Person 类，不用比较
14.     Person p=(Person)o;//强制类型转换成 Person
15.     //字符串默认比较规则就是升序
16.     return this.name.compareTo(p.getName());
17. }
18.
19.
20. //重写 equals(Object obj)方法
21. @Override
22. public boolean equals(Object obj) {
23.     if (this == obj)//判断是否是同一个元素
24.         return true;//如果是，直接返回同一个元素
25.     if (obj == null)//如果比较对象是空值
26.         return false;//直接返回不是同一个元素
27.     if (getClass() != obj.getClass())//如果不是同一个 Person 类
28.         return false;//直接返回不是同一个元素
29.     Person other = (Person) obj;//将对象强制转换为 Person
30.     if (name == null) {//如果有一个对象的 name 为空直接返回 false
31.         if (other.name != null)
32.             return false;
33.     } else if (!name.equals(other.name))
```

```
34.         return false;//如果名字不相同
35.     if (phone == null) {//如果有一个对象的phone为空
36.         if (other.phone != null)
37.             return false;
38.     } else if (!phone.equals(other.phone))
39.         return false;//如果phone不相同
40.     return true;
41. }
```

再次执行以下代码，可以看到 HashSet 中没有出现重复的 Person 信息，如图 8-17 所示。

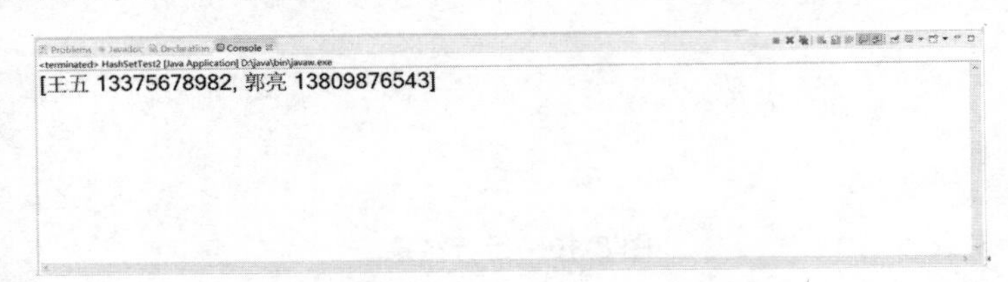

图 8-17　运行结果

3. TreeSet

TreeSet 是一个有序的 Set，其底层采用二叉树的原理进行排序，因此放到 TreeSet 集合中的对象是可以排序的，并用到了集合框架提供的另外两个实用接口 Comparable 和 Comparator。一个类是可排序的，它就应该实现 Comparable 接口。如果需要多个类具有相同的排序算法，那就不需要为每个类分别重复定义相同的排序算法，只要实现 Comparator 接口即可。

```
1.  public class TreeSetTest {
2.
3.      public static void main(String[] args) {
4.          Set set=new TreeSet();
5.          Person p=new Person("小丽", “13590876543”);
6.          set.add(p);
7.      }
8.  }
```

运行结果如图 8-18 所示。

```
<terminated> TreeSetTest [Java Application] D:\java\bin\javaw.exe
Exception in thread "main" java.lang.ClassCastException: com.nle.demo2.Person cannot be cast to java.lang.Comparable
	at java.util.TreeMap.compare(TreeMap.java:1294)
	at java.util.TreeMap.put(TreeMap.java:538)
	at java.util.TreeSet.add(TreeSet.java:255)
	at com.nle.demo2.TreeSetTest.main(TreeSetTest.java:10)
```

图 8-18　运行结果

从运行结果来看，当往 TreeSet 集合中存放一个自定义的 Person 类对象时，程序运行出错了，报了 ClassCastException 类型转换出错，原因是 Person 类没有实现 Comparable 接口，从而没有实现比较规则，而 TreeSet 因为有排序功能，当把一个没有比较规则的对象加入 TreeSet 集合时就会报错。

要解决上述问题，就要让 Person 类实现 Comparable 接口，重写 compareTo()方法。接下来改写 Person 类，重写 compareTo()方法，比较规则是按 name 升序排序。

让 Person 类实现 comparable 接口，重写 compareTo()方法：

```
1.  public class Person implements Comparable{
2.      …
3.      @Override
4.      public int compareTo(Object o) {
5.          if(!(o instanceof Person))
6.              return -1;//如果不是 Person 类，不用比较
7.          Person p=(Person)o;//强制类型转换成 Person
8.          //字符串默认比较规则就是升序
9.          return this.name.compareTo(p.getName());
10.         }
11. …
```

接着把 Person 对象放入 TreeSet 中，可以看到 Person 对象因为有了比较规则，能存放进 TreeSet 集合中了，同时也实现了按名字升序排序。

```
1.  public class TreeSetTest2 {
2.      public static void main(String[] args) {
3.          Set set=new TreeSet();
4.          set.add(new Person("王二", "18609876543"));
5.          set.add(new Person("王一", "13509876543"));
6.          System.out.println(set);
7.      }
8.  }
```

运行结果如图 8-19 所示。

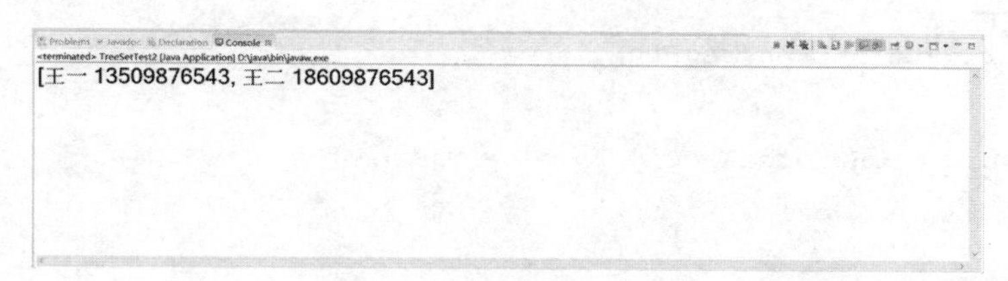

图 8-19　运行结果

【任务实施】

1. 任务分析

本任务要完成智慧园区系统用户注册功能。

(1) 完成注册界面并创建 controller 类。

(2) 创建用户数据模型类 User，并创建保存用户的 HashSet。

(3) 获取用户填写信息经过验证后保存到集合。

2. 任务实施

(1) 新建工程 project8_task2，新建包 com.nle.task2，并准备好程序入口类。

(2) 新建 RegView.fxml 文件，在 SceneBuilder 中设计好界面，并创建相应的 controller 类，如图 8-20 所示。

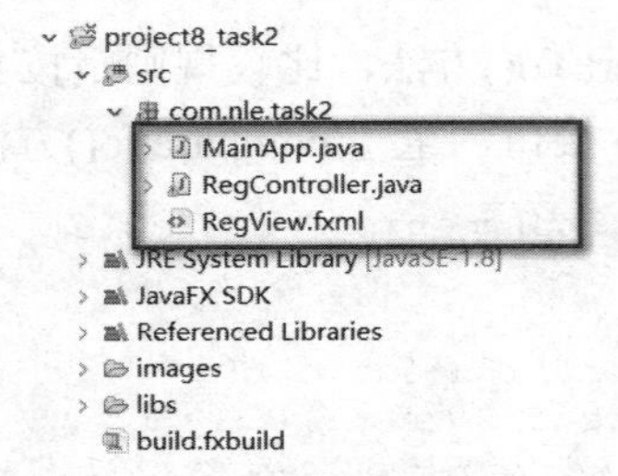

图 8-20　创建 FXML 文件和 controller 类

(3) 在 controller 中添加控件映射和控件监听方法。

```
1.   public class RegController {
2.       @FXML
3.       private TextField phone;
4.
5.       @FXML
6.       private TextField nickName;
7.
8.       @FXML
9.       private TextField confirmPwd;
10.
11.      @FXML
12.      private Button backBtn;
13.
14.      @FXML
15.      private TextField pwd;
16.
17.      @FXML
18.      private Button regBtn;
19.
20.      @FXML
21.      void regHandler(ActionEvent event) {
22.
23.      }
24.  }
```

(4) 声明存放注册用户的 HashSet 集合变量。

```
1.   public static HashSet<User> userLists;
2.       static
3.       {
4.           userLists=new HashSet();
5.           userLists.add(new User("13975091234", "abc123"));
6.           userLists.add(new User("15989076543", "def456"));
7.       }
```

(5) 创建用户类 User，重写 hashCode 方法和 equals 方法。

```
1.   public class User {
2.       private String phone;
3.       private String pwd;
4.       public User() {
5.       }
6.       public User(String phone, String pwd) {
7.           this.phone = phone;
8.           this.pwd = pwd;
9.       }
```

```
10.     public String getPhone() {
11.        return phone;
12.     }
13.     public void setPhone(String pwd) {
14.        this.pwd = pwd;
15.     }
16.     public String getPass() {
17.        return pwd;
18.     }
19.     public void setPass(String pwd) {
20.        this.pwd = pwd;
21.     }
22.     @Override
23.     public String toString() {
24.        return "电话号码: " + phone + ", 密码: " + pwd;
25.     }
26.     @Override
27.     public int hashCode() {
28.        final int prime = 31;
29.        int result = 1;
30.        result = prime * result + ((pwd == null) ? 0 : pwd.hashCode());
31.        result = prime * result + ((phone == null) ? 0 : phone.hashCode());
32.        return result;
33.     }
34.     @Override
35.     public boolean equals(Object obj) {
36.        if (this == obj)
37.           return true;
38.        if (obj == null)
39.           return false;
40.        if (getClass() != obj.getClass())
41.           return false;
42.        User other = (User) obj;
43.        if (pwd == null) {
44.           if (other.pwd != null)
45.              return false;
46.        } else if (!pwd.equals(other.pwd))
47.           return false;
48.        if (phone == null) {
49.           if (other.phone != null)
50.              return false;
51.        } else if (!phone.equals(other.phone))
52.           return false;
53.        return true;
54.     }
55.  }
```

(6) 创建验证工具类，添加表单验证方法，包括验证电话、密码等。

```
1.   public class CheckUtil {
2.      public static String check(String nickName,String phone,String pwd,
        String confirmPwd)
3.      {
4.         if(nickName==null||"".equals(nickName)) {
5.            return "昵称不能为空，请重输";
6.         }
7.         boolean flag=checkCellphone(phone);
8.         if(!flag)//手机号码不正确
9.         {
10.           return "手机号码不正确，请重输!";
```

```
11.         }
12.         flag=checkPass(pwd);
13.         if(!flag)//密码不正确
14.         {
15.             return "密码不正确，请重输!";
16.         }
17.
18.         if(!pwd.equals(confirmPwd)) {
19.             return "两次密码不一致";
20.         }
21.
22.         //将用户添加到集合中，添加不成功则说明用户重复
23.         User user=new User(phone,pwd);
24.         flag=RegController.userList.add(user);
25.         if(!flag)//用户添加不成功
26.         {
27.             return "用户重复";
28.         }
29.         else
30.         {
31.             return "注册成功";
32.         }
33.     }
34.
35.     /**
36.      * 验证手机号码
37.      *
38.      * 移动号码段:139、138、137、136、135、134、150、151、152、157、158、159、
   182、183、187、188、147、182
39.      * 联通号码段:130、131、132、136、185、186、145
40.      * 电信号码段:133、153、180、189、177
41.      *
42.      * @param phone
43.      * @return true    false
44.      */
45.    public static boolean checkCellphone(String phone) {
46.        String regex = "^((13[0-9])|(14[5|7])|(15([0-3]|[5-9]))|(18[0,1,2,
            5-9])|(177))\\d{8}$";
47.        Pattern pattern=Pattern.compile(regex);
48.        Matcher matcher=pattern.matcher(phone);
49.        return matcher.matches();
50.    }
51.
52.    /**
53.     * 验证密码(6-16位数字和字母的组合)
54.     * @param pass
55.     * @return
56.     */
57.    public static boolean checkPass(String pass) {
58.        String regex = "^(?![0-9]+$)(?![a-zA-Z]+$)[0-9A-Za-z]{6,16}$";
59.        Pattern pattern=Pattern.compile(regex);
60.        Matcher matcher=pattern.matcher(pass);
61.        return matcher.matches();
62.    }
63. }
```

(7) 完成 controller 类中注册按钮监听方法 regHandler。

```
1.  @FXML
2.  void regHandler(ActionEvent event) {
3.      // 获取输入的手机号与密码、昵称
4.      String nickName = nickNameField.getText().trim();
5.      String phone =  phoneField.getText().trim();
6.      String pwd = pwdField.getText().trim();
7.      String confirmPwd = confirmPwdField.getText().trim();
8.      //调用校验方法检查手机号和密码输入是否正确，无误后添加进用户集合中
9.      String message=CheckUtil.check(nickName, phone, pwd, confirmPwd);
10.     alert.setContentText(message);
11.     alert.show();
12.     if(!message.equals("注册成功"))//如果输入出错，清除所有的输入
13.     {
14.         phoneField.setText("");
15.         pwdField.setText("");
16.         confirmPwdField.setText("");
17.     }
18. }
```

3. 运行结果

运行程序，结果如图 8-21 所示。

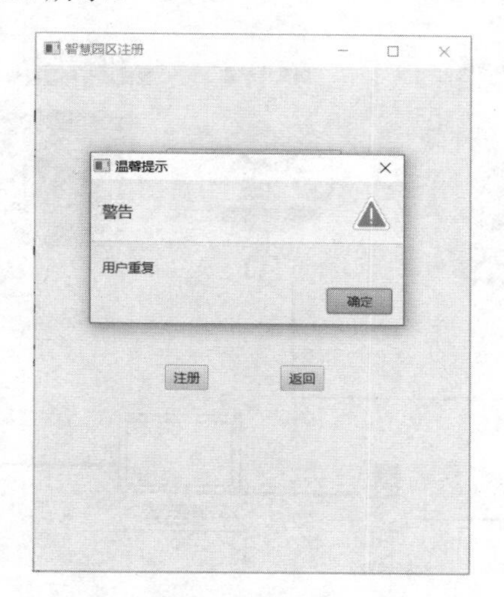

图 8-21　运行结果

任务 3　使用 Map 存储采集器数据

【任务描述】

物联网应用的开发主要是围绕着设备数据进行的，在开发过程中经常需要获取设备数据，并且需要保证设备数据的即时性，如果每次在需要数据时再去设备串口获取则会严重影响应用效率。处理方法是，通过开启线程定时获取各个设备的数据，并将数据保存到一个容器中，需要设备数据的时候就去容器中获取。设备与设备数据就是一个键值对的映射关系，这时使用 Map 集合存储再合适不过。通过本任务将向读者演示 Map 集合的特点和使用。任务清单如表 8-6 所示。

表 8-6　任务清单

任务课时	4 课时	任务组员数量	建议 3 人
任务组采用设备	1 个人体传感器 1 个烟雾传感器 1 个火焰传感器(可选) 1 个 ADAM-4150 1 个 485 转 232 1 个 ZigBee 温湿度传感器 1 个 ZigBee 光照传感器 1 个 ZigBee 协调器 相关电源、导线、工具		

【拓扑图】

本任务拓扑图如图 8-22 所示。

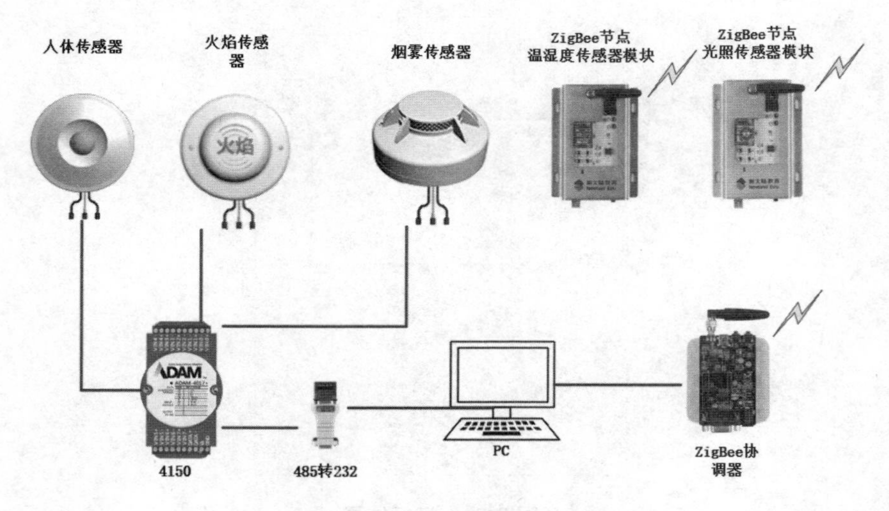

图 8-22　拓扑图

备注：本任务人体传感器接 DI0 口，火焰传感器接 DI1 口，烟雾传感器接 DI2 口，COM 口根据实际情况确定。

【知识解析】

1. Map 接口

集合框架中的 Map 指 Map 接口和 Map 接口的实现类。Map 接口没有继承自 Collection，Map 用于保存具有“映射关系”的“键值对”数据。键是 key，值是 value，key 和 value 可以是任何引用类型的数据。Map 的 key 不允许重复，即同一个 Map 对象的任何两个 key 通过 equals()比较的结果总是返回 false。Map 接口中常用的实现类有 HashMap 和 TreeMap，如表 8-7 所示。

表 8-7 Map 接口的常用方法

方法	描述
void clear()	从该地图中删除所有的映射(可选操作)
boolean containsKey(Object key)	如果此映射包含指定键的映射，则返回 true
Set<Map.Entry<K,V>> entrySet()	返回此地图中包含的映射的 Set 视图
V get(Object key)	返回到指定键所映射的值，如果此映射还包含该键的映射，则返回 null
boolean isEmpty()	如果此地图不包含键值映射，则返回 true
Set<K> keySet()	返回此地图中包含的键的 Set 视图
V put(K key,V value)	将指定的值与该映射中的指定键相关联
V remove(Object key)	如果存在(可选的操作)，从该地图中删除一个键的映射
Collection<V> values()	返回此地图中包含的值的 Collection 视图

2. HashMap

HashMap 是 Map 接口的一个实现类，其底层用哈希表的方式存储键值映射关系，不能出现重复的键。

```
1.  public class HashMapTest {
2.      public static void main(String[] args) {
3.          // 创建对象，键是 String 类型，值是 Integer 类型
4.          Map<String, Object> map = new HashMap();
5.          map.put("名字", "小红");//存储键和值
6.          map.put("年龄", 23);
7.          //根据键获取值
8.          System.out.println("名字="+map.get("名字"));
9.          System.out.println("年龄="+map.get("年龄"));
10.         //放入相同的键
11.         map.put("名字", "李四");//存储键和值
12.         System.out.println("名字="+map.get("名字"));
13.     }
14. }
```

运行结果如图 8-23 所示。

图 8-23 运行结果

上面的程序先通过 put 的方式向集合中加入 2 个元素，然后通过 get 的方法获取键对应的值。当向 Map 中存储相同的 key 时，后添加的值会替换前面的值，这也证实了 Map 中的键必须是唯一的，不能重复。

在程序的开发中，经常需要获取 Map 中所有的值，有四种方法可以实现，具体代码如下：

```
1.   public class HashMapTest2 {
2.
3.       public static void main(String[] args) {
4.           Map<Integer, Person> map =new HashMap();
5.           map.put(1, new Person("张三", "13509876541"));
6.           map.put(2, new Person("李四", "15278909876"));
7.           map.put(3, new Person("王五", "13912345678"));
8.           map.put(4, new Person("赵六", "13709341256"));
9.           //遍历 Map 集合中的数据。方法一 ：直接用集合对象名
10.          System.out.println("方法一：");
11.          System.out.println(map);
12.
13.          System.out.println("方法二：");
14.          //方法二：map 中的值是 person 对象，map.values()把所有的 person 放到
               Collection 中
15.          for(Person p:map.values())
16.          {
17.             System.out.println(p);
18.          }
19.
20.          System.out.println("方法三：");
21.          //方法三：map.keySet()方法 把 Map 集合中所有的 key 放在一个 Set 集合中
22.          Set<Integer> set=map.keySet();
23.          //对 set 进行迭代，取出所有的 key
24.          Iterator<Integer> it=set.iterator();
25.          while(it.hasNext())
26.          {
27.             Integer key=it.next();//取出每个 key
28.             Person p=map.get(key);//凭着 key 得到对应的值
29.             System.out.println(key+"="+p);
30.          }
31.
32.          System.out.println("方法四：");
33.          //方法四：map.entrySet() 把放映射关系的 Entry 对象放在一个 Set 集合中
34.          Set<Entry<Integer, Person>> set2=map.entrySet();
35.          Iterator<Entry<Integer, Person>> it2=set2.iterator();
36.          while(it2.hasNext())
37.          {
38.              //从 Set 中取出一个个的 Entry 对象
39.              Entry<Integer, Person> entry=it2.next();
40.              Integer key=entry.getKey();//取出 key
41.            Person p= entry.getValue();//取出 value
42.            System.out.println(key+"="+p);
43.          }
44.      }
45.  }
```

运行结果如图 8-24 所示。

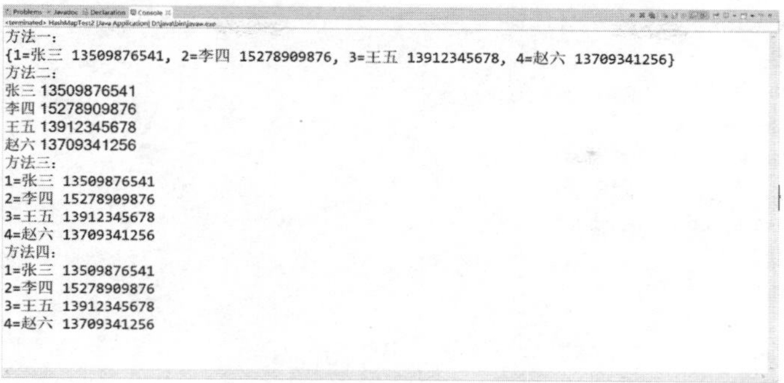

```
方法一：
{1=张三 13509876541, 2=李四 15278909876, 3=王五 13912345678, 4=赵六 13709341256}
方法二：
张三 13509876541
李四 15278909876
王五 13912345678
赵六 13709341256
方法三：
1=张三 13509876541
2=李四 15278909876
3=王五 13912345678
4=赵六 13709341256
方法四：
1=张三 13509876541
2=李四 15278909876
3=王五 13912345678
4=赵六 13709341256
```

图 8-24　运行结果

【任务实施】

1. 任务分析

(1) 通过串口服务器连接 ZigBee 协调器和 4150 采集数据。

(2) ZigBee 协调器采集数据：温湿度、光照数值。

(3) 4150 采集器采集数据：人体、火焰、烟雾。

(4) 封装一个对象保存设备数据。

(5) 将配置参数统一保存在一个工具类中。

2. 任务实施

(1) 新建工程 project8_task3，新建 com.nle.task3 包，根据连线准备好工具类，保存配置参数；定义好存储设备数据的 Map 集合。

```
1.   public class Config {
2.       //4150COM 口
3.       public static String comAdam = "COM201";
4.       //ZigBee 协调器 COM 口
5.       public static String comZigBee = "COM200";
6.       //4150 波特率
7.       public static int baudrateAdam = 9600;
8.       //ZigBee 波特率
9.       public static int baudrateZigBee = 38400;
10.      //4150 采集指令
11.      public static byte[] collectCmd = new byte[] {0x01, 0x01, 0x00, 0x00,
           0x00, 0x07, 0x7D, (byte) 0xC8};
12.      //人体 DI 口
13.      public static String human = "DI0";
14.      //火焰 DI 口
15.      public static String fire = "DI1";
16.      //烟雾 DI 口
17.      public static String smoke = "DI2";
18.      //设备数据存储 Map 集合
19.     public static Map<String, DeviceValue> deviceValues = new HashMap<String,
          DeviceValue>();
20.  }
```

(2) 创建设备数据保存模型对象，包含设备名称、设备类型、采集时间和数据。

```
1.   public class DeviceValue {
2.       //采集时间
3.       private String time;
4.       //设备名称
5.       private String deviceName;
6.       //1，有线；2，无线
7.       private int deviceType;
8.       //采集数值
9.       private String value;
10.
11.      public String getTime() {
12.          return time;
13.      }
14.      public void setTime(String time) {
15.          this.time = time;
16.      }
17.      public String getDeviceName() {
18.          return deviceName;
19.      }
20.      public void setDeviceName(String deviceName) {
21.          this.deviceName = deviceName;
22.      }
23.      public int getDeviceType() {
24.          return deviceType;
25.      }
26.      public void setDeviceType(int deviceType) {
27.          this.deviceType = deviceType;
28.      }
29.      public String getValue() {
30.          return value;
31.      }
32.      public void setValue(String value) {
33.          this.value = value;
34.      }
35.      @Override
36.      public String toString() {
37.          return time + " " + deviceName + " " + deviceType + " " + value;
38.      }
39.
40.  }
```

(3) 创建 4150 串口监听器，并编写数据解析代码，将解析得到的结果通过设备数据模型对象保存，将设备名称与模型对象当作键值对存入 Config 中定义的 Map 集合。

```
1.   public class AdamListener implements SerialPortEventListener{
2.
3.       private SerialPort port;
4.
5.       public AdamListener(SerialPort port) {
6.           this.port = port;
7.       }
8.
9.
10.      @Override
11.      public void serialEvent(SerialPortEvent arg0) {
12.
13.          switch (arg0.getEventType()) {
14.          case SerialPortEvent.DATA_AVAILABLE:
15.              if(port!=null)
```

```
16.             {
17.                 byte[] res=SerialPortManager.readFromPort(port);
18.                 String s= ByteUtils.byteArrayToHexString(res);
19.                 //System.out.println("从串口收到的数据是="+s);
20.                 //如果是采集数据指令响应数据
21. 
22.                 if(s.substring(2, 4).equals("01")) {
23.                     //截取应答数据
24.                     String msg = s.substring(6, 8);
25.                     DeviceValue valueHuman = new DeviceValue();
26.                     //获取时间
27.                     DateFormat df = new SimpleDateFormat("yyyy-mm-dd
                          hh:MM:ss");
28.                     valueHuman.setTime(df.format(new Date()));
29.                     valueHuman.setDeviceType(1);
30.                     //获取每个通道数值
31.                     int result = Integer.parseInt(msg, 16);
32.                     String str = ByteUtils.toBinary7(result);
33.                     //判断人体 DI0 口
34.                     valueHuman.setDeviceName("人体");
35.                     if(str.charAt(DiToCharat(Config.human))=='0') {
36.                         valueHuman.setValue("有人");
37.                     }else {
38.                         valueHuman.setValue("无人");
39.                     }
40.                     //将结果存入 Map 集合
41.                     Config.deviceValues.put("人体", valueHuman);
42.                     DeviceValue valueSmoke = new DeviceValue();
43.                     valueSmoke.setTime(df.format(new Date()));
44.                     valueSmoke.setDeviceType(1);
45.                     //判断烟雾 DI2 口
46.                     valueSmoke.setDeviceName("烟雾");
47.                     if(str.charAt(DiToCharat(Config.smoke))=='1') {
48.                         valueSmoke.setValue("有烟");
49.                     }else {
50.                         valueSmoke.setValue("无烟");
51.                     }
52.                     //将结果存入 Map 集合
53.                     Config.deviceValues.put("烟雾", valueSmoke);
54.                 }
55. 
56.             }
57.             break;
58.         }
59. 
60.     }
61. 
62.     /**
63.      * DI 口对应应答数据位
64.      * @param DI
65.      * @return
66.      */
67.     private int DiToCharat(String DI) {
68.         int charat=6-Integer.parseInt(DI.charAt(2)+"");
69.         return charat;
70.     }
71. 
```

(4) 创建 ZigBee 串口监听器，并编写数据解析代码，将解析得到的结果通过设备数据模型对象保存，将设备名称与模型对象当作键值对存入 Config 中定义的 Map 集合。

```
1.  public class ZigBeeListener implements SerialPortEventListener {
2.      private SerialPort port ;
3.
4.      public ZigBeeListener(SerialPort port ) {
5.          this.port = port;
6.      }
7.
8.      @Override
9.      public void serialEvent(SerialPortEvent arg0) {
10.         //确定日期格式
11.         DateFormat df = new SimpleDateFormat("yyyy-mm-dd hh:MM:ss");
12.         switch (arg0.getEventType()) {
13.         case SerialPortEvent.DATA_AVAILABLE:
14.             byte[] res = SerialPortManager.readFromPort(port);
15.             String hexStr = ByteUtils.byteArrayToHexString(res);
16.             String[] strArr = hexStr.split("FE");
17.
18.             for (int i = 1; i < strArr.length; i++) {
19.
20.                 byte[] resData = ByteUtils.hexStrToByte(strArr[i]);
21.                 //温湿度传感器
22.                 if(resData[16]==1) {
23.                     //获取湿度十六进制字符串
24.                     String humiHexStr = ByteUtils.byteToHex(resData[20]) +
                          ByteUtils.byteToHex(resData[19]);
25.                     int humi = Integer.parseInt(humiHexStr, 16);
26.                     //获取温度十六进制字符串
27.                     String tempHexStr = ByteUtils.byteToHex(resData[18]) +
                          ByteUtils.byteToHex(resData[17]);
28.                     int temp = Integer.parseInt(tempHexStr, 16);
29.                     //将温度存入 Map
30.                     DeviceValue valueTemp = new DeviceValue();
31.                     valueTemp.setTime(df.format(new Date()));
32.                     valueTemp.setDeviceType(2);
33.                     valueTemp.setValue(temp/10+"");
34.                     valueTemp.setDeviceName("温湿度");
35.                     Config.deviceValues.put("温度", valueTemp);
36.                     //将湿度存入 Map
37.                     DeviceValue valueHumi = new DeviceValue();
38.                     valueHumi.setTime(df.format(new Date()));
39.                     valueHumi.setDeviceType(2);
40.                     valueHumi.setValue(humi/10+"");
41.                     valueHumi.setDeviceName("温湿度");
42.                     Config.deviceValues.put("湿度", valueHumi);
43.                 }
44.                 //光照传感器
45.                 if(resData[16]==33) {
46.                     String beamHexStr = ByteUtils.byteToHex(resData[18]) +
                          ByteUtils.byteToHex(resData[17]);
47.                     int beam = Integer.parseInt(beamHexStr, 16);
48.                     double resbeam = beam / 100.0;
49.                     double val =Math.pow(10, ((1.78 - Math.log10(33 / resbeam
                            - 10)) / 0.6));
50.                     DecimalFormat numDf = new DecimalFormat("#.00");
```

```
51.                     DeviceValue valueBeam = new DeviceValue();
52.                     valueBeam.setTime(df.format(new Date()));
53.                     valueBeam.setDeviceType(2);
54.                     valueBeam.setValue(numDf.format(val));
55.                     valueBeam.setDeviceName("光照");
56.                     Config.deviceValues.put("光照", valueBeam);
57.                 }
58.
59.             }
60.             break;
61.         }
62.
63.     }
64.
65. }
```

(5) 定义 4150 数据采集指令发送线程。每隔一秒发送采集指令。

```
1.  public class CollectTask implements Runnable {
2.
3.      private SerialPort port;
4.
5.      private boolean collectStart;
6.
7.      public CollectTask(SerialPort port) {
8.          this.port = port;
9.          collectStart = true;
10.     }
11.
12.     @Override
13.     public void run() {
14.
15.         while(collectStart) {
16.             try {
17.                 //发送采集指令
18.                 SerialPortManager.sendToPort(port, Config.collectCmd);
19.                //每间隔 1000 毫秒发送一次采集指令
20.                 Thread.sleep(1000);
21.                //输出 Map 中的数据查看
22.                 System.out.println(Config.deviceValues);
23.
24.             } catch (InterruptedException e) {
25.                 // TODO Auto-generated catch block
26.                 e.printStackTrace();
27.             } catch (SendDataToSerialPortFailure e) {
28.                 // TODO Auto-generated catch block
29.                 e.printStackTrace();
30.             } catch (SerialPortOutputStreamCloseFailure e) {
31.                 // TODO Auto-generated catch block
32.                 e.printStackTrace();
33.             }
34.         }
35.
36.     }
37.
38. }
```

(6) 创建程序入口，创建串口对象并添加监听、开启采集指令发送线程。

```
1.  public class MainApp {
2.
3.      public static void main(String[] args) {
4.
5.          try {
6.              //有线串口对象
7.              SerialPort portAdam = SerialPortManager.openPort(Config.comAdam,
                  Config.baudrateAdam);
8.              //无线串口对象
9.              SerialPort portZigBee = SerialPortManager.openPort
                  (Config.comZigBee, Config.baudrateZigBee);
10.             //添加串口监听
11.             SerialPortManager.addListener(portAdam, new AdamListener
                  (portAdam));
12.             SerialPortManager.addListener(portZigBee, new ZigBeeListener
                  (portZigBee));
13.             //开启采集线程
14.             Thread collectThread = new Thread(new CollectTask(portAdam));
15.             collectThread.start();
16.
17.         } catch (SerialPortParameterFailure e) {
18.             // TODO Auto-generated catch block
19.             e.printStackTrace();
20.         } catch (NotASerialPort e) {
21.             // TODO Auto-generated catch block
22.             e.printStackTrace();
23.         } catch (NoSuchPort e) {
24.             // TODO Auto-generated catch block
25.             e.printStackTrace();
26.         } catch (PortInUse e) {
27.             // TODO Auto-generated catch block
28.             e.printStackTrace();
29.         } catch (TooManyListeners e) {
30.             // TODO Auto-generated catch block
31.             e.printStackTrace();
32.         }
33.
34.     }
35.
36. }
```

3. 运行结果

运行结果如图 8-25 所示。

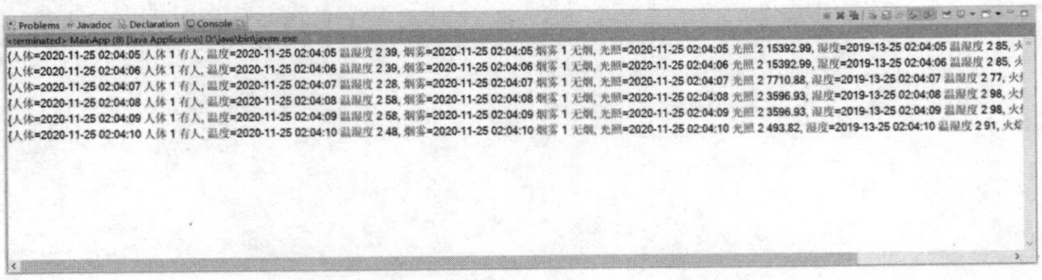

图 8-25　运行结果

思考与练习

1. 简述 ArrayList 和 LinkedList 的区别与应用场景。
2. 简述 List、Set、Map 各集合的存储结构。
3. 写出 HashMap 不同遍历方式的代码。

项目 9 使用 IO 流

【项目描述】

应用程序的编写过程都要处理数据流、序列化和文件系统的输入和输出工作。Java 的 IO 包就是一套用来读写数据的 API。当需要处理来自设备、硬盘、程序、网络的数据，或将数据写出到其他目标位置时就必须使用 IO 流。本项目将完成智慧园区项目中的读取目录、保存用户信息到硬盘、保存应用配置信息等功能。通过案例演示介绍了 File 类对象的使用以及 Java IO 流的概念和分类，主要围绕字节流、字符流、输入流、输出流、节点流和包装流等展开讲解，使读者能够全方位掌握 Java 中用于处理输入/输出的各种流。具体任务列表如图 9-1 所示。

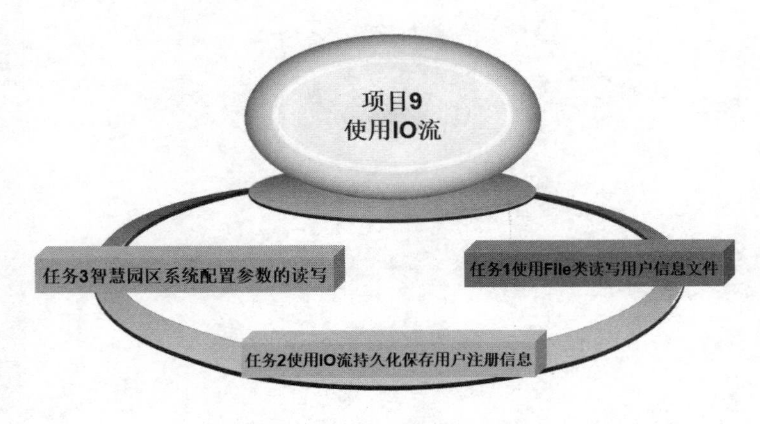

图 9-1　项目 9 任务列表

【学习目标】

知识目标： 掌握 File 类对象；掌握字节流；掌握字符流；掌握缓冲流。

技能目标： 能使用 File 类读取文件列表；能使用字节流、字符流等完成数据读写操作；能使用文件保存应用配置信息。

任务 1 使用 File 类读写用户信息文件

扫码观看视频讲解

【任务描述】

在前面的任务中，实现注册功能时使用集合来保存用户数据，但是不论使用何种容器，当程序进程结束后，数据就会被系统清理。要想持久地保存用户注册的信息，必须将信息保存到硬盘中。本任务要求在用户注册时，将注册用户的信息保存到集合中，并为每个注册用户创建一个信息文件，以用户名命名文件。每次程序运行时将所有用户信息读取保存到集合中，以避免用户名重复。任务清单如表 9-1 所示。

表 9-1 任务清单

任务课时	4 课时	任务组员数量	建议 1 人
任务组采用设备	无		

【知识解析】

File 类是 java.io 包下用来操作文件的类。在应用程序中利用 File 类对象可以对文件本身进行操作，例如新建、删除、重命名文件，判断文件是否可读可写，列出目录列表等操作。但是 File 类本身不能读取文件内容，因为操作文件内容需要输入流和输出流。

注意：相对路径名必须根据从其他路径名获取的信息进行解释。java.io 包中的类始终会根据当前用户目录解析相对路径名，如图 9-2 所示。

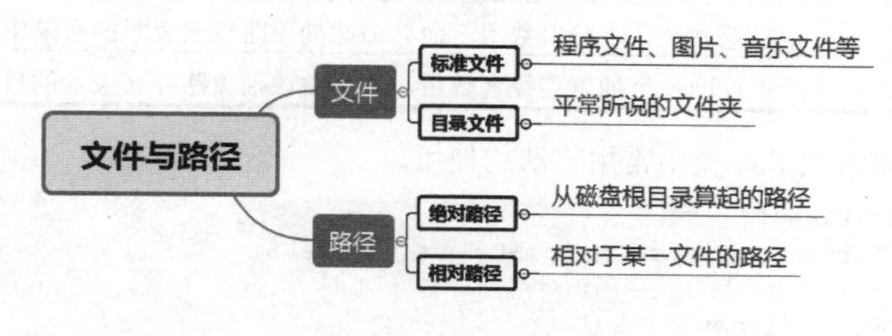

图 9-2 文件与路径

(1) File 类中常用的方法。

File 类常用的构造方法如表 9-2 所示。

通常来说，如果知道该目录或文件的路径，用 File(String pathname)比较合适；如果操作的是网络路径，用 File(URI uri)比较合适；如果知道要处理的文件中有多少子目录或文件，用其他两个比较合适。

表 9-2　File 类常用的构造方法

方　法	描　述
File(File parent,String child)	从父抽象路径名和子路径名字符串创建新的 File 实例
File(String pathname)	通过将给定的路径名字符串转换为抽象路径名来创建新的 File 实例
File(String parent,String child)	从父路径名字符串和子路径名字符串创建新的 File 实例
File(URI uri)	通过将给定的 file: URI 转换为抽象路径名来创建新的 File 实例

(2)　File 类的常用方法。

File 类的常用方法如表 9-3 所示。

表 9-3　File 类的常用方法

方　法	描　述
boolean exists()	测试此抽象路径名表示的文件或目录是否存在
boolean createNewFile()	当且仅当具有该名称的文件不存在时，创建一个由该抽象路径名命名的新的空文件
boolean delete()	删除由此抽象路径名表示的文件或目录
boolean isDirectory()	测试由此抽象路径名表示的文件是否为目录
boolean isFile()	测试由此抽象路径名表示的文件是否为普通文件
boolean mkdir()	创建由此抽象路径名命名的目录
boolean mkdirs()	创建由此抽象路径名命名的目录，包括任何必需但不存在的父目录
String getParent()	返回父目录
String getAbsolutePath()	返回此抽象路径名的绝对路径名字符串
long length()	返回文件的大小
boolean canRead()	文件是否可读，返回真为可读
boolean canWrite()	文件是否可写，返回真为可写
String[] list()	返回一个字符串数组，命名由此抽象路径名表示的目录中的文件和目录
File[]listFiles()	返回一个抽象路径名数组，表示由该抽象路径名表示的目录中的文件

下面程序用来说明 File 类中常用方法的使用。

```
1.   public class FileTest {
2.       public static void main(String[] args) {
3.           //创建一个目录，使用相对路径
4.           File f1=new File("abc");
5.           if(!f1.exists())//如果文件不存在
6.           {
7.               f1.mkdir();//创建目录
8.           }
9.           // 创建一个文件，名为config.ini，使用的是相对路径
10.          File file = new File("abc/config.ini");
11.      if (!file.exists())// 如果文件不存在
12.          {
13.              try {
```

```
14.                  // 创建文件，该方法会抛出异常，要处理异常
15.                  file.createNewFile();
16.                  //获取文件名称
17.                  System.out.println("文件名称："+file.getName());
18.                  //获取文件的相对路径
19.                  System.out.println("文件的相对路径："+file.getPath());
20.                  //获取文件的绝对路径
21.                  System.out.println("文件的绝对路径："+file.getAbsolutePath());
22.                  //获取文件的父路径
23.                  System.out.println("文件的父路径："+file.getParent());
24.                  //判断文件是否可读
25.                  boolean flag=file.canRead();
26.                  System.out.println(flag?"文件可读":"文件不可读");
27.                  //判断文件是否可写
28.                  flag=file.canWrite();
29.                  System.out.println(flag?"文件可写":"文件不可写");
30.                  //判断文件是否是一个文件
31.                  flag=file.isFile();
32.                  System.out.println(flag?"是一个文件":"不是一个文件");
33.                  //判断文件是否是一个目录
34.                  flag=file.isDirectory();
35.                  System.out.println(flag?"是一个目录":"不是一个目录");
36.                  flag=file.isFile();
37.                  System.out.println("文件大小是："+file.length());
38.              } catch (IOException e) {
39.                  e.printStackTrace();
40.              }
41.          } else // 如果文件存在，则删除
42.          {
43.              System.out.println("是否成功删除已存在的文件："+file.delete());
44.          }
45.      }
46.  }
```

程序通过 mkdir()方法创建新目录，用 new File("abc/config.ini")在构造方法中传入要操作的文件名 config.ini，使用的是相对路径，当程序第一次运行时用 file.exists()判断，如果文件不存在会返回假，取非则条件满足，所以当文件不存在时用 file.createNewFile()创建文件，再次运行则文件已存在，用 file.delete()删除文件，程序的中间则是测试 File 类的各种方法。

运行结果如图 9-3 所示。

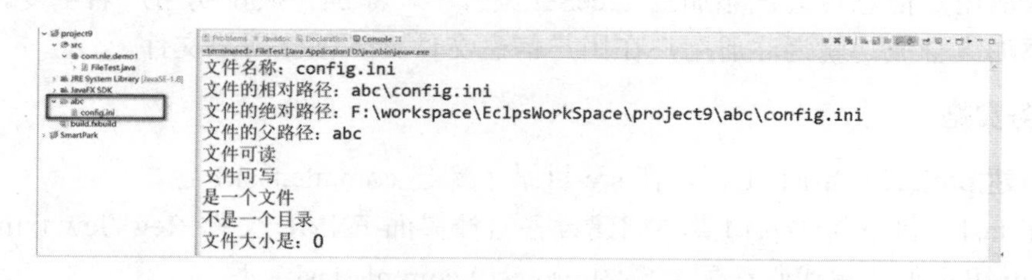

图 9-3　运行结果

下面的程序是利用 listFiles 方法遍历目录下的内容。

```
1.  public class FileTest2 {
2.
```

```
3.      public static void main(String[] args) {
4.          File file = new File("c:/java_book9/chapter9");
5.          // 遍历目录下的文件，返回的是字符串数组
6.          String[] str = file.list();
7.          for (String s : str) {
8.              System.out.println(s);// 输出文件名
9.          }
10.         System.out.println("********");
11.         // 遍历目录下的文件，返回的是 File 类数组
12.         File[] files = file.listFiles();
13.         for (File f : files) {
14.             System.out.println(f);// 输出文件名
15.         }
16.     }
17. }
```

运行结果如图 9-4 所示。

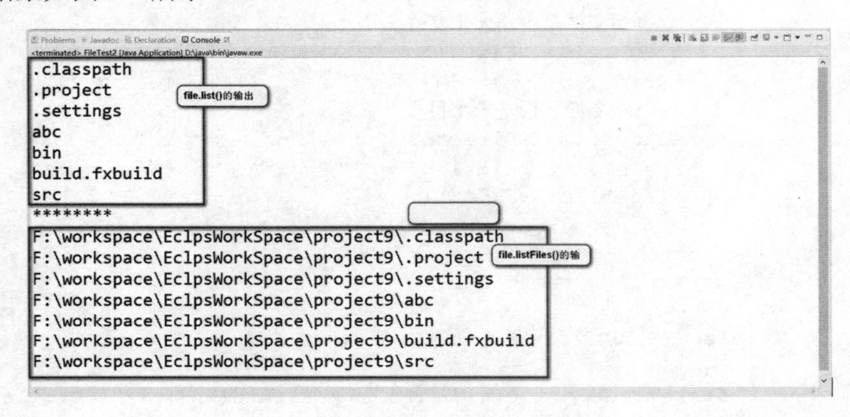

图 9-4　运行结果

【任务实施】

1. 任务分析

(1) 用户注册界面。

(2) 程序启动时获取用户信息文件夹下所有文件名作为用户名存入 TreeSet 集合。

(3) 获取用户信息后尝试添加到 TreeSet 集合中，添加失败说明用户名重复。

(4) 添加集合成功就为注册用户在用户信息文件夹下创建用户文件。

2. 任务实施

(1) 新建 project9_task1 工程，在 src 目录下新建 com.nle.task1 包。

(2) 在 task1 包下新建入口类，创建设备监控界面 FXML 文件 RegView.fxml，并创建对应的 Controller 类，源代码在配套资源\project9\com\nle\task1 下。

(3) 创建用户 User 类。

(4) 创建程序入口 MainApp 类。

工程结构如图 9-5 所示。

```
project9_task1
  src
    com.nle.task1
      MainApp.java
      RegViewController.java
      User.java
      RegView.fxml
  JRE System Library [JavaSE-1.8]
  JavaFX SDK
  abc
  users
  build.fxbuild
```

图 9-5　工程结构

(5) 在 RegViewController 中的 initialize 方法中初始化控件，并读取用户信息。

```
1.  private void initialize() {
2.          alert = new Alert(AlertType.WARNING);
3.          alert.setTitle("温馨提示");
4.          //读取用户
5.          readAllUser();
6.      }
7.     /**
8.   * 读取所有用户
9.   */
10. private void readAllUser() {
11.     File file = new File("users");
12.     if(!file.exists()) {
13.        file.mkdirs();
14.     }
15.     String[] userNames = file.list();
16.     for (int i = 0; i < userNames.length; i++) {
17.        User user = new User();
18.        user.setUsername(userNames[i]);
19.        users.add(user);
20.     }
21. }
```

(6) 完成 RegViewController 中的注册按钮监听，判断用户是否重复，并为用户创建信息保存文件。

```
1.  @FXML
2.  void regHandler(ActionEvent event) {
3.     String username = userNameField.getText();
4.     String pwd1 = pwd1Field.getText();
5.     String pwd2 = pwd2Field.getText();
6.     if(checkNotNull(username)&&checkNotNull(pwd1)) {
7.        if(pwd1.trim().equals(pwd2.trim())) {
8.           User user = new User();
9.           user.setUsername(username);
10.          user.setPwd(pwd1);
11.          if(users.add(user)) {
12.             alert.setContentText("注册成功");
13.             saveUserFile(user);
14.          }else {
15.             alert.setContentText("用户名已存在");
16.          }
17.       }else {
18.          alert.setContentText("两次密码不一致");
```

```
19.         }
20.     }else {
21.         alert.setContentText("用户名或密码不能为空");
22.     }
23.     alert.show();
24. }
25.
26. /**
27.  * 创建用户信息文件
28.  * @param user
29.  */
30. private void saveUserFile(User user) {
31.     File file = new File("users/"+user.getUsername());
32.     try {
33.         file.createNewFile();
34.     } catch (IOException e) {
35.         // TODO Auto-generated catch block
36.         e.printStackTrace();
37.     }
38. }
```

3. 运行结果

运行程序，结果如图 9-6 和图 9-7 所示。

图 9-6　运行结果

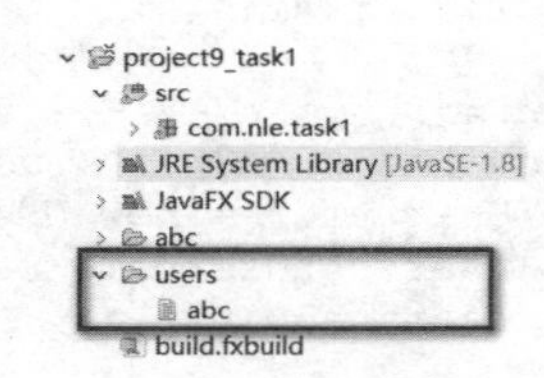

图 9-7　工程结构

任务 2　使用 IO 流持久化保存用户注册信息

扫码观看视频讲解

【任务描述】

在任务 1 中，已经为每个注册用户以其用户名创建了用户信息文件，本任务将在此基础上把用户的密码写入用户所属的文件中，在登录的时候可以将文件中的信息读取出来做比对。要实现持久保存注册用户的用户名和密码，完成智慧园区应用的注册和登录功能，就必须用到 Java 的 IO 包，完成对文件的读写。任务清单如表 9-4 所示。

表 9-4　任务清单

任务课时	6 课时	任务组员数量	建议 1 人
任务组采用设备	无		

【知识解析】

1. Java 的 IO 包

在程序中，输入和输出都是相对当前程序而言的。例如，从硬盘上读取一个配置文件的内容到程序中，相当于将文件的内容输入程序内部，因此输入和“读”对应，而将程序中的内容保存到硬盘上，则相当于将文件的内容输出到程序外部，因此输出和“写”对应。熟悉输入和输出的对应关系，将有助于后续内容的学习。

在 Java 语言中，输入和输出的概念要比其他语言输入和输出的概念涵盖的内容广泛，不仅包含文件的读写，也包含网络数据的发送，甚至内存数据的读写以及控制台数据的接收等都由 IO 来完成。为了使输入和输出的结构保持统一，从而方便程序员使用 IO 相关的类，在 Java 语言的 IO 类设计中引入了一个新的概念——Stream(流)。

由于在进行 IO 操作时，需要操作的内容很多，例如文件、内存和网络连接等，这些都被称作数据源(data source)。不同数据源的处理方式是不一样的，如果直接交给程序员进行处理，对于程序员来说处理起来则会比较复杂。所以在所有的 IO 类设计中，在读数据时，JDK API 将数据源的数据转换为一种固定的数据序列，在写数据时，将需要写的数据以一定的格式写入数据序列，由 JDK API 将数据序列中的数据写入对应的数据源中。这样由系统完成复杂的数据转换以及不同数据源之间的变换，从而简化程序员的编码。Java 的 IO 包层次如图 9-8 所示。

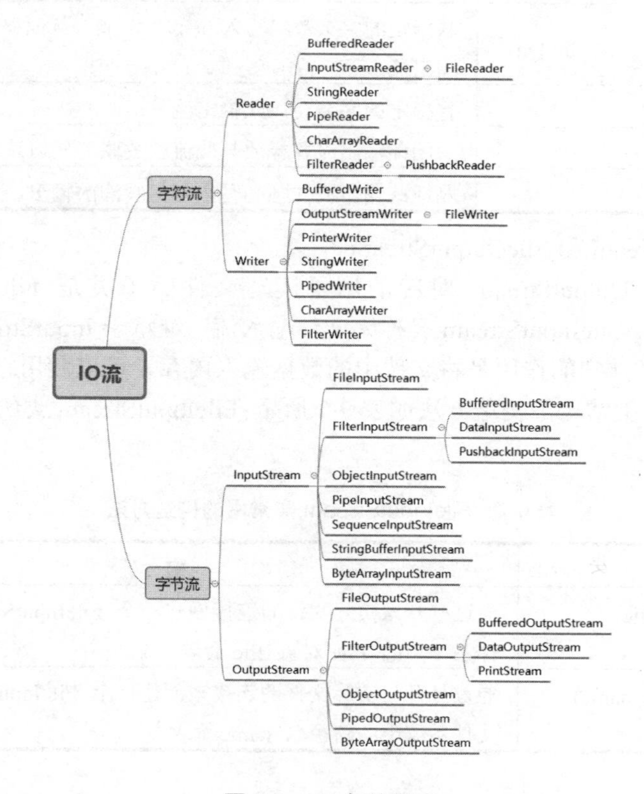

图 9-8　IO 包的层次

2. 字节流

字节流和字符流是依据所处理数据的类型来分类的。字符流通常处理文本文件，而字节流则可以处理所有的文件。本任务重点介绍字节流的使用，字符流可通过查询 API 来使用。InputStream 是所有字节输入流的抽象基类，OutputStream 是所有字节输出流的抽象基类，它们都无法直接创建实例对象，只能通过其子类的实现类创建对象。

InputStream 类的常用方法如表 9-5 所示；OutputStream 类的常用方法如表 9-6 所示。

表 9-5　InputStream 类的常用方法

方　法	描　述
abstract int read()	从输入流读取数据的下一个字节
int read(byte[] b)	从输入流读取一些字节数，并将它们存储到缓冲区 b
int read(byte[] b，int off，int len)	从输入流读取最多 len 字节的数据到一个字节数组
void close()	关闭此输入流并释放与流相关联的任何系统资源
abstract int read()	从输入流读取数据的下一个字节

表 9-6　OutputStream 类的常用方法

方　法	描　述
void write(byte[] b)	将 b.length 字节从指定的字节数组写入此输出流
void write(byte[] b，int off，int len)	从指定的字节数组写入 len 个字节，从偏移 off 开始输出到此输出流
abstract write(int b)	将指定的字节写入此输出流
void close()	关闭此输出流并释放与此流相关联的任何系统资源
void flush()	刷新此输出流并强制写出任何缓冲的输出字节，清空缓冲区

(1) FileInputStream 和 FileOutputStream。

InputStream 和 OutputStream 对应的用于读写文件的子类是 FileInputStream 类和 FileOutputStream 类。FileInputStream 类称为文件输入流，继承于 InputStream 类，是进行文件读操作的最基本类，它的作用是将文件中的数据输入内存，可以利用它来读文件。

FileInputStream 类常用的构造方法如表 9-7 所示；FileInputStream 类的常用方法如表 9-8 所示。

表 9-7　FileInputStream 类常用的构造方法

方　法	描　述
FileInputStream(File file)	通过打开与实际文件的连接创建一个 FileInputStream，该文件由文件系统中的 File 对象 file 命名
FileInputStream(String name)	通过打开与实际文件的连接来创建一个 FileInputStream，该文件由文件系统中的路径名 name 命名

表 9-8　FileInputStream 类的常用方法

方　法	描　述
int read()	从该输入流读取一个字节的数据
int read(byte[] b)	从该输入流读取最多 b.length 个字节的数据到字节数组中
int read(byte[] b，int off，int len)	从该输入流读取最多 len 个字节的数据到字节数组中
void close()	关闭此输出流并释放与此流相关联的任何系统资源

FileOutputStream 类称为文件输出流，继承于 OutputStream 类，是进行文件写操作的最基本类；它的作用是将内存中的数据输出到文件中，可以利用它来写文件。

FileOutputStream 类常用的构造方法如表 9-9 所示；FileOutputStream 类的常用方法如表 9-10 所示。

表 9-9　FileOutputStream 类常用的构造方法

方　法	描　述
FileOutputStream(File file)	创建文件输出流以写入由指定的 File 对象表示的文件
FileOutputStream(File file, boolean append)	创建文件输出流以写入由指定的 File 对象表示的文件，append 为 true，在文件尾部进行追加
FileOutputStream(String name)	创建文件输出流以指定的名称写入文件
FileOutputStream(String name, boolean append)	创建文件输出流以指定的名称写入文件，append 为 true，在文件尾部进行追加

表 9-10　FileOutputStream 类的常用方法

方　法	描　述
void write(byte[] b)	将 b.length 个字节从指定的字节数组写入此文件输出流
void write(byte[] b，int off，int len)	将 len 个字节从位于偏移量 off 的指定字节数组写入此文件输出流
void write(int b)	将指定的字节写入此文件输出流
void close()	关闭此输出流并释放与此流相关联的任何系统资源

接下来通过一些例子来演示如何用字节流的方式进行文件的读写。

例：下面程序是用字节流把字符串“I Love Java”写到文件 test.txt 中。

```
1.   public class FileOutputStreamDemo
2.   {
3.     //在方法内部不进行异常处理，将异常抛给 JVM
4.     public static void main(String[] args) throws Exception
5.     {
6.       String str = "I Love Java";
7.       //通过文件名创建文件输出流对象，同时会创建文件 test.txt
8.      FileOutputStream fos = new FileOutputStream("test.txt");
9.       //将字符串转换为字节数组
10.      byte[] buffer = str.getBytes();
11.      //将字节数组中包含的数据一次性写入文件中
12.      fos.write(buffer);
13.      fos.close();//关闭流
```

```
14.     }
15.  }
```

例：下面程序是用字节流读取文件 test.txt 的内容并输出到控制台。

```
1.   public class FileInputStreamDemo {
2.       public static void main(String[] args) {
3.           File file = null;                    //声明一个 File 对象
4.           FileInputStream fis = null;          //声明一个文件输入流对象
5.           try {
6.               file = new File("test.txt");   // 创建文件对象
7.               // 使用文件对象创建文件输入流对象，相当于打开文件
8.               fis = new FileInputStream(file);
9.               //变量 buf 用于存储从文件中收到的字节数据
10.              byte[] buf = new byte[1024];
11.              //变量 str 用于把字节数据转换成字符串以方便输出到控制台
12.              String str = null;
13.              int n = 0;
14.              // 循环从文件中读取字节数据存放到 buf，n 为-1 代表读到文件尾部
15.              while ((n = fis.read(buf)) != -1) {
16.             // 把读到的字节数据转换成 String，输出到控制台上
17.                  str = new String(buf, 0, n);
18.                  System.out.print(str);
19.              }
20.          } catch (Exception e) {
21.              e.printStackTrace();
22.          } finally {
23.              try {
24.                  //放在 finally 块中，保证无论如何都能关流
25.                  fis.close();
26.              } catch (IOException e) {
27.                  e.printStackTrace();
28.              }
29.          }
30.      }
31.  }
```

上面的第一个例子因为要写的内容比较少，所以只用 fos.write(buffer)进行了一次读写，而第二个例子在读取的时候用了循环多次读取判断真正读到了多少数据。

在读取和写出字节操作时，增大读取次数或者写出次数无疑会降低程序的运行效率。因此，可以使用缓冲输出流一次性批量写出若干数据或者一次性批量读取若干数据来提高程序的效率，Java 中可以通过缓冲流来解决上述问题。

(2) 字节缓冲流。

这里重点介绍两个字节缓冲流，分别是 BufferedInputStream 和 BufferedOutputStream。BufferedInputStream 是字节缓冲输入流，其内部维护着一个缓冲区(字节数组)。当使用该流读取数据时，会尽可能多地一次性读取若干字节并存入缓冲区，然后逐一地将字节返回，直到缓冲区中的数据被全部读取完毕，之后再次读取若干字节，不断重复上述过程，这样就减少了读取的次数，从而提高了读取的效率。

BufferedOutputStream 缓冲输出流内部维护着一个缓冲区，向该流写入的数据都会先存入缓冲区，当缓冲区满时，缓冲流就会将数据一次性地全部写出。

例：使用缓冲流把文件 text.txt 中的内容复制到 dest.dat 中。

```
1.  public class BufferedDemo {
2.      public static void main(String[] args) throws IOException {
3.          File srcFile = new File("text.txt"); // 源文件对象
4.          File destFile = new File("dest.dat"); // 目标文件对象
5.          if (!(destFile.exists())) { // 判断目标文件是否存在
6.              destFile.createNewFile();// 如果不存在则创建新文件
7.          }
8.      // 使用源文件对象创建文件输入流对象
9.      FileInputStream fis = new FileInputStream(srcFile);
10.     // 使用目标文件对象创建文件输出流对象
11.     FileOutputStream fos = new FileOutputStream(destFile);
12.     //使用字节输入缓冲流以提高效率
13.     BufferedInputStream bin=new BufferedInputStream(fis);
14.     //使用字节输出缓冲流以提高效率
15.     BufferedOutputStream bout=new BufferedOutputStream(fos);
16.     // 创建字节数组，作为临时缓冲
17.         byte[] buf = new byte[1024];
18.         System.out.println("开始复制文件...");
19.         // 循环从输入缓冲流中读取数据
20.         int n=0;
21.         // 循环从输入缓冲流中读取数据
22.         while ((n=bin.read(buf)) != -1) {
23.              // 写入到缓冲输出流中
24.             bout.write(buf, 0, n); //刷新缓冲区，不必等缓冲区满就可以及时把缓冲区
                                       中的数据输出到文件中
25.             bout.flush();
26.         }
27.         System.out.println("文件复制成功！");
28.         // 关闭流
29.         fis.close();
30.         fos.close();
31.         bin.close();
32.         bout.close();
33.     }
34. }
```

3. 字符流

FileInputStream 类和 FileOutputStream 类虽然可以高效率地读/写文件，但对于 Unicode 编码的文件，使用它们有可能出现乱码。考虑到 Java 是跨平台的语言，要经常操作 Unicode 编码的文件，使用字符流操作文件是有必要的。Java 的字符流的父类是 Reader 和 Writer。

(1) Reader 和 Writer。

Reader 是所有字符输入流的抽象基类，Writer 类是所有字符输出流的抽象基类，它们本身不能创建对象，只能通过其子类创建对象，如图 9-9 所示。

Reader 和 Writer 类的常用方法如表 9-11 和表 9-12 所示。

(2) FileReader 和 FileWriter。

FileReader 类称为文件读取流，允许以字符流的形式对文件进行读操作，其构造方法有 3 种重载方式，常用的两种如表 9-13 所示。

FileReader 类将从文件中逐个地读取字符，效率比较低，因此一般将该类对象包装到缓冲流中进行操作。

FileWriter 类称为文件写入流，以字符流的形式对文件进行写操作，其构造方法有 5 种

重载方式，常用的 4 种如表 9-14 所示。

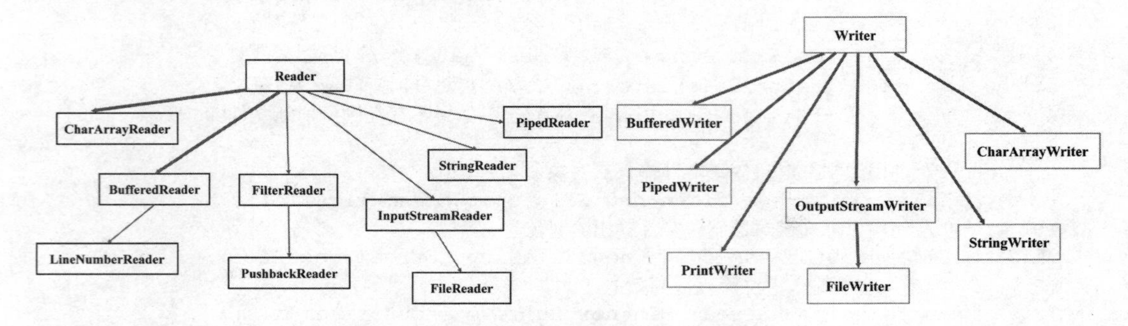

图 9-9　字符输入流和输出流的继承关系

表 9-11　Reader 类的常用方法

方　法	描　述
int read()	读一个字符
int read(char[] cbuf)	将字符读入数组
int read (char[] cbuf, int off, int len)	将字符读入数组的一部分
void close()	关闭输出流并释放与此流相关联的任何系统资源

表 9-12　Writer 类的常用方法

方　法	描　述
void write (char[] cbuf)	写入一个字符数组
void write(char[] cbuf, int off, int len)	写入字符数组的一部分
void write(int b)	写一个字符
void write(String str)	写一个字符串
void flush()	刷新流
void close()	关闭流，先刷新

表 9-13　FileReader 类常用的构造方法

方　法	描　述
FileReader(File file)	创建一个新的 FileReader，给出 File 读取
FileReader(String fileName)	创建一个新的 FileReader，给定要读取的文件的名称

表 9-14　FileWriter 类常用的构造方法

方　法	描　述
FileWriter(File file)	给一个 File 对象构造一个 FileWriter 对象
FileWriter(File file, boolean append)	给一个 File 对象构造一个 FileWriter 对象并指定是追加还是覆盖
FileWriter(String fileName)	构造一个给定文件名的 FileWriter 对象
FileWriter(String fileName, boolean append)	构造一个 FileWriter 对象，给出一个带有布尔值的文件名，表示是否附加写入的数据

下面通过例子讲解字符流操作文件的用法。

例：使用 FileReader 和 FileWriter 完成文件的复制。

```
1.  public class FileReaderAndWriterTest {
2.      public static void main(String[] args) throws IOException {
3.          File srcFile = new File("src.dat"); // 源文件对象
4.          // 判断源文件是否存在，不存在则退出程序
5.          if(!srcFile.exists())
6.          {
7.              System.out.println("源文件不存在");
8.              return;
9.          }
10.         File destFile = new File("dest.dat"); // 目标文件对象
11.         // 判断目标文件是否存在
12.         if (!(destFile.exists())) {
13.             // 如果不存在则创建新文件
14.             destFile.createNewFile();
15.         }
16.         // 使用源文件对象创建文件输入字符流对象
17.         FileReader fin = new FileReader(srcFile);
18.         // 使用目标文件对象创建文件输出字符流对象
19.         FileWriter fout = new FileWriter(destFile);
20.         // 创建字符数组
21.         char[] buf = new char[1024];
22.         System.out.println("开始复制文件...");
23.         int n=0;
24.         // 循环从输入字符流中读取数据
25.         while ((n=fin.read(buf)) != -1) {
26.             fout.write(buf,0,n);   // 写入到输出流中
27.             fout.flush();          //刷新，输出到文件中
28.         }
29.         System.out.println("文件复制成功！");
30.         fin.close();// 关闭流
31.         fout.close();
32.     }
33. }
```

运行程序时要先在工程下创建 src.dat 文件，并把一些字符放进文件中，执行程序后即可看到生成了 dest.dat 文件，完成了文件的复制。

(3) 字符缓冲流。

BufferedReader 类主要为字符流提供缓冲，从字符输入流读取文本，缓冲字符，以提供字符、数组和行的高效读取。其构造方法有两种重载方式，表 9-15 中第一行是常用的一种。

表 9-15 BufferedReader 类的构造方法和常用方法

方 法	描 述
BufferedReader(Reader in)	将字符读取流对象包装成缓冲读取流对象
String readLine()	读一行文字
int read()	读一个字符
int read(char[] cbuf, int off, int len)	将字符读入数组的一部分

BufferedWriter 类可以为 FileWriter 类提供缓冲，将文本写入字符输出流，缓冲字符，以实现单个字符、数组和字符串的高效写入。其构造方法有两种重载方式，表 9-16 中常用的

一种。

表 9-16　BufferedWriter 类常用的构造方法

方　法	描　述
BufferedWriter(Writer out)	创建使用默认大小的输出缓冲区的缓冲字符输出流

例： 使用 BufferedReader 和 BufferedWriter 完成文件的复制。

```
1.   public class BufferedReaderAndWriterTest {
2.       public static void main(String[] args) throws IOException {
3.           //源文件
4.           String sfileName="src.dat";
5.           // 目标文件对象
6.           File destFile = new File("dest.dat");
7.           // 判断目标文件是否存在
8.           if (!(destFile.exists())) {
9.               // 如果不存在则创建新文件
10.              destFile.createNewFile();
11.          }
12.          // 使用源文件对象创建文件输入字符缓冲流对象
13.          BufferedReader bin = new BufferedReader(new FileReader(new File
               (sfileName)));
14.          // 使用目标文件对象创建文件输出字符缓冲流对象
15.          BufferedWriter bout = new BufferedWriter(new FileWriter(destFile));
16.          // 接收一行字符串
17.          String line=null;
18.          System.out.println("开始复制文件...");
19.          // 循环从源文件中读取一行行的数据
20.          while ((line=bin.readLine()) != null) {
21.              bout.write(line);// 写入目标文件中
22.              bout.newLine();//写入换行符
23.              bout.flush();//刷新缓冲区，输出到文件中
24.          }
25.          System.out.println("文件复制成功！");
26.          bin.close();    // 关闭流
27.          bout.close();
28.      }
29.  }
```

(4) 转换流。

InputStreamReader 是字符输入流，它可以将字节输入流转换为字符输入流。每次调用 InputStreamReader 的 read()方法目的是从底层字节输入流读取一个或多个字节。为了使字节有效地转换为字符，可以从底层流读取比满足当前读取操作所需的更多字节。因此为了获得最高的效率，请考虑在 BufferedReader 中包装一个 InputStreamReader。

OutputStreamWriter 是字符输出流，它可以将字节输出流转换为字符输出流。每次调用 write()方法都会使编码转换器在给定字符上被调用。所得到的字节在写入底层输出流之前累积在缓冲区中。请注意，传递给 write()方法的字符不会缓冲。为了最高的效率，请考虑在 BufferedWriter 中包装一个 OutputStreamWriter，以避免频繁地调用转换器。

InputStreamReader 类中常用的方法如表 9-17 所示。

OutputStreamWriter 类中常用的方法如表 9-18 所示。

表 9-17　InputStreamReader 类中常用的方法

方　法	描　述
InputStreamReader(InputStream in)	创建一个使用默认字符集的 InputStreamReader
InputStreamReader(InputStream in, String charsetName)	创建一个使用命名字符集的 InputStreamReader
InputStreamReader(InputStream in)	创建一个使用默认字符集的 InputStreamReader

表 9-18　OutputStreamWriter 类中常用的方法

方　法	描　述
OutputStreamWriter(OutputStream out)	创建一个使用默认字符编码的 OutputStreamWriter
OutputStreamWriter(OutputStream out, String charsetName)	创建一个使用命名字符集的 OutputStreamWriter

例：从标准输入控制台输入一行行数据并保存到文件 a.txt 中。

```
1.  public class ChangeIOTest {
2.      public static void main(String[] args) throws Exception {
3.          //创建标准输入流
4.          InputStream in=System.in;
5.          //把标准输入的字节流转换成字符流
6.          InputStreamReader reader=new InputStreamReader(in);
7.          //为了提高效率，一行行读取标准输入控制台的数据，创建字符输入缓冲流
8.          BufferedReader input=new BufferedReader(reader);
9.  
10.         //创建目的地的文件输出流
11.         FileOutputStream out=new FileOutputStream(new File("a.txt"));
12.         //把标准输出的字节流转换成字符流
13.         OutputStreamWriter writer=new OutputStreamWriter(out);
14.         //为了提高效率，一行行输出字符数据到文件中，创建字符输出缓冲流
15.         BufferedWriter output=new BufferedWriter(writer);
16. 
17.         String line=null;
18.         //循环从控制台接收一行行字符数据
19.         while((line=input.readLine())!=null)
20.         {
21.             output.write(line);     //一行行地写到文件中
22.             output.newLine();       //写入换行符
23.             output.flush();         //刷新数据
24.         }
25.         in.close();
26.         out.close();
27.         input.close();
28.         output.close();
29.     }
30. }
```

(5)　PrintWriter 类。

PrintWriter 类是具有自动刷新功能的字符缓冲输出流，不需要 flush()。

PrintWriter 类中常用的方法如表 9-19 所示。

表 9-19　PrintWriter 类中常用的方法

方　法	描　述
PrintWriter(File file)	使用指定的文件创建一个新的 PrintWriter，而不需要自动执行刷新
PrintWriter(OutputStream out)	从现有的 OutputStream 创建一个新的 PrintWriter，而不需要自动执行刷新
PrintWriter(String fileName)	使用指定的文件名创建一个新的 PrintWriter，而不需要自动执行刷新
PrintWriter(Writer out)	创建一个新的 PrintWriter 而不需要自动执行刷新

例：往文件 a.txt 中写入一行数据：你好，中国。

```
1.   public class PrintWriterTest {
2.       public static void main(String[] args) throws Exception {
3.           // 创建字符输出流
4.           OutputStreamWriter out=new OutputStreamWriter(
5.                   new FileOutputStream(new File("a.txt")));
6.           //对字符流进行包装，创建带有自动刷新的 PrintWriter
7.           PrintWriter pwriter=new PrintWriter(out);
8.           //向文件 a.txt 中写入一行数据：你好，新大陆教育
9.           pwriter.println("你好，中国");
10.          pwriter.close();
11.      }
12.  }
```

【任务实施】

1. 任务分析

(1)　注册界面在任务 1 中已经完成，可直接使用。

(2)　注册按钮的监听完成了注册用户文件的创建，添加使用 IO 写入注册信息代码。将用户名存入 Set 集合，同时将用户名和用户信息对象作为键值对存入 Map 集合中。

(3)　程序初始化时读取所有用户注册信息文件中的内容，用户名存入 Set 与 Map 集合中。

(4)　注册成功或取消注册可返回到登录界面。

2. 任务实施

(1)　新建工程 project9_task2，新建包 com.nle.task2，并将 task1 中的内容复制到 task2，注意修改 fxml 文件的 Controller 类名，如图 9-10 所示。

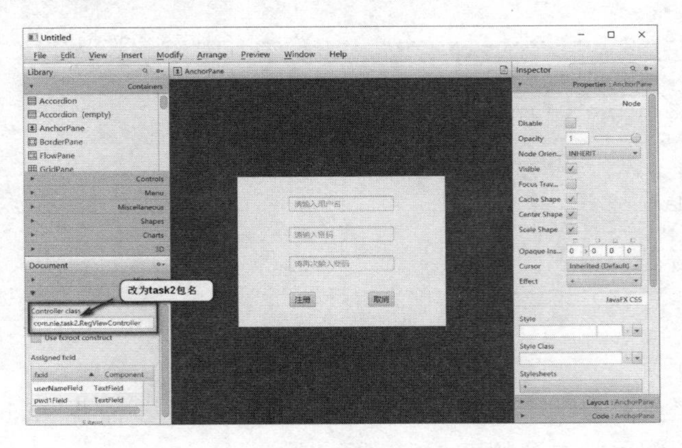

图 9-10　修改文件的 Controller 类名

(2) 新建 LoginView.fxml 文件，在 SceneBuilder 中设计好界面，并创建相应的 Controller，如图 9-11 所示。

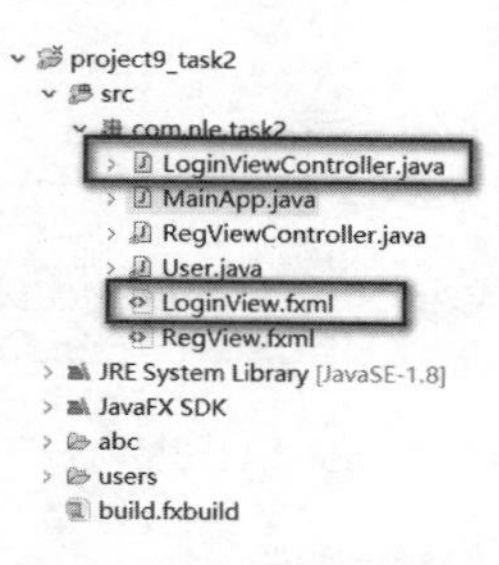

图 9-11 创建界面文件

(3) 在 Controller 中添加控件映射和控件监听方法，将原来 RegViewController 中读取的用户代码迁移至 LoginController，并添加读取用户信息的代码，在 initialize()方法中调用。

```
1.  public class LoginViewController {
2.      @FXML
3.      private TextField usernameField;
4.
5.      @FXML
6.      private TextField pwdField;
7.
8.      @FXML
9.      private Button logBtn;
10.
11.     @FXML
12.     private Button regBtn;
13.
14.     private Alert alert;
15.
16.     @FXML
17.     private void initialize() {
18.         alert = new Alert(AlertType.WARNING);
19.         alert.setTitle("温馨提示");
20.         //读取所有用户
21.         readAllUser();
22.     }
23.
24.     @FXML
25.     void loginHandler(ActionEvent event) {
26.
27.     }
28.
29.     @FXML
30.     void regHandler(ActionEvent event) {
31.         MainApp.gotoView("RegView", "智慧园区注册界面");
32.     }
33.
34.     /**
35.      * 读取所有用户
36.      */
37.     private void readAllUser() {
38.         File file = new File("users");
39.         if (!file.exists()) {
40.             file.mkdirs();
```

```
41.         }
42.         String[] userNames = file.list();
43.         for (int i = 0; i < userNames.length; i++) {
44.             User user = new User();
45.             user.setUsername(userNames[i]);
46.             //读取用户信息，目前只有密码一个信息
47.             File userFile = new File("users/"+userNames[i]);
48.             try {
49.                 FileInputStream fis = new FileInputStream(userFile);
50.                 BufferedReader br = new BufferedReader(new InputStreamReader
                      (fis));
51.                 user.setPwd(br.readLine());
52.                 br.close();
53.             } catch (FileNotFoundException e) {
54.                 // TODO Auto-generated catch block
55.                 e.printStackTrace();
56.             } catch (IOException e) {
57.                 // TODO Auto-generated catch block
58.                 e.printStackTrace();
59.             }
60.             RegViewController.users.add(user);
61.             RegViewController.userMsgs.put(userNames[i], user);
62.         }
63.     }
64.
65.
66.     /**
67.      * 字符串检测
68.      *
69.      * @param str
70.      * @return
71.      */
72.     private boolean checkNotNull(String str) {
73.         if (str == null || "".equals(str.trim())) {
74.             return false;
75.         }
76.         return true;
77.     }
78. }
```

(4) 完成登录按钮的监听方法。

```
1.    @FXML
2.    void loginHandler(ActionEvent event) {
3.        String username = usernameField.getText().trim();
4.        String pwd = pwdField.getText().trim();
5.        if(checkNotNull(username)&&checkNotNull(pwd)) {
6.            if(RegViewController.userMsgs.containsKey(username)) {
7.
      if(RegViewController.userMsgs.get(username).getPwd().equals(pwd)) {
8.                    alert.setContentText("登录成功！");
9.                }else {
10.                   alert.setContentText("密码错误，请重试！");
11.               }
12.           }else {
13.               alert.setContentText("用户名不存在，请先注册！");
14.           }
15.       }else {
16.           alert.setContentText("请填写用户名和密码！");
```

```
17.     }
18.     alert.show();
19. }
```

(5) 修改注册界面控制器中保存用户信息的方法，添加写入用户信息的代码。

```
1.  private void saveUserFile(User user) {
2.      File file = new File("users/" + user.getUsername());
3.      try {
4.          file.createNewFile();
5.          //将用户密码写入用户信息文件
6.          FileOutputStream fos = new FileOutputStream(file);
7.          fos.write(user.getPwd().getBytes());
8.          fos.close();
9.          //将用户信息添加到 Map 集合
10.         userMsgs.put(user.getUsername(), user);
11.     } catch (IOException e) {
12.         // TODO Auto-generated catch block
13.         e.printStackTrace();
14.     }
15. }
```

3. 运行结果

运行程序，结果如图 9-12 所示。

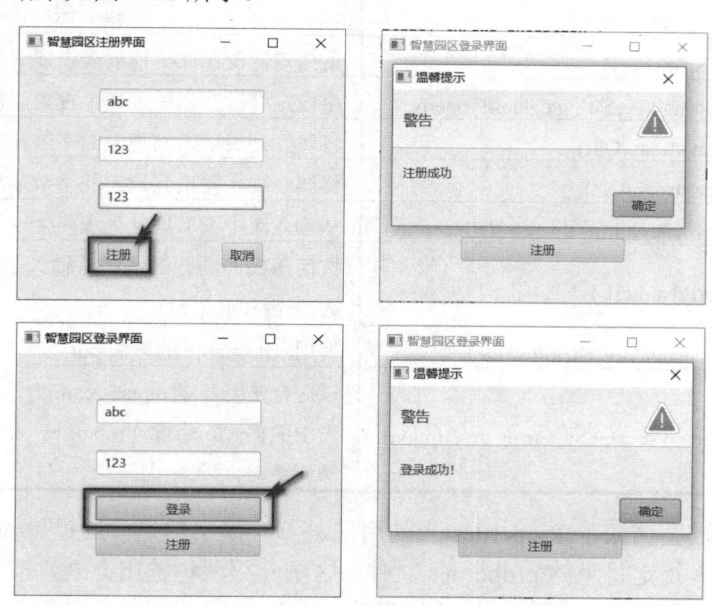

图 9-12 运行结果

任务 3 智慧园区系统配置参数的读写

扫码观看视频讲解

【任务描述】

在前面的任务中，通过 Java 的 IO 实现了对文件的读写，每次调用 read()或 write()的方法可以分别实现对文件固定长度的读取或写入。但是，在应用编写的过程中，经常需要将应用的配置保存到文件中以便用户设置。配置信息通常是以键值对的形式存在，若使用普通的读写 API 则需要对字符串进行一定的设计。Properties 类是 Java 为此提供的一个解决方案，它是 properties 文件和程序间的桥梁，通常用来实现对键值对数据的读写。本任务要求通过 Properties 类完成项目配置文件的读写。任务清单如表 9-20 所示。

表 9-20　任务清单

任务课时	4 课时	任务组员数量	建议 1 人
任务组采用设备	无		

【知识解析】

Properties 类主要用于读取 Java 的配置文件，不同的编程语言有自己所支持的配置文件，配置文件中的很多变量是经常改变的，为了方便用户的配置，需要让用户能够脱离程序本身去修改相关的变量设置。就像在 Java 中，其配置文件为.properties 文件，是以键值对的形式进行参数配置的。Properties 类的常用方法如表 9-21 所示。

表 9-21　Properties 类的常用方法

方　法	描　述
String getProperty(String key)	用指定的键在此属性列表中搜索属性
String getProperty(String key, String defaultProperty)	用指定的键在属性列表中搜索属性
void list(PrintStream streamOut)	将属性列表输出到指定的字节输出流
void list(PrintWriter streamOut)	将属性列表输出到指定的字节输出流
void load(InputStream streamIn) throws IOException	从输入流中读取属性列表(键和元素对)
Enumeration propertyNames( )	按简单的面向行的格式从输入字符流中读取属性列表(键和元素对)
Object setProperty(String key, String value)	调用 Hashtable 的方法 put
void store(OutputStream streamOut, String description)	以适合使用 load(InputStream)方法加载到 Properties 表中的格式，将此 Properties 表中的属性列表(键和元素对)写入输出流

接下来，通过案例演示 Properties 类的用法。首先演示的是，通过 Properties 类将需要保存的键值对存入指定目录的 properties 文件。这里介绍两种输出方式，第一种是通过 store 方法，先创建目标文件 File 对象，然后创建 Properties，将需要保存的键值对设置给对象，调用 store()方法即可将信息存入文件中。第二种是使用 list 方法，也可以通过同样的步骤将数据存入文件。

Properties 类向文件写入键值对：

```
1.  public class PropertiesDemo {
2.
3.      public static void main(String[] args) throws IOException {
4.          File proFile = new File("abc/db.properties");
5.          if(!proFile.exists()) {
6.              proFile.createNewFile();
```

```
7.          }
8.
9.          //创建 Properties 类对象
10.         Properties properties = new Properties();
11.         //存储键值对
12.         //第一种方法
13.         //properties.setProperty("username", "zhangsan");
14.         //properties.setProperty("pwd", "123");
15.         //数据存储到文件
16.         //properties.store(new FileOutputStream(proFile), "test");
17.         //第二种方法
18.         properties.put("username", "tom");
19.         properties.put("pwd", "123");
20.         //数据存储到文件
21.         properties.list(new PrintStream(proFile));
22.     }
23.
24. }
```

下面通过案例演示 Properties 类对文件内容的读取。

Properties 类从文件读取键值对：

```
1.  public class PropertiesDemo2 {
2.
3.      public static void main(String[] args) throws IOException {
4.          File proFile = new File("abc/db.properties");
5.          if(!proFile.exists()) {
6.              proFile.createNewFile();
7.          }
8.          //创建 Properties 类对象
9.          Properties properties = new Properties();
10.         //加载文件流
11.         properties.load(new FileInputStream(proFile));
12.         System.out.println(properties.getProperty("username"));
13.         System.out.println(properties.getProperty("pwd"));
14.     }
15.
16. }
```

【任务实施】

1. 任务分析

(1) 参数配置界面采用标签结合下拉列表框的形式让用户进行选择。

(2) 界面初始化时，将配置文件中的信息读取出来，并回填到界面。

(3) 用户修改设置后，单击“保存”按钮，覆盖配置文件中的信息。

2. 任务实施

(1) 新建工程 project9_task3，新建 com.nle.task3 包，创建参数配置界面的 fxml 文件，并创建该文件对应的 Controller 类，源代码在配套资源\project9_task3\com\nle\task3 下。

(2) 在 Controller 类中新建读取配置文件的方法 readConfig()，并在 initialize()方法中调用。

```
1.   /**
2.    * 回填配置参数
3.    */
4.   private void readConfig() {
5.       try {
6.           File config = new File("config.properties");
7.           if (!config.exists()) {
8.               config.createNewFile();
9.           }
10.          Properties properties = new Properties();
11.          properties.load(new FileInputStream(config));
12.          alarmRed.setValue(properties.getProperty("alarmRed"));
13.          alarmYellow.setValue(properties.getProperty("alarmYellow"));
14.          light.setValue(properties.getProperty("light"));
15.          blower.setValue(properties.getProperty("blower"));
16.          alarmGreen.setValue(properties.getProperty("alarmGreen"));
17.          human.setValue(properties.getProperty("human"));
18.          smoke.setValue(properties.getProperty("smoke"));
19.      } catch (FileNotFoundException e) {
20.          // TODO Auto-generated catch block
21.          e.printStackTrace();
22.      } catch (IOException e) {
23.          // TODO Auto-generated catch block
24.          e.printStackTrace();
25.      }
26.  }
```

(3) 给保存按钮添加监听方法 saveConfig()。

```
1.   void saveConfig(ActionEvent event) {
2.       Properties properties = new Properties();
3.       properties.setProperty("alarmRed", alarmRed.getValue().toString());
4.       properties.setProperty("alarmYellow", alarmYellow.getValue().toString());
5.       properties.setProperty("light", light.getValue().toString());
6.       properties.setProperty("blower", blower.getValue().toString());
7.       properties.setProperty("alarmGreen", alarmGreen.getValue().toString());
8.       properties.setProperty("human", human.getValue().toString());
9.       properties.setProperty("smoke", smoke.getValue().toString());
10.      try {
11.          properties.store(new PrintStream(new File("config.properties")),
               "system config");
12.          alert.setContentText("保存成功");
13.      } catch (IOException e) {
14.          e.printStackTrace();
15.          alert.setContentText("保存失败");
16.      }
17.      alert.show();
18.  }
```

3. 运行结果

运行结果如图 9-13 所示。

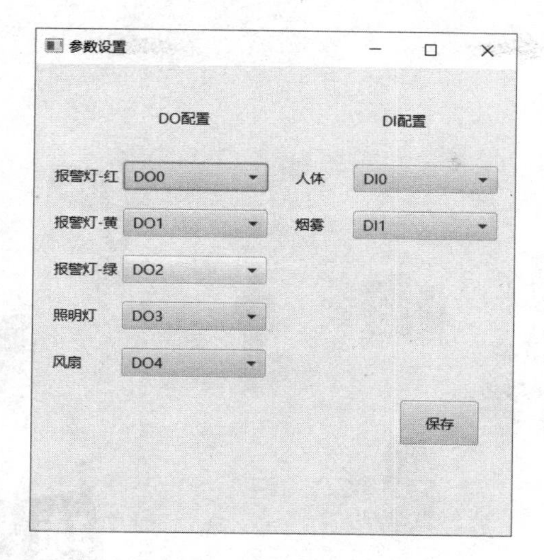

图 9-13　运行结果

思考与练习

1. 什么是 IO?
2. 简述字节流与字符流的区别。
3. 通过 IO 流实现记住用户密码和上次登录时间的功能。

项目10 实时更新数据

【项目描述】

本项目实现智慧园区中的门禁监测和重大火警报警功能，利用多线程技术实时更新可用串口。在串口采集到传感数据后，根据采集到的红外对射传感数据判断是否有人经过，有人用亮灯来模拟开门从而实现门禁监测功能；根据采集到的火焰传感数据判断是否有火警，有则三色灯轮流闪烁模拟火警报警功能。通过本项目的学习，掌握在物联网系统的应用开发中如何创建线程、启动线程、停止线程等，同时了解线程同步和互斥的实际应用。具体任务列表如图 10-1 所示。

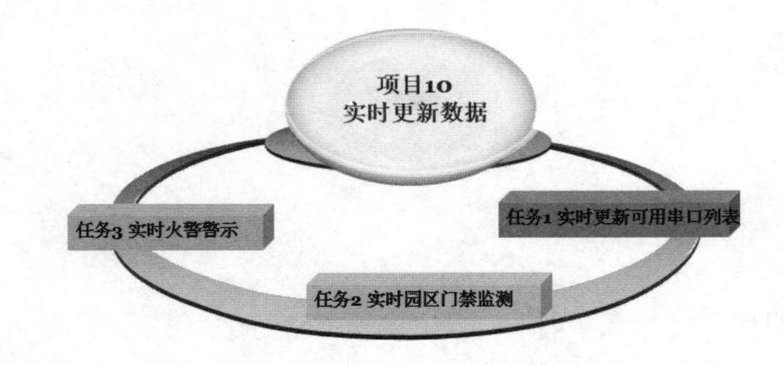

图 10-1　项目 10 任务列表

【学习目标】

知识目标： 会描述进程与线程的区别；会创建、启动、停止、合并线程；会辨识线程的状态和设置线程的优先级；会使用线程的同步和互斥技术。

技能目标：能用多线程技术实现可用串口的实时更新；能用多线程技术实现园区门禁监测；能用多线程技术实现火警警示信号。

任务1　实时更新可用串口列表

扫码观看视频讲解

【任务描述】

在物联网的应用系统中，传感器数据经过采集后，可以通过串口传输，也可以经由网关采集后传输到物联网云平台。如果是传输到串口，则物联网的应用程序应该知道有哪些COM口可用，以便可以选择对应的串口进行数据收发。

本任务要求实现当检测到可用串口发生变化时，程序能收集到这些变化信息。实现的功能是为后面智能园区的综合应用开发做准备。任务清单如表10-1所示。

表10-1　任务清单

任务课时	4课时	任务组员数量	建议3人
任务组采用设备	1台PC 一根USB口转串口线		

【拓扑图】

本任务的拓扑图如图10-2所示。

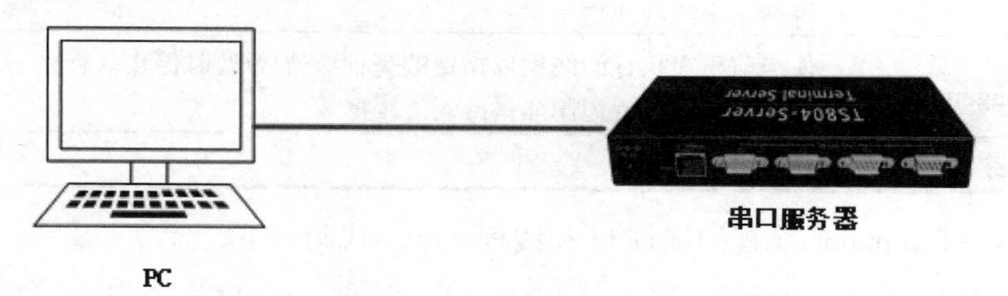

图10-2　任务1的拓扑图

【知识解析】

1. 进程与线程

当一个应用程序启动时，就有一个进程被操作系统创建，与此同时，一个主线程也立刻运行。运行中的程序称为一个**进程(Progress)**，每个进程中又可以包含多个顺序执行的流程，每个顺序执行的流程就是一个**线程(Thread)**。

线程是程序执行流的最小单元，是**进程**中的一个实体，一个标准的线程由自己的线程ID、当前指令指针、寄存器集合和堆栈组成，是被系统独立调度和分配的基本单元。线程本身不拥有系统资源，只拥有一些在运行时必不可少的资源，同一个进程内的所有线程共享进程所拥有的全部资源，所以进程ID是全局唯一的，线程ID是进程内唯一的，只有进程竞争到CPU执行权，进程内的线程才有可能被执行。

通常计算机的一个 CPU 在任意时刻只能执行一条机器指令，每个线程只有获得 CPU 的使用权才会被执行，当同一个程序中同时运行多个执行顺序完成不同的工作时，操作系统通过将 CPU 时间划分为时间片的方式，让进入就绪状态的线程轮流获得 CPU 的使用权，只不过这个时间片很短，使用户觉得多个线程在同时执行。

Java 的设计思想是建立在当前大多数操作系统都支持线程调度的基础上。Java 虚拟机的很多任务都依赖于线程调度，而且所有的类库都是为多线程设计的，所以多线程的开发是必须掌握的。

2. Thread 类

Thread 类是一个线程类，其中常用的方法包括 start()方法、run()方法等，具体如表 10-2 所示。

表 10-2　Thread 类的常用方法

方　法	描　述
void start()	使线程开始执行，Java 虚拟机调用此线程的 run()方法
void run()	如果线程使用单独的 Runnable 运行对象构造，则调用该 Runnable 对象的 run()方法；否则，此方法不执行任何操作并返回
long getId()	返回此线程的标识符
String getName()	返回此线程的名称
void interrupt()	中断这个线程
void join()	等待这个线程死亡
void sleep(long millis)	使当前正在执行的线程以指定的毫秒数暂停(暂时停止执行)，具体取决于系统定时器和调度程序的精度和准确性
currentTread()	获取当前线程

例 10-1：打印 main()线程的线程 id 和线程名字，代码如下。

```
1.   public class Demo1 {
2.       public static void main(String[] args) {
3.           System.out.println(Thread.currentThread().getId());
4.           System.out.println(Thread.currentThread().getName());
5.       }
6.   }
```

当程序运行后，main()方法就是程序的主线程，可以用 getId()获取线程号，用 getName()获取线程名。

运行结果如图 10-3 所示。

图 10-3　main 线程的 id 和名字

3. 创建线程的两种方式

(1) 继承 Thread 类，并重写 run()函数。

将一个类声明为 Thread 的子类，重写 run()方法，把线程执行的代码写在 run()中，代码如下：

```
1.   public class MyThread extends Thread{
2.   …
3.     @Override
4.     public void run() {
5.   //把该线程要执行的代码写在这里
6.     }
7.   }
```

建立子类的实例对象，调用 start()启动线程，代码如下：

```
1.   MyThread thread=new MyThread();
2.   thread.start();
```

例 10-2：创建一个线程，与主线程一起交替运行，代码如下：

```
1.   public class Demo2 {
2.       public static void main(String[] args) {
3.           MyThread m = new MyThread();//创建一个新线程对象
4.           m.start();//启动新线程
5.           //主线程循环一次 线程休眠 1s，让出 CPU 资源
6.           for(int i=0;i<10;i++)
7.           {
8.               try {
9.       System.out.println(Thread.currentThread().getName() +"  "+i);
10.                  Thread.sleep(1000);
11.              } catch (InterruptedException e) {
12.                  e.printStackTrace();
13.              }
14.          }
15.      }
16.  }
17.
18.   class MyThread extends Thread{
19.      @Override
20.      public void run() {
21.          for(int i=0;i<10;i++)
22.          {
23.              try {
24.      System.out.println(Thread.currentThread().getName() +"  "+i);
25.                  Thread.sleep(1000);
26.              } catch (InterruptedException e) {
27.                  e.printStackTrace();
28.              }
29.          }
30.      }
31.  }
```

运行结果如图 10-4 所示。

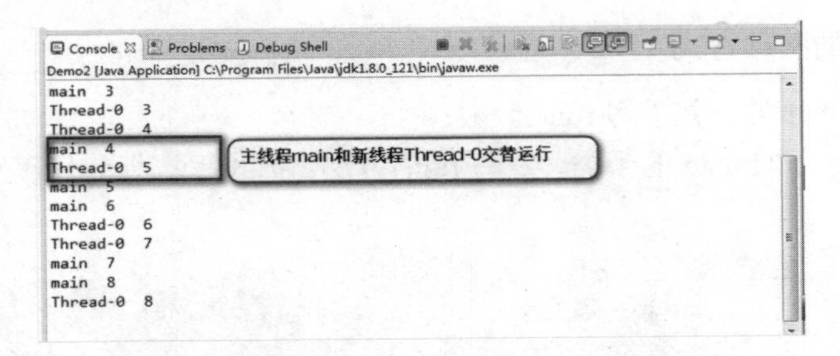

图 10-4　用继承方式产生线程的运行结果

上述代码用继承 Thread 类的方式创建了一个新线程，并且调用 start()方法启动线程，线程启动后自动执行线程类中的 run()方法。新线程和 main()线程各执行 10 次，每执行一次线程就调用 sleep(1000)休眠 1s 让出 CPU 资源，两个线程在竞争 CPU 资源，谁竞争到资源谁先运行。

需要注意的是，用 m.start()的方式启动新线程，如果改成 m.run()，程序的输出结果如图 10-5 所示，程序并没有产生新的线程。

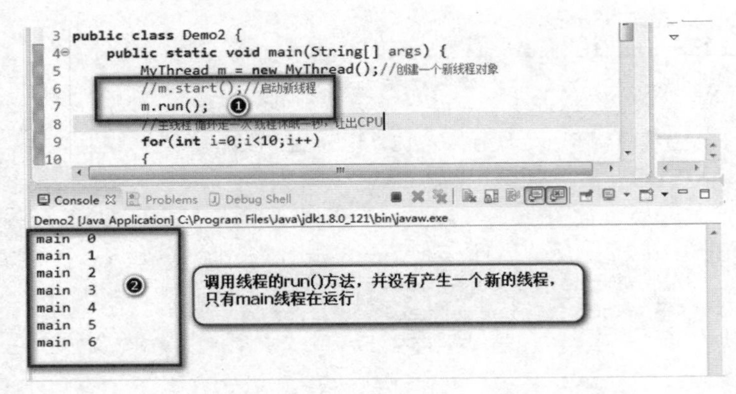

图 10-5　直接执行 run 方法的运行结果

(2) 实现 Runnable 接口，并重写 run()函数。

因为 Java 是单继承的，在某些情况下一个类可能已经继承了某个父类，这时若该类想成为线程类，则用继承 Thread 类的方法来创建线程显然违反了 Java 的单继承规则，所以 Java 的设计者们提供了另外一种方式创建线程，就是通过实现 Runnable 接口来创建线程。

通过实现 Runnable 接口来创建线程的步骤如下。

① 先创建一个类，该类实现 Runnable 接口并重写 run()方法。

② 把线程执行的代码写在 run()方法中。

③ 在创建 Thread 对象时把实现 Runnable 接口的类的实例对象作为参数传递给 Thread 的构造方法。

④ 通过 Thread 对象的 start()方法启动线程。

例 10-3：使用 Runnable 接口的实现类创建线程，代码如下。

```
1.   public class Demo3 {
2.       public static void main(String[] args) {
3.           //1. 创建一个 Runnable 对象
```

```
4.          MyRunnable r = new MyRunnable();
5.          //2. 创建一个线程对象，把 Runnable 当构造参数传递进去
6.          Thread t = new Thread(r);
7.          //3. 启动新线程
8.          t.start();
9.          //4. 主线程循环一次 线程休眠一秒，让出 CPU
10.         for (int i = 0; i < 10; i++) {
11.             try {
12.                 System.out.println(Thread.currentThread().getName() + "   " +
    i);
13.                 Thread.sleep(1000);
14.             } catch (InterruptedException e) {
15.                 e.printStackTrace();
16.             }
17.         }
18.     }
19. }
20. class MyRunnable implements Runnable {
21.     @Override
22.     public void run() {
23.         for (int i = 0; i < 10; i++) {
24.             try {
25.                 System.out.println(Thread.currentThread().getName() + "   " +
    i);
26.                 Thread.sleep(1000);
27.             } catch (InterruptedException e) {
28.                 e.printStackTrace();
29.             }
30.         }
31.     }
32. }
```

上述代码先写了一个类 MyRunnable 实现 Runnable 接口，重写了 run()方法，在 run()方法里打印当前线程的名字，线程每隔 1s 休眠，让出 CPU 资源。在 main()中创建 MyRunnable 对象，把它当成 Thread 的构造参数进行传递，然后调用线程类的 start()方法启动线程。

程序运行结果如图 10-6 所示。

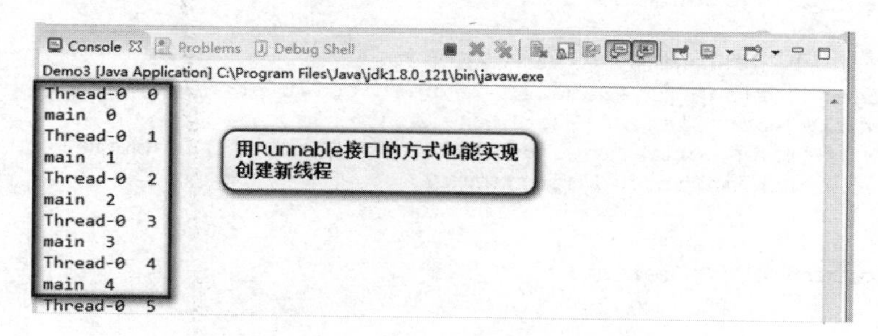

图 10-6 实现 Runnable 接口的线程的运行结果

【任务实施】

1. 任务分析

(1) 创建工程，添加串口通信包。

(2) 编写获取所有串口的方法。

(3) 用线程实时获取可用串口。

2. 任务实施

(1) 搭建工程，添加串口通信包。

新建工程 project10_task1，由于用到了串口通信包，所以要先在项目中添加 RXTXcomm.jar 包(放在项目中的 libs 目录下，并添加到 build Path 中)，同时还需要将解压后的 rxtxParallel.dll 和 rxtxSerial.dll 两个文件放在%JAVA_HOME%\jre\bin 目录下，这样该包才能被正常加载和调用。参考项目 4 任务 1 把之前写好的串口管理类、异常、工具类添加到工程中，如图 10-7 所示。

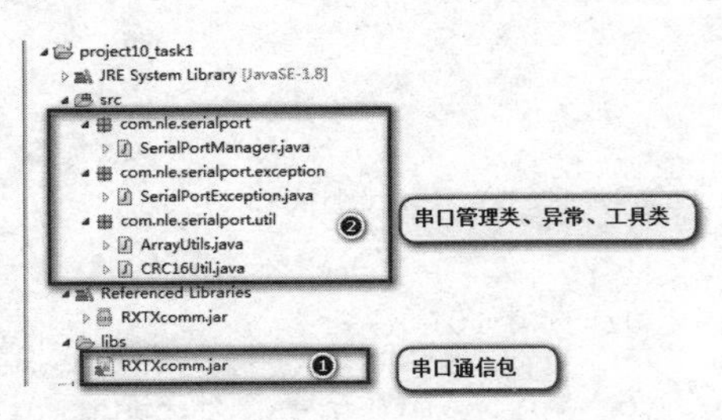

图 10-7 创建工程并添加 jar 包

(2) 编写获取所有串口的方法。

在类 SerialPortManager.java 里添加方法 findPort()用于查找所有的串口。

```
1.    /**
2.    * 查找所有可用串口
3.    * @return 可用串口名称列表
4.    */
5.    public static final List<String> findPort() {
6.           // 获得当前所有可用串口
7.       @SuppressWarnings("unchecked")
8.       Enumeration<CommPortIdentifier> portList = CommPortIdentifier.
           getPortIdentifiers();
9.
10.      // 将可用串口名添加到 List 并返回该 List
11.      List<String> portNameList = new ArrayList<String>();
12.      while (portList.hasMoreElements()) {
13.         String portName = portList.nextElement().getName();
14.         portNameList.add(portName);
15.      }
16.
17.      return portNameList;
18.  }
```

上述代码使用 RXTX 包中提供的 CommPortIdentifier.getPortIdentifiers()方法获取所有可用的串口 CommPortIdentifier 对象，该方法的返回值是一个枚举接口 CommPortIdentifier，实现枚举接口的对象生成一系列元素，一次一个，连续调用 nextElement()方法返回系列的连续元素，hasMoreElements()用于判断是否有更多的元素，如果有返回真。getName()用于获取所有的串口名字，通过遍历所有的可用串口，把串口的名字添加到集合 portNameList 中。

(3) 用线程实时获取可用串口。

新建包 com.nle.thread，在包里新建线程类 GetComsThread，代码如下：

```
1.  public class GetComsThread extends Thread {
2.      private SerialPortManager manager;
3.      private List<String> coms;
4.
5.      public GetComsThread() {
6.          manager = new SerialPortManager();
7.          coms = new ArrayList();
8.      }
9.
10.     @Override
11.     public void run() {
12.         while (true) {
13.
14.             coms = manager.findPort();
15.             System.out.println("可用的串口是: "+coms);
16.             try {
17.                 Thread.sleep(1000);
18.             } catch (InterruptedException e) {
19.                 // TODO Auto-generated catch block
20.                 e.printStackTrace();
21.             }
22.         }
23.     }
24.
25. }
```

上述代码在线程类的构造方法中对串口管理工具类对象manager进行了初始化，在run()方法中用manager. findPort()方法获取所有的串口，每次间隔1s获取一次，并打印出所有可用的串口。

新建包com.nle.main，在包里新建类Test，在main()入口方法中调用线程并启用线程：

```
1.  public class Test {
2.      public static void main(String[] args) {
3.          GetComsThread thread=new GetComsThread();
4.          thread.start();
5.      }
6.  }
```

3. 运行结果

把USB转串口线接到PC的USB口，先验证计算机设备管理器中有哪些串口，接着运行程序，可以看到COM3口被列出来了，如图10-8所示。

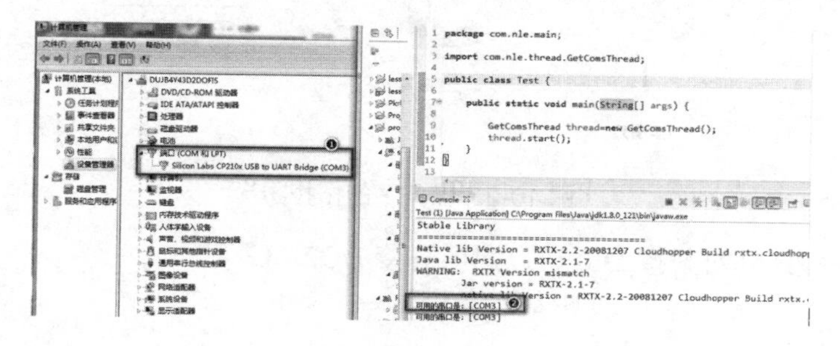

图10-8 程序搜索到了可用的COM3口

拔掉USB转串口线，可以看到程序输出的结果是没有可用的串口，至此，可用串口列表的实时更新已实现了。

任务 2 实时园区门禁监测

【任务描述】

本任务利用红外对射传感器、4150 数字量采集器、继电器、灯组成一个园区门禁监测模块，利用多线程技术实时监控园区门口是否有人进入，如果有人则联动控制灯亮模拟有人进来了。联动可以撤销，如果撤销了，则不再监控。任务清单如表 10-3 所示。

表 10-3 任务清单

任务课时	4 课时	任务组员数量	建议 3 人
任务组采用设备	1 个红外对射传感器 1 个 4150 数字量采集器 1 个继电器 1 个照明灯 1 台 PC 电源、USB 转串口线、连接线若干		

【拓扑图】

本任务拓扑图如图 10-9 所示。

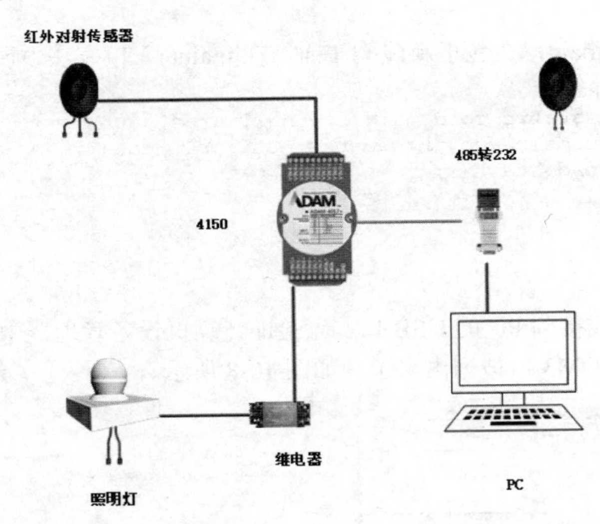

图 10-9 任务 2 的拓扑图

备注：本任务红外对射接 DI0 口，照明灯接 DO2 口。

【知识解析】

1. 线程状态的转换

在使用线程的过程中，常常伴随着线程状态的转换。Java 多线程状态的转换如图 10-10

所示。

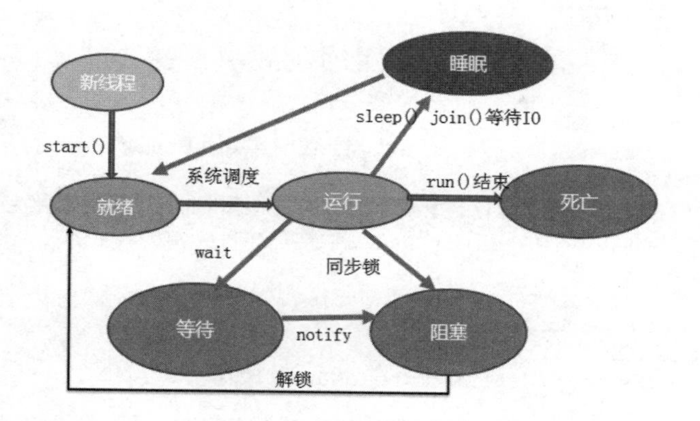

图 10-10　Java 多线程的转换

线程有以下几种状态。

(1) 初始态(new)：一个线程在调用了 new()方法之后，并在调用 start()方法之前所处的状态。在初始态中，可以调用 start()和 stop()方法(但不建议这样做)。

(2) 可运行(Runnable)：一旦线程调用了 start()方法，线程就转到 Runnable 状态。注意，如果线程处于 Runnable 状态，它也有可能不在运行，这是因为还有优先级和调度问题。

(3) 阻塞(Blocked)：线程处于阻塞(Blocked)状态，这是由两种可能性造成的，要么是因为挂起而暂停的，要么是由于某些原因而阻塞的。

阻塞(Blocked)状态是最复杂的，也是最需要关注的。应该密切监控线程的状态，避免发生阻塞现象，如果发生了，需要将阻塞状态转换为可运行或退出状态。打个比方，这就像交通警察的工作一样，要尽量协调多辆车运行，避免阻塞，保证顺畅。

发生阻塞的原因可能是多样的，一种常见的情况是线程共享资源造成冲突，将在后面进行详细介绍。另外一种常见的情况是多个线程互相干扰，比如某个线程强行加到一个正在运行的线程之前(调用 join()方法)，或者强行打断一个正在运行的线程(调用 interrupt()方法)。

(4) 退出：线程转到退出状态，这有两种可能性，要么是 run()方法执行结束，要么是调用了 stop()方法(该方法已声明作废)。

(5) 死亡状态(Dead)。

例 10-4：测试线程阻塞、线程加入及线程的优先级。使用情况的代码如下。

```
1.  public class ThreadJoinAndIsAlive {
2.      public static void main(String[] args) {
3.          //启动了两个线程 a、b，main 线程是主线程，a、b 是子线程
4.          Counter a = new Counter(" Counter a");
5.          Counter b = new Counter(" Counter b");
6.          try {
7.              System.out.println("wait for thd child thread of finish");
8.              //a、b 线程分别调用 join()，子线程会强行加到父线程之前，而父线程暂时阻塞
9.              a.join();
10.             b.join();
11.             if(!a.isAlive())//判断 a 是否存活
12.             {
13.                 System.out.println("Counter a is not alive.");
```

```
14.            }
15.            if(!b.isAlive())
16.            {
17.                System.out.println("Counter b is not alive.");
18.            }
19.        } catch (Exception e) {
20.            System.out.println("main "+e.getStackTrace());
21.        }
22.        //当a、b子线程运行结束并退出后，被中断的main线程才会继续运行
23.        System.out.println("Exit from  main thread");
24.    }
25. }
26. class Counter extends Thread{
27.     private int currentValue;
28.     public Counter(String threadName)
29.     {
30.         super(threadName);
31.         currentValue=0;
32.         System.out.println(Thread.currentThread().getName());
33.         setPriority(10);//设置线程的优先级为10，最高为10，最低为1，不设为5
34.         start();//启动线程
35.     }
36.     @Override
37.     public void run() {
38.         try {
39.             while(currentValue<5)
40.             {
41.                 System.out.println(getName()+":"+ currentValue++ );
42.                 Thread.sleep(1000);
43.             }
44.         } catch (Exception e) {
45.             System.out.println(getName()+" "+e.getStackTrace());
46.         }
47.         System.out.println("Exit from  "+ getName());
48.     }
49.     public int getValue()
50.     {
51.         return currentValue;
52.     }
53. }
```

运行结果如图 10-11 所示。

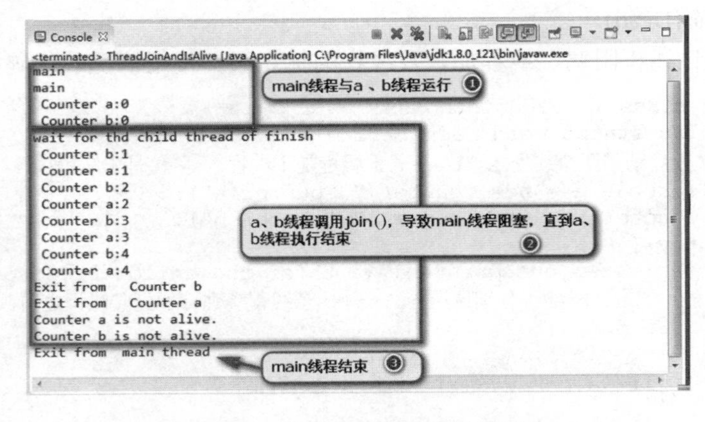

图 10-11　线程的其他方法使用结果

2. 守护线程

当 main()方法运行时，main()方法所在的线程是用户线程，守护线程是为用户线程运行提供服务的线程。当用户线程运行结束时，守护线程也跟着结束。当守护线程结束时，用户线程不一定结束。

要把线程设置为守护线程，可以用 setDaemon(true)，参数为 true 代表是守护线程。

例 10-5：测试守护线程的代码如下。

```
1.  public class DaemonThreadDemo {
2.      public static void main(String[] args) throws InterruptedException {
3.          Thread thread = new DaemonThread();
4.          thread.setDaemon(true);
5.          thread.start();
6.          for (int i = 0; i < 2; i++) {
7.              System.out.println("" + Thread.currentThread().getName());
8.              Thread.sleep(1000);
9.          }
10.     }
11. }
12.
13. class DaemonThread extends Thread {
14.     @Override
15.     public void run() {
16.
17.         for (int i = 0; i < 10; i++) {
18.             System.out.println("" + Thread.currentThread().getName());
19.             try {
20.                 Thread.sleep(1000);
21.             } catch (InterruptedException e) {
22.
23.                 e.printStackTrace();
24.             }
25.         }
26.     }
27. }
```

上述代码中，用户线程 main 中的 for 循环执行 2 次，守护线程执行 10 次。

运行程序，结果如图 10-12 所示。

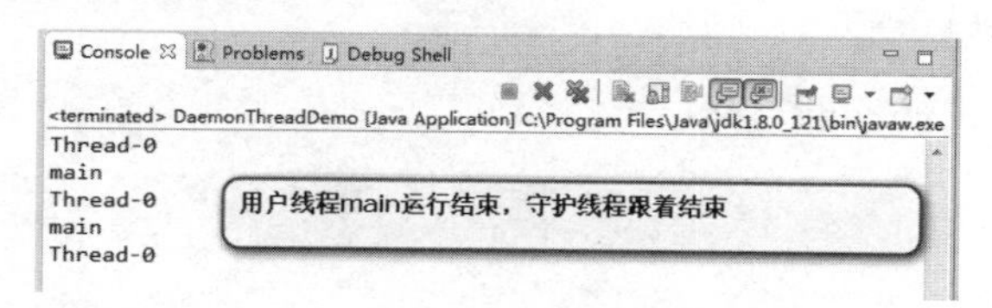

图 10-12 守护线程(1)

把第 6 行代码和第 17 行代码进行互换，结果如图 10-13 所示。

3. 退出/停止线程

终止正在运行的线程有三种方法。

(1) 使用退出标志，使线程正常执行完 run()方法终止。

(2) 使用 interrupt 方法，使线程异常，线程进行捕获或抛异常，然后正常执行完 run()方法终止。

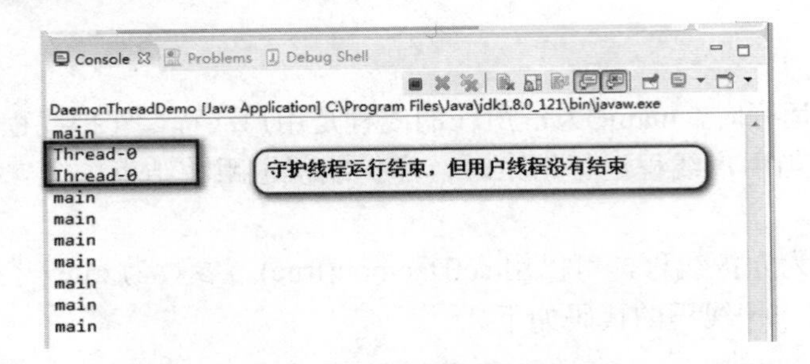

图 10-13　守护线程(2)

(3)　使用 stop()方法强制退出。

因为 stop()方法已声明作废，所以只要能让 run()执行结束，则线程就会停止。

调用 Thread 类的 interrupt()方法可以中断线程，并且会给线程打上中断标志，如果线程被中断时该线程正在被 wait、join、sleep 阻塞，则线程的中断状态会被清除，并且会抛出 InterruptedException 异常。

调用 Thread 类的 isInterrupted()方法可以判断线程是否被中断。

当线程被中断时，会被打上中断标志，interrupted()方法可以检测这个标志位。如果线程被中断，interrupted()返回真；如果线程没有被中断，interrupted()方法返回假。

例 10-6：使用线程中断标志实现停止线程的代码如下。

```
1.  public class StopThread {
2.
3.      public static void main(String[] args) {
4.
5.          Mythread thread=new Mythread();
6.          thread.start();
7.          for(int i=0;i<10;i++)
8.          {
9.              System.out.println("main i="+i);
10.             try {
11.                 Thread.sleep(1000);
12.             } catch (InterruptedException e) {
13.                 e.printStackTrace();
14.             }
15.             if(i==3)
16.             {
17.                 thread.interrupt();
18.             }
19.         }
20.     }
21.
22. }
23. class Mythread extends Thread{
24.     @Override
25.     public void run() {
26.         int i=0;
27.         while(!Thread.interrupted())
28.         {
29.             System.out.println("MyThread i="+i++);
30.             try {
31.                 Thread.sleep(1000);
32.             } catch (InterruptedException e) {
33.                 e.printStackTrace();
```

```
34.                 }
35.             }
36.         }
37.  }
```

程序运行结果如图 10-14 所示。

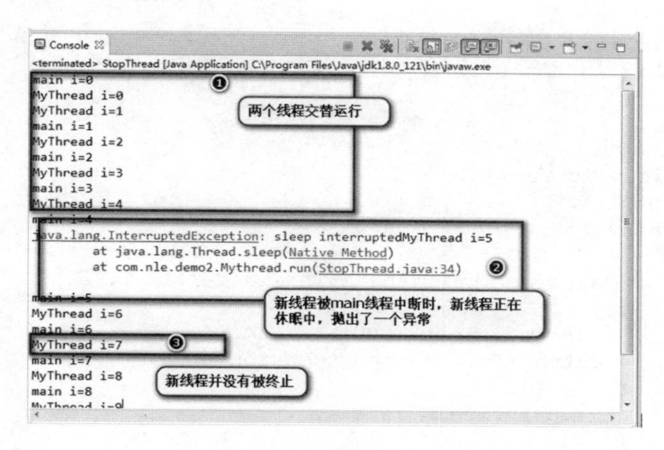

图 10-14　线程中止

从运行结果看，当运行到第 20 行代码，main 线程将新线程中断后，新线程并没有被结束。因为当线程在 sleep 时如果被中断了，会抛出一个线程被中断的异常，同时中断标志会被清除，所以单靠线程中断标志不能实现终止线程。如果一定要靠中断标志来实现，可以在代码第 33 行前加上 break 语句，当抛出中断异常时，同时退出 while()循环，从而结束 run()方法。

上面的程序用线程中断标志的方式去停止线程不是很可靠，在实际应用中，可以用一个成员变量作为中断标志，控制着线程中 run()方法的执行，想停止线程时，只需改变中断标志的值让 run()方法结束即可。

例 10-7：使用停止标志实现停止线程的代码如下。

```
1.   public class StopThread2 {
2.
3.       public static void main(String[] args) {
4.
5.           Mythread2 thread= new Mythread2();
6.           thread.start();
7.           for(int i=0;i<10;i++)
8.           {
9.               System.out.println("main i="+i);
10.              try {
11.                  Thread.sleep(1000);
12.              } catch (InterruptedException e) {
13.                  e.printStackTrace();
14.              }
15.              if(i==3)
16.              {
17.                  thread.setStop(true);
18.              }
19.          }
20.      }
21.  }
```

```
22.  class Mythread2 extends Thread{
23.      private boolean isStop;
24.
25.      public boolean isStop() {
26.          return isStop;
27.      }
28.
29.      public void setStop(boolean isStop) {
30.          this.isStop = isStop;
31.      }
32.
33.      @Override
34.      public void run() {
35.          int i=0;
36.          while(!isStop)
37.          {
38.              System.out.println("MyThread i="+i++);
39.              try {
40.                  Thread.sleep(1000);
41.              } catch (InterruptedException e) {
42.
43.                  e.printStackTrace();
44.              }
45.          }
46.      }
47.  }
```

上述代码在新线程类 MyThread2 里添加了一个 boolean 类型的 isStop 变量，并提供了操作这个变量的 get()和 set()方法，在代码 36 行用 isStop 控制 run()中的循环，当线程没有被停止时，isStop 值为假，!isStop 值为真，所以线程一直在运行。当想让线程停止时，如代码 17 行，就调用 thread.setStop(true)设置 isStop 的值为真，!isStop 的值就为假，循环结束，run()方法结束，从而达到停止线程的目的。

程序运行如图 10-15 所示。

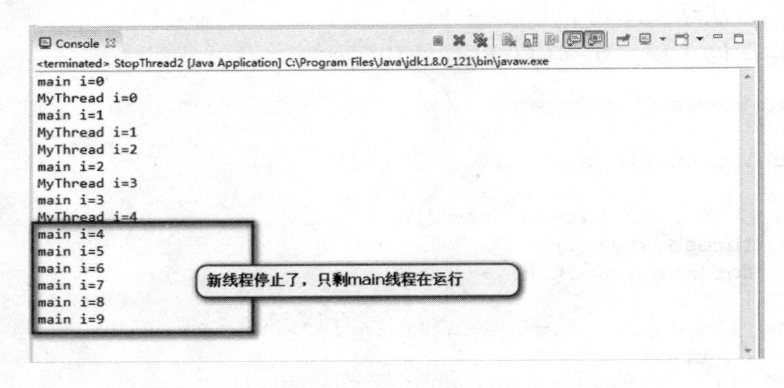

图 10-15　停止线程

【任务实施】

1. 任务分析

(1) 创建工程，添加串口通信包和串口管理工具包。

(2) 采集传感器数据并分析。

(3) 联动控制指令分析。

(4) 使用线程发送采集指令。

(5) 撤销联动后，不再检测。

(6) 编写测试类，实现有人灯亮，无人灯灭。

2. 任务实施

1) 创建工程，添加串口通信包和串口管理工具包

参考任务 1，创建工程 project10_task2，把随书配套资料中提供的串口通信包 RXTX 和串口管理工具类等添加到工程中，如图 10-16 所示。

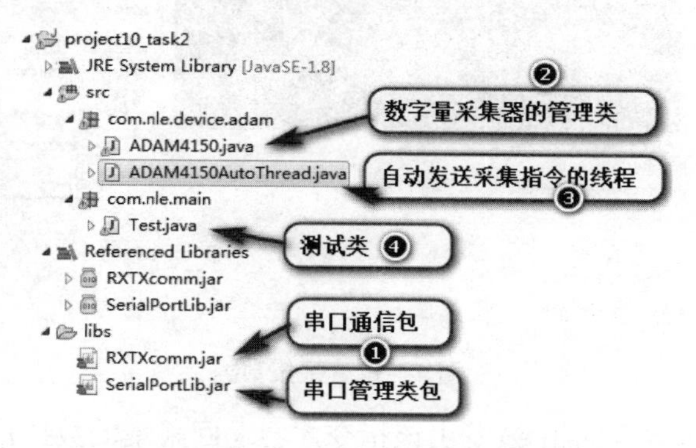

图 10-16 创建工程并导入 jar 包

2) 采集数字量传感器数据并分析

本任务用到了 ADAM-4150 数字量采集器，它有 7 个输入口(DI0～DI6)，8 个输出口(DO0～DO7)，为了获取采集到的传感器数据并进行分析，在类 ADAM4150 里进行相关数据的处理。

第一步：在 ADAM-4150 类里添加以下成员变量，并通过构造方法传入串口管理类和串口对象：

```
1.  public class ADAM-4150 {
2.      /* 有 7 个输入口 */
3.      public static final int DI_COUNT = 7;
4.      /* 有 8 个输出口 */
5.      public static final int DO_COUNT = 8;
6.
7.      /* 串口管理类*/
8.      private SerialPortManager manager;
9.      /* 串口对象*/
10.     private SerialPort serialPort;
11.
12.
13.
14.     public ADAM-4150(SerialPortManager manager, SerialPort serialPort) {
15.         this.manager = manager;
16.         this.serialPort = serialPort;
17.         actionStatus = new Boolean[7];// 输入口只有 7 个: DI0 ～ DI6;
18.     }
19. }
```

第二步：测试串口都读到了什么数据。

在分析数据前，先连接好设备并检查无误后通电，把 485 转 232 口接到 PC 的 COM 口，假设现在使用的是 COM200 口，打开串口助手，发送 4150 模块的采集指令，如图 10-17 所示。

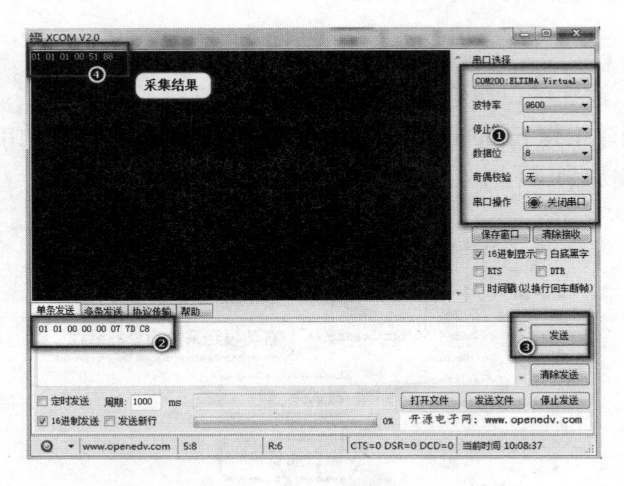

图 10-17　采集数据

数字量 4150 采集数据的指令(发出该指令才会返回采集的数据)：

```
0x01，0x01，0x00，0x00，0x00，0x07，0x7D，0xC8
```

数字量 4150 响应格式分析：

01	01	01	0c(代表 7 个通道值)	51	8d
地址码	功能码	位数	转换成二进制数不够补 0	CRC 低位	CRC 高位

从响应格式分析中得知，第 4 位返回值才是 4150 输入口采集到的数据。

接下来，可以试着发送一条控制设备的指令，结果发现控制指令也会返回到串口，如图 10-18 所示。

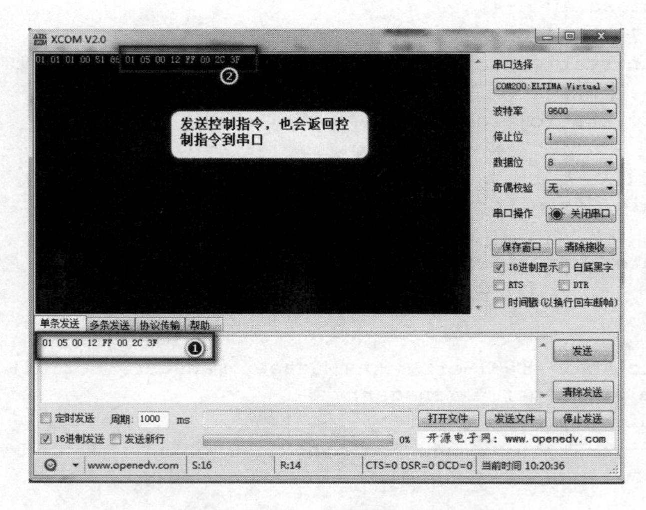

图 10-18　返回控制指令

第三步：按照第二步的分析结果，串口读取的数据不单只有采集到的传感器数据，还有控制指令等其他指令，因此要从串口中取出传感器数据进行分析，应该在从串口中读取到的所有数据 01 01 01 0c(代表 7 个通道值) 51 8d 中截取 7 个通道值。

在 ADAM-4150 的类里添加 ADAM-4150 查询状态指令 SEARCH_COMMAND 和方法 flashStatus(byte[] data)，对采集的传感器数据进行分析并存放到数组 actionStatus 中，详细说明如下：

```
1.  /** ADAM-4150 查询状态指令*/
2.  public static final String SEARCH_COMMAND = "01 01 00 00 00 07 7D C8";
3.
4.
5.  /**
6.   * 分析 ADAM-4150 采集到的数据并存放在 actionStatus 中
7.   * @param data 从串口读取到的数据
8.   * @return 是否有采集到结果
9.   */
10. private boolean flashStatus(byte[] data) {
11.
12.     if (data == null)
13.        return false;
14.     // 本采集结果是：01 01 01 00 51 88，用 toHexString 方法处理了其中的空格
15.     String result = ArrayUtils.toHexString(data);
16.     System.out.println("采集结果："+result);
17.     // result = 010101005188
18.     // RETURN_HEAD = "010101";//状态在此标志之后，所以 status=00
19.     // status = 00 二进制数 0000 0000
20.     // status = 04 二进制数 0000 0100 表示 DI2 动作了
21.     // status = 08 二进制数 0000 1000 表示 DI3 动作了
22.     // status = 09 二进制数 0000 1001 表示 DI3、DI0 动作了
23.
24.        // 查找回应值数据头的下标
25.        int statusIndex = result.indexOf(RETURN_HEAD);
26.        // 如果找到了
27.        if (statusIndex != -1) {
28.           int beginIndex = RETURN_HEAD.length();
29.           // 截取数据值
30.           String status = result.substring(statusIndex+beginIndex,
              statusIndex+beginIndex + 2);
31.           // 转成十六进制
32.           int statusNum = Integer.valueOf(status, 16);
33.           System.out.println("statusNum="+statusNum);
34.           int bitVerify = 1;
35.           for (int i = 0; i < actionStatus.length; i++) {
36.              // 取出每一个位的 0 或 1 值，存放到 actionStatus 中
37.              actionStatus[i] = (statusNum & bitVerify) != 0;
38.              // 通过位移的方式取每一位的值
39.              bitVerify <<= 1;
40.           }
41.        }
42.
43.
44.     return true;
45. }
```

第四步：在 ADAM-4150 类里添加一个方法，用于获取指定输入通道的传感器的值。

```
1.   /**
2.    * 获取指定输入口的状态
3.    *
4.    * @param channelNo 输入口的通道号 0 ～ 6
5.    * @return
6.    */
7.   public Boolean getStatus(int channelNo) {
8.       return actionStatus[channelNo];
9.   }
```

3) 联动控制指令分析

当 ADAM-4150 采集到人体红外传感器的数据变化时，用灯的亮灭来表示，要做到这一点，可以对采集到的数据进行分析后，记在数组 actionStatusOld 中，当再一次采集到新数据时，放在 actionStatus 数组中。通过比较两个数组中对应位的值是否发生变化，从而得知传感器的数值是否变了，如果有了变化，要控制对应的继电器动作，就需要把变化的位和对应的传感器的值回调给调用者，这个可以用接口 Controller 来实现。在 ADAM-4150 的类里添加以下代码：

```
1.    /**回调采集传感数据的接口*/
2.   private Controller controller;
3.   public interface Controller {
4.       /**
5.        * 传送 ADAM-4150 对应输入口采集到的数字量传感器的数据
6.        * @param index ADAM-4150 对应的 DI 口  DI0 对应下标 0
7.        * @param flag  ADAM-4150 对应的 DI 口采集到的传感器数据
8.        */
9.       void comm(Integer index, Boolean flag);
10.  }
11.
12.  /**
13.   * 获取 ADAM-4150  7 个输入口的状态改变值
14.   * @param data 采集到的传感器数据
15.   * @param controller 回调接口 用于把数据传给调用者
16.   */
17.  public void getChangeStatus(byte[] data, Controller controller) {
18.      // 1.处理采集回来的传感器数据并存放在 actionStatus 中
19.      flashStatus(data);
20.      // 2.存放 ADAM-4150  7 个输入口传感器的值
21.      Boolean[] actionStatusOld = new Boolean[7];
22.
23.      // 3.判断各个输入口传感器的值是否改变了
24.
25.      for (int i = 0; i < actionStatus.length; i++) {
26.
27.          if (actionStatus[i] != actionStatusOld[i]) {
28.
29.       //4.如果传感器的数值发生了改变，就将新的传感器值传出去供外部调用，外部负责处理数据
30.              controller.comm(i, actionStatus[i]);
31.          }
32.      }
33.      //5.把传感器的值记录下来
34.      actionStatusOld = Arrays.copyOf(actionStatus, actionStatus.length);
35.  }
```

通过接口 Controller 把值传递出去后，如果要控制继电器动作，需要发送相应的指令，如图 10-19 所示。

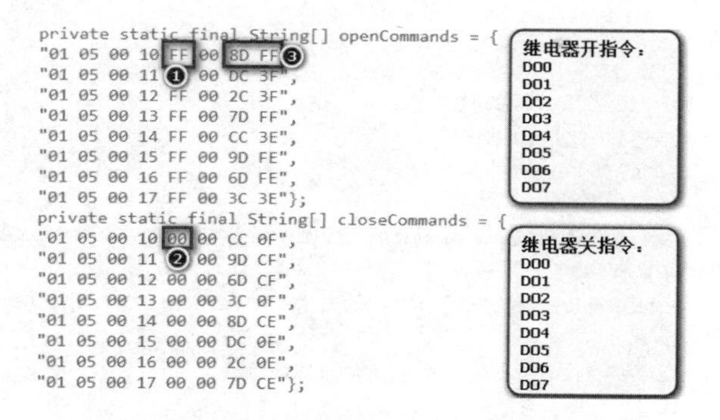

图 10-19 ADAM-4150 控制指令

对比以上指令，可以看出开与关的指令差别为图 10-19 中的①②③点，其中③用公式生成，因为在 ADAM-4150 类里添加方法 getDOCommand()可快速生成控制指令。

```
1.   /**
2.    * 生成 ADAM-4150 输出口控制命令
3.    *
4.    * @param channelId DO 口编号 0～7
5.    * @param isOpen    是否要动作
6.    * @return
7.    */
8.   public static String getDOCommand(int channelId, boolean isOpen) {
9.
10.      StringBuilder command = new StringBuilder(HEAD);
11.      command.append(" ").append(Integer.toHexString(0x10 + channelId));
12.      command.append(" ").append(isOpen ? OPEN : CLOSE);
13.
14.      String readData = "00";
15.      command.append(" ").append(readData);
16.
17.      byte[] bAry = ArrayUtils.toByteAry(command.toString());
18.
19.      String crc = Integer.toHexString(CRC16Util.calcCrc16(bAry));
20.
21.      command.append(" ").append(crc.substring(0, 2));
22.      command.append(" ").append(crc.substring(2, 4));
23.
24.      return command.toString();
25.  }
```

添加 sendCommand()方法和 controllOut()方法用于发送控制指令：

```
1.   /**
2.    * 往串口发送命令
3.    *
4.    * @param command
5.    * @return
6.    */
7.   public boolean sendCommand(String command) {
8.      boolean isOk = manager.sendToPort(serialPort, ArrayUtils.toByteAry
          (command));
```

```
9.      return isOk;
10.  }
11.
12.  /**
13.   * 控制输出口
14.   *
15.   * @param channelId 数据口编号 0 ～ 6
16.   * @param isActive 是否动作
17.   * @return 是否成功
18.   */
19.  public boolean controllOut(int channelId, boolean isActive)  {
20.      // 生成控制指令
21.      String command = getDOCommand(channelId, isActive);
22.      // 发控制指令
23.      boolean isOk = sendCommand(command);
24.      return isOk;
25.  }
```

4) 使用线程发送采集指令

编写线程类 ADAM-4150AutoThread，用停止标志位 isStop 控制线程是否停止，在 run() 方法中每隔 1s 发送一个采集指令。

```
1.   /**
2.    * ADAM-4150 自动采集指令的发送线程
3.    *
4.    */
5.   public class ADAM4150AutoThread extends Thread {
6.
7.      private boolean isStop;
8.      public boolean isStop() {
9.         return isStop;
10.     }
11.     public void setStop(boolean isStop) {
12.        this.isStop = isStop;
13.     }
14.
15.     private ADAM4150 adam4150;
16.
17.     // 每当输入口的状态改变时，就把当前状态值传入 controller，供外部调用
18.
19.     public ADAM4150AutoThread(ADAM4150 adam4150) {
20.
21.        this.adam4150 = adam4150;
22.
23.     }
24.
25.     @Override
26.     public void run() {
27.
28.        while (!isStop) {
29.           try {
30.              adam4150.sendCommand(ADAM4150.SEARCH_COMMAND);
31.               Thread.sleep(1000);
32.           } catch (Exception e) {
33.
34.           }
35.        }
36.     }
```

```
37.
38. }
```

当需要停止线程时，调用 setStop(true)就可以实现线程的停止。

5) 编写测试类，实现有人灯亮，无人灯灭

在测试类里，获取串口管理工具类对象，打开串口并给串口添加监听。当串口有数据可读时，根据回调出来的哪个口有数据发生了改变，从而联动控制灯亮灯灭。因为红外对射传感器有人时返回值为 0，所以注意 41 行代码处控制指令要取非。

```
1.  public class Test {
2.      // 串口管理类
3.      public static SerialPortManager manager;
4.      // 串口对象
5.      public static SerialPort serialPort;
6.
7.      public static void main(String[] args) throws Exception {
8.          //1.获取串口管理对象
9.          manager = new SerialPortManager();
10.
11.         try {
12.             //2.打开串口
13.             serialPort = manager.openPort("COM200", 9600);
14.         } catch (SerialPortException e) {
15.             e.printStackTrace();
16.         }
17.
18.         ADAM4150 adam4150 = new ADAM4150(manager, serialPort);
19.         //3.开启自动发送采集指令的线程
20.         ADAM4150AutoThread thread = new ADAM4150AutoThread(adam4150);
21.         thread.start();
22.         //4.给串口添加事件侦听
23.         manager.addListener(serialPort, new SerialPortEventListener() {
24.
25.             @Override
26.             public void serialEvent(SerialPortEvent arg0) {
27.                 switch (arg0.getEventType()) {
28.                 case SerialPortEvent.DATA_AVAILABLE: // 5.如果串口存在可用数据
29.                     if (serialPort == null) {
30.                         System.out.println("串口对象为空！监听失败！");
31.                     } else {
32.                         // 6.读取串口数据
33.                         byte[] data = manager.readFromPort(serialPort);
34.                         //7.调用相关方法进行数据分析，并判断对应的 DI 口是否有数据改变，
                              如果有联动控制灯亮灭
35.                         adam4150.getChangeStatus(data, new Controller() {
36.
37.                             @Override
38.                             public void comm(Integer index, Boolean flag) {
39.                                 if (index == 0)// 红外对射接在 DI0 口
40.                                 { // 灯接在 DO2 口
41.                                     adam4150.controllOut(2, !flag);
42.                                 }
43.                             }
44.                         });
45.                     }
46.                 }
```

```
47.             }
48.
49.         });
50.
51.     }
52.
53. }
```

3. 运行结果

再次确认设备连接无误后，运行程序，用物体挡住红外对射口，可以看到有人时，灯亮了；无人时，灯灭了。

任务 3　实时火警警示

【任务描述】

园区内的 ZigBee 火焰传感器如果监测到有火警情况，点亮三色灯，并让三色灯按指定顺序轮流闪烁模拟火警提示功能。本任务仅完成三色灯按指定顺序闪烁部分，ZigBee 火焰传感器检测火警部分请参考智慧园区的综合实现篇。任务清单如表 10-4 所示。

表 10-4　任务清单

任务课时	4 课时	任务组员数量	建议 3 人
任务组采用设备	1 个 ADAM-4150 模块 3 个继电器 1 个 485 转 232 USB 1 个三色灯		

【拓扑图】

本任务拓扑图如图 10-20 所示。

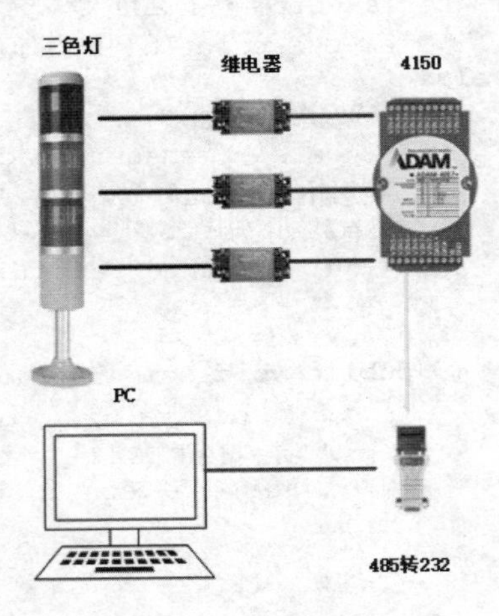

图 10-20　任务 3 拓扑图

备注：本任务红灯接 DO0 口，绿灯接 DO1 口，黄灯接 DO2 口，COM 口根据实际情况确定。

【知识解析】

1. 为什么要使用线程同步

在支持多线程的系统中，多个线程在并发运行时，会有同步(Synchronization)的需求。同步包含两个方面：互斥(Mutex)与协作(Cooperation)。接下来，写一个模拟下载的程序，用 3 个线程同时进行下载，以此来探讨多个线程交叉访问临界资源时可能遇到的问题和解决方法。

把将要下载的资源划分为 3 个部分，线程 1 负责 1～30 部分进度数据的下载，线程 2 负责 31～60 部分进度数据的下载，线程 3 负责 61～100 部分进度数据的下载，这就相当于三个线程在同时下载数据。

例 10-8：用 3 个线程模拟下载同一个资源的代码如下。

```
1.   //在线程中执行下载
2.   public class MyDownload {
3.
4.      private int currentProgress;// 总下载进度量
5.
6.      public static void main(String[] args) {
7.
8.          MyDownload frame = new MyDownload();
9.
10.     }
11.
12.     public MyDownload() {
13.         currentProgress = 0;// 初始为 0
14.         Thread t1 = new Thread(new MyDownloadThread(0, 30));
15.         Thread t2 = new Thread(new MyDownloadThread(31, 60));
16.         Thread t3 = new Thread(new MyDownloadThread(61, 100));
17.         t1.start();// 启动线程 1，下载 0～30 部分的数据
18.         t2.start();// 启动线程 2，下载 31～60 部分的数据
19.         t3.start();// 启动线程 3，下载 61～100 部分的数据
20.
21.     }
22.
23.     class MyDownloadThread implements Runnable {
24.         private int begin;     // 本线程开始下载的位置
25.         private int end;       // 本线程结束下载的位置
26.         private int nowDownloadSize;// 本线程当前已下载的进度量
27.         // 通过构造方法控制每个线程开始和结束的位置
28.
29.         public MyDownloadThread(int begin, int end) {
30.             this.begin = begin;
31.             this.end = end;
32.             this.nowDownloadSize = 0;
33.         }
34.
35.         @Override
36.         public void run() {
37.             for (int i = begin; i <= end; i++) {
38.                 this.nowDownloadSize++;// 当前线程下载进度量
39.                 currentProgress++;// 三个线程共享全局的总下载进度量
```

```
40.                 if (currentProgress > 100) {
41.                     System.out.println("当前进度：下载完成");
42.
43.                     break;
44.                 }
45.                 try {
46.                     Thread.sleep(1000);// 让线程休眠 1s
47.                 } catch (InterruptedException e1) {
48.                     e1.printStackTrace();
49.                 }
50.                 System.out.println(" 总下载量=" + currentProgress + "%  线程"
                      + Thread.currentThread().getId() + " 下载了各自的："
51.                         + this.nowDownloadSize);
52.
53.
54.             }
55.         }
56.     }
57.
58. }
```

运行程序，结果如图 10-21 所示。

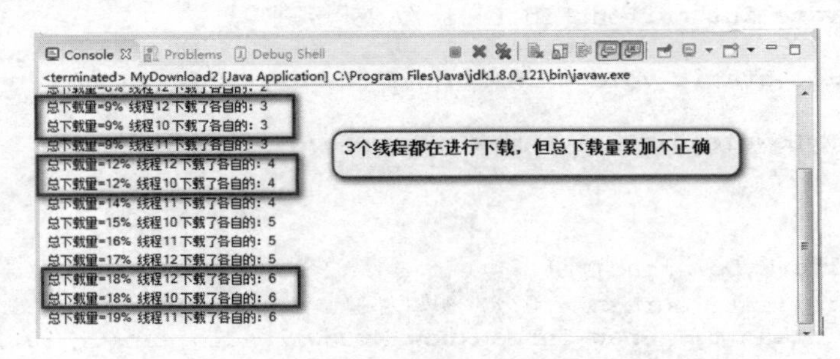

图 10-21　没有控制住共享变量的线程执行结果

观察程序的输出，发现不同的线程都下载了，但总下载量累计出错。

这是因为多个线程间共享了成员变量 currentProgress。当线程 12 执行到代码 39 行处时，对变量 currentProgress 进行了加 1 的操作，然后执行到 46 行，该线程休眠了，此时线程没有输出下载总量的值。这时另外一个线程也执行了 39 行，也做了加 1 的操作。当线程 12 休眠回来再次读取值时，读到了不正确的 currentProgress 数值。也就是说，该程序是非线程安全的程序。要解决这一问题，需保证一个线程在访问和修改 currentProgress 值的过程中，不会被另一个线程“打扰”，实现对共享资源 currentProgress 变量的互斥访问。

2. 同步代码块与同步方法

从上面的程序可以看出，多个线程在访问共享资源的时候，如果不对共享资源进行互斥管理，执行的结果可能不是读者预想的结果，原因是多个线程在操作共享的同一个数据的多条语句时，一个线程对多条语句只执行了一部分，还没有执行完，另一个线程就参与进来了，导致共享数据出现问题。

那么应该如何解决呢？

应该让操作共享数据的多条语句，保证它们在一个线程中都执行完，在执行过程中，其他线程不能参与执行，这个叫作互斥访问控制。Java 语言通过 Synchronized 同步代码块和同步方法实现对共享资源的互斥访问控制。Java 线程在进入这些同步方法或同步代码块语句标识的关键资源区时申请被保护对象的对象锁；离开关键资源区的时候(包括出现异常而离开的时候释放该对象锁)；如果该对象锁已经被别的线程锁定，则当前进入的线程被挂起等待。

1) 使用同步代码块

```
1.  synchronized (锁对象){
2.  //关键资源代码
3.  }
```

此时需要一把锁，这把锁可以是一个任意对象。它起到一个标志的功能，拥有这把锁的线程可以执行同步代码块中的代码。

2) 使用同步方法

```
1.  public synchronized void 方法名{
2.  //关键资源代码
3.  }
```

任何线程进入同步代码块、同步方法之前，必须先获得对对象锁的锁定。持有锁的线程可以在同步中执行，没有持有锁的线程即使获得了 CPU 的执行权，也进不去，因为没有获取锁对象。

注意同步的前提。

- 必须有两个或两个以上的线程；
- 必须是多个线程使用同一把锁；
- 必须保证同步中只能有一个线程在运行。

同步解决了线程安全的问题，但是每次都要判断锁，增加了开销，消耗了资源，但是这种消耗在允许的范围。

例 10-9：用同步代码块实现线程安全的下载程序，代码如下。

```
1.  @Override
2.  public void run() {
3.      for (int i = begin; i <= end; i++) {
4.          synchronized (obj) {
5.
6.              this.nowDownloadSize++;// 当前线程下载量
7.              currentProgress++;// 三个线程共享全局的总下载量
8.              if (currentProgress > 100) {
9.                  System.out.println("当前进度：下载完成");
10.                 break;
11.             }
12.             try {
13.                 Thread.sleep(1000);// 让线程休眠 1s
14.             } catch (InterruptedException e1) {
15.                 e1.printStackTrace();
16.             }
17.             System.out.println(" 总下载量=" + currentProgress + "% 线程" +
                 Thread.currentThread().getId()
18.                 + " 下载了各自的：" + this.nowDownloadSize);
19.         }
20.     }
21. }
```

关键资源就是 run 方法中进行变量加 1 和访问变量的代码，用同步代码块来实现线程安全，锁对象用类内定义的 Object obj=new Object()。

运行程序，结果如图 10-22 所示。

```
Console  Problems  Debug Shell
MyDownload2 [Java Application] C:\Program Files\Java\jdk1.8.0_121\bin\javaw.exe
总下载量=1% 线程10下载了各自的：1
总下载量=2% 线程10下载了各自的：2
总下载量=3% 线程10下载了各自的：3
总下载量=4% 线程10下载了各自的：4
总下载量=5% 线程10下载了各自的：5
总下载量=6% 线程10下载了各自的：6
总下载量=7% 线程10下载了各自的：7
总下载量=8% 线程10下载了各自的：8
总下载量=9% 线程10下载了各自的：9
总下载量=10% 线程10下载了各自的：10
总下载量=11% 线程10下载了各自的：11
```

用同步锁进行控制后，总下载量统计正确了

图 10-22　使用同步锁的结果

用普通同步方法实现线程安全的下载程序，代码如下。

```
1.   @Override
2.   public void run() {
3.       for (int i = begin; i <= end; i++) {
4.           show();
5.       }
6.   }
7.
8.   public synchronized void show() {
9.       this.nowDownloadSize++;// 当前线程下载量
10.      currentProgress++;// 三个线程共享全局的总下载量
11.      if (currentProgress > 100) {
12.          System.out.println("当前进度：下载完成");
13.          return;
14.      }
15.      try {
16.          Thread.sleep(1000);// 让线程休眠 1 秒
17.      } catch (InterruptedException e1) {
18.          e1.printStackTrace();
19.      }
20.      System.out.println(" 总下载量=" + currentProgress + "%  线程" +
           Thread.currentThread().getId() + " 下载了各自的：" 
21.              + this.nowDownloadSize);
22.  }
```

用普通同步方法(非静态方法)来实现线程安全，同步方法的锁对象是 this。

执行程序，采用同步代码块进行互斥后，三个线程同时正确下载并且总下载量的读数正确了。

同步方法锁对象分两种，普通实例方法的锁对象是 this，静态方法的锁对象要求类在它就在，所以可以用当前类的.class 当作锁对象。

当多个线程不是用同一把锁时，无法保证线程安全。

用不同的锁不能实现线程安全，代码如下。

```
1.   @Override
2.   public void run() {
3.       for (int i = begin; i <= end; i++) {
4.           if(i%2==0)
```

```
5.          {
6.              synchronized (obj) {
7.                  this.nowDownloadSize++;// 当前线程下载量
8.                  currentProgress++;// 三个线程共享全局的总下载量
9.                  if (currentProgress > 100) {
10.                     System.out.println("当前进度：下载完成");
11.                     return;
12.                 }
13.                 try {
14.                     Thread.sleep(1000);// 让线程休眠 1s
15.                 } catch (InterruptedException e1) {
16.                     e1.printStackTrace();
17.                 }
18.                 System.out.println(" 总下载量=" + currentProgress + "% 线程"
                      + Thread.currentThread().getId() + " 下载了各自的："
19.                         + this.nowDownloadSize);
20.             }
21.         }
22.         else
23.         {
24.             show();
25.         }
26.     }
27. }
28.
29. public synchronized void show() {
30.
31.     this.nowDownloadSize++;// 当前线程下载量
32.     currentProgress++;// 三个线程共享全局的总下载量
33.     if (currentProgress > 100) {
34.         System.out.println("当前进度：下载完成");
35.         return;
36.     }
37.     try {
38.         Thread.sleep(1000);// 让线程休眠 1s
39.     } catch (InterruptedException e1) {
40.         e1.printStackTrace();
41.     }
42.     System.out.println(" 总下载量=" + currentProgress + "% 线程" +
      Thread.currentThread().getId() + " 下载了各自的："
43.       + this.nowDownloadSize);
44. }
```

上述代码利用变量 i 为奇偶数时执行同步方法或同步代码块，运行程序，结果如图 10-23 所示。

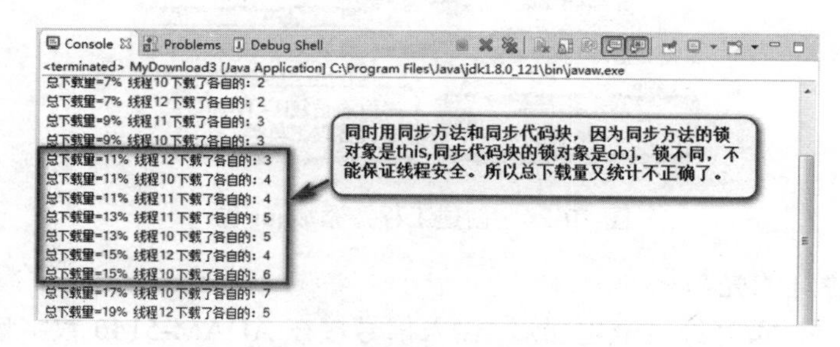

图 10-23 锁不同，线程不能同步

3. wait 与 notify

除了 synchronized 机制以外，Java 还使用 wait、notify 信号机制在线程之间进行通信，以保证它们的正常运行顺序。

wait()、notify()、notifyAll()都不属于 Thread 类，而是属于 Object 基础类，也就是每个对象都有 wait()、notify()、notifyAll()的功能，因为每个对象都有锁，锁是每个对象的基础，当然操作锁的方法也是最基础的了。

当需要调用以上的方法时，一定要对竞争资源进行加锁，如果不加锁，则会出现 IllegalMonitorStateException 异常。当想要调用 wait()进行线程等待时，必须取得所加锁对象的控制权(对象监视器)，一般是将代码放到 synchronized(obj)代码中。

wait()和 notify()在两个独立的线程之间使用，而 notifyAll()可以通知所有正处于等待状态的线程。

一旦程序被分成几个逻辑线程，就必须清晰地知道这些线程之间如何相互通信。Java 提供了 wait 和 notify 等功能来使线程之间相互交谈。一个线程可以进入某个对象的 synchronized 方法并进入等待状态，直到其他线程显式地将它唤醒。可以有多个线程进入同一个方法并等待同一个唤醒消息。

【任务实施】

1. 任务分析

(1) 创建工程，添加串口通信包和串口管理工具包。

(2) 控制三色灯的亮与灭。

(3) 用多线程技术控制三色灯按指定顺序亮灭。

(4) 测试三色灯是否按指定顺序轮流闪烁。

2. 任务实施

1) 创建工程，添加串口通信包和串口管理工具包

创建工程 project10_task3，把串口通信包和串口管理工具类添加进来，并且把任务 2 的 ADAM-4150 管理类添加进来，如图 10-24 所示。

图 10-24　创建工程并添加 jar 包

2) 控制三色灯的亮与灭

三色灯的红、绿、黄灯经过继电器后，输入信号接在 ADAM-5140 数字量采集器的 DO0、DO1、DO2 口。

在Test类里添加常量RED_CHANNELID、GREEN_CHANNELID、YELLOW_CHANNELID分别代表三个灯所接的DO口，通过方法lampFlick(char color)让三个灯按传入的颜色亮与灭。

```
1.  public class Test {
2.      private static final int RED_CHANNELID=0;//红灯
3.      private static final int GREEN_CHANNELID=1;//绿灯
4.      private static final int YELLOW_CHANNELID=2;//黄灯
5.      private static SerialPortManager manager;
6.      private static SerialPort serialPort ;
7.      private static ADAM4150 adam4150;
8.      public static void main(String[] args) throws Exception {
9.          //待后面处理
10.
11.     }
12.     /**
13.      * 三色灯亮与灭
14.      * @param color 三色灯中指定颜色的灯
15.      * @throws Exception
16.      */
17.     public static void lampFlick(char color) throws Exception {
18.         switch (color) {
19.         case 'r':
20.             adam4150.controllOut(RED_CHANNELID, true);
21.             Thread.sleep(500);
22.             adam4150.controllOut(RED_CHANNELID, false);
23.             break;
24.         case 'g':
25.             adam4150.controllOut(GREEN_CHANNELID, true);
26.             Thread.sleep(500);
27.             adam4150.controllOut(GREEN_CHANNELID, false);
28.             break;
29.         case 'y':
30.             adam4150.controllOut(YELLOW_CHANNELID, true);
31.             Thread.sleep(500);
32.             adam4150.controllOut(YELLOW_CHANNELID, false);
33.             break;
34.         }
35.     }
36.
37. }
```

3) 用多线程技术控制三色灯按指定顺序亮灭

在线程FlickThread 的构造方法中传入三色灯的颜色、等待锁对象和唤醒锁对象，在run()方法中用锁对象控制线程的顺序。

```
1.  public class FlickThread extends Thread{
2.
3.      private char color;//三色灯的颜色
4.
5.      private Object waitObj;//等待锁
6.
7.      private Object notifyObj;//唤醒锁
8.      private boolean isStop;
9.
10.
11.     public boolean isStop() {
```

```
12.          return isStop;
13.      }
14.
15.      public void setStop(boolean isStop) {
16.          this.isStop = isStop;
17.      }
18.
19.      public FlickThread(char color, Object waitObj, Object notifyObj) {
20.          this.color = color;
21.          this.waitObj = waitObj;
22.          this.notifyObj = notifyObj;
23.
24.      }
25.
26.      @Override
27.      public void run() {
28.          while(!isStop) {
29.              try {
30.                  //1.让某个颜色的灯亮
31.                  Test.lampFlick(color);
32.                  Thread.sleep(500);
33.                  //2.如果获取到 notifyObj 锁，唤醒所有线程
34.                  synchronized (notifyObj) {
35.                      notifyObj.notifyAll();
36.                  }
37.                  //3.如果获取到 waitObj 锁，则线程等待
38.                  synchronized(waitObj){
39.                      waitObj.wait();
40.                  }
41.              } catch (InterruptedException e) {
42.                  e.printStackTrace();
43.              } catch (Exception e) {
44.                  e.printStackTrace();
45.              }
46.          }
47.      }
48.
49.  }
```

4) 测试三色灯是否按指定顺序轮流闪烁

在 Test 类的 main()方法中初始化串口管理对象，打开串口，初始化 ADAM-4150 对象，初始化三个锁对象、初始化三条闪烁线程后启动线程。

```
1.   public static void main(String[] args) throws Exception {
2.       try {
3.
4.            manager=new SerialPortManager();
5.            serialPort = manager.openPort("COM3", 9600);
6.           adam4150 =new ADAM4150(manager, serialPort);
7.           //初始化三个锁对象
8.           Object obj1 = new Object();
9.           Object obj2 = new Object();
10.          Object obj3 = new Object();
11.          //初始化三条闪烁线程
12.          FlickThread thread1 = new FlickThread('r', obj1, obj2);
13.          FlickThread thread2 = new FlickThread('g', obj2, obj3);
14.          FlickThread thread3 = new FlickThread('y', obj3, obj1);
15.          //开启线程
```

```
16.          thread1.start();
17.          Thread.sleep(1000);
18.          thread2.start();
19.          Thread.sleep(1000);
20.          thread3.start();
21.
22.      } catch (Exception e) {
23.          e.printStackTrace();
24.      }
25.
26.  }
```

代码分析如图 10-25 所示。

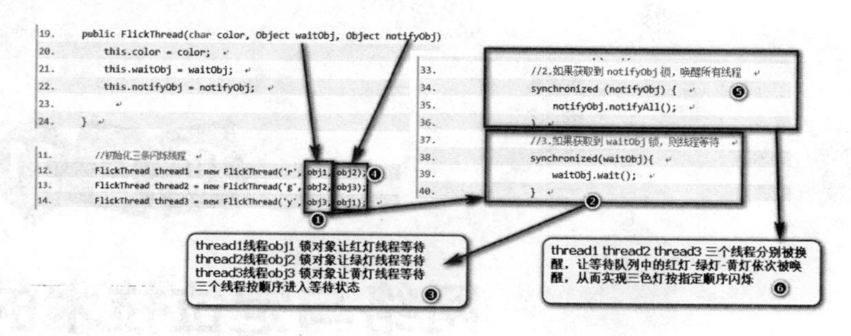

图 10-25 三色灯执行过程

3. 运行结果

运行程序，可以看到三色灯按指定的顺序亮和灭。在实际运行过程中，如果火焰报警器检测到火警信号，就启用线程，如果想让线程停止，调用 setStop(false)即可。

思考与练习

1. 简述创建线程的两种方式。
2. 简述停止线程的方法。
3. 简述同步的前提。

项目 11 网络与定位技术的使用

【项目描述】

现如今，网络技术已成为人们工作、生活中必不可少的一项服务。它把互联网上分散的资源融为一个有机的整体，实现资源的全面共享和有机协作，使人们能够透明地使用资源并按需获取信息。北斗导航系统拥有导航定位功能，天上的北斗卫星负责广播自己的精确位置和时间，地面上的接收机只要能看到 4 颗以上的北斗卫星，就能确定自己的位置和时间。北斗导航系统可对全球任何一个点进行定位、导航、测速、授时和短语通信。

结合网络与定位技术，便可以实现获取物体当前位置信息、跟踪物体行动轨迹、地理围栏等功能，本项目通过北斗定位模块获取经纬度数据，利用百度地图 API 获取地理位置信息，最后将经纬度数据上报到云平台，为用户跟踪物体或为应用开发提供数据，如图 11-1 所示。项目任务列表如图 11-2 所示。

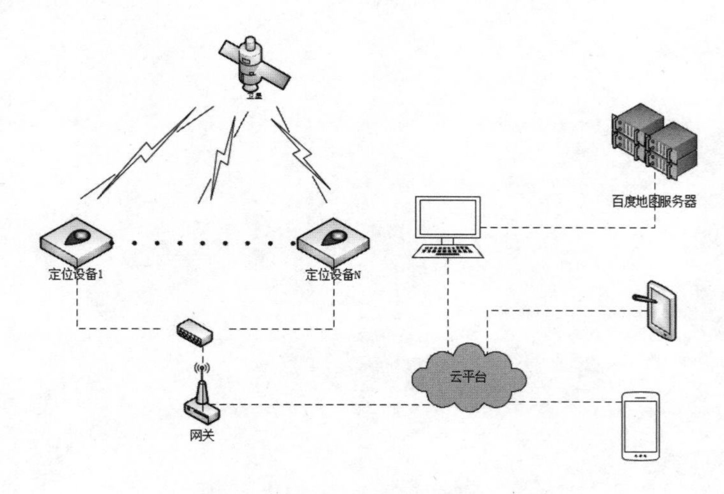

图 11-1　网络与定位结合使用

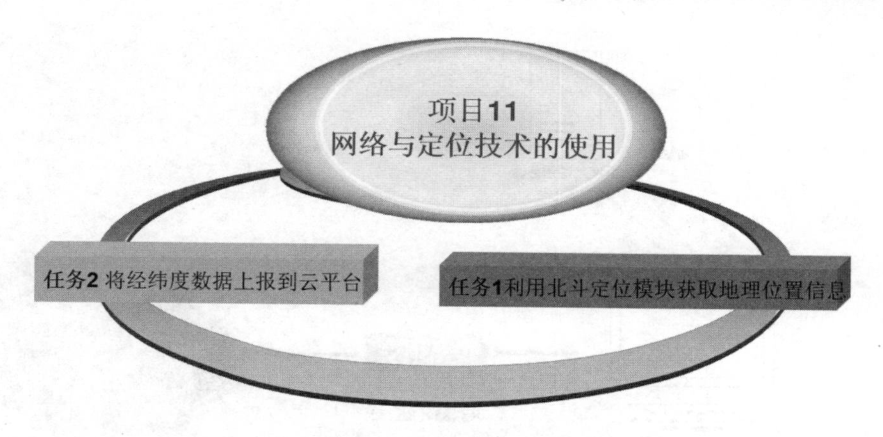

图 11-2 项目任务列表

【学习目标】

知识目标： 了解网络通信基本概念；会使用 URLConnection；会 TCP 协议的 Socket 套接字编程；会 UDP 协议的数据报编程；知道物联网云平台的接入流程；会采集和解析定位数据；会使用地图 API。

技能目标： 能解析北斗定位模块的数据帧格式；能使用 Socket 套接字和数据报文开发网络应用程序；能上报传感数据到物联网云平台；能使用百度地图 API 获取地理位置信息。

任务 1 利用北斗定位模块获取地理位置信息

【任务描述】

本任务要求编程实现——通过串口采集北斗定位模块的定位数据，再对采集的数据进行解析，解析得到经纬度数据后，通过百度地图提供的全球逆地理编码服务，将坐标点(经纬度)转换为对应地理位置信息。任务清单如表 11-1 所示。

表 11-1 任务清单

任务课时	2 课时	任务组员数量	建议 3 人
任务组采用设备	1 个北斗定位模块 1 个 RS485 转 RS232 转接头 1 台 PC 电源、USB 转串口、连接线若干		

【拓扑图】

本任务的拓扑图如图 11-3 所示。

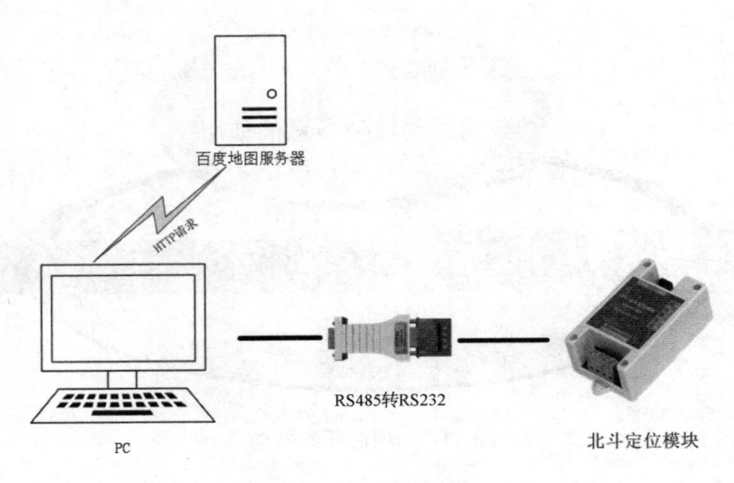

图 11-3　拓扑图

【知识解析】

1. 网络通信基础知识

在 Java 中，开发网络通信程序首先要了解以下基础知识。

1)　计算机网络

计算机网络是把分布在不同地理区域的计算机与专门的外部设备用通信线路互联成一个规模大、功能强的系统，并使用某些通信协议，使众多计算机可以方便地互相传递信息，共享硬件、软件、数据信息等资源。

2)　通信协议

在计算机网络中进行通信，需要遵从一定的约定，这些约定被称为通信协议。通信协议定义了在网络设备间交换消息的内容、格式、定时、顺序和错误控制。

3)　TCP

TCP(Transmission Control Protocol，传输控制协议)是一种面向连接的、可靠的、基于字节流的传输层通信协议，由 IETF 的 RFC 793 定义。在因特网协议(Internet protocol suite)中，TCP 层是位于 IP 层之上、应用层之下的中间层。不同主机的应用层之间经常需要可靠的、像管道一样的连接，但是 IP 层不提供这样的流机制，而是提供不可靠的包交换。

TCP 主要负责数据的分组和重组，它与 IP 组合使用称为 TCP/IP。TCP 负责将数据包按次序传送，接收端收到后再将其正确地还原。TCP 采用三次握手机制建立可靠连接，丢包会重传，乱序会重排，所以 TCP 能够保证通信的可靠性。

4)　UDP

UDP(User Datagram Protocol，用户数据报协议)是一种不可靠的非持续连接的通信协议，它不能保证数据的安全传送，数据丢失不会重传，也不保证发送和接收数据的顺序。相比 TCP，它少了各种验证机制，所以传送速度较快。在目前正常的网络环境中，数据都可以送达目的地，所以 UDP 适合那些对可靠性要求不高的应用场景，如网络聊天等。

5)　IP 地址

IP(Internet Protocol)地址是计算机在网络中的唯一标识，基于 IP 协议网络传输的数据包，都必须使用IP地址来标识。IP地址是一个32位(IPv4)或128位(IPv6)的无符号整数。

IPv4通常分成4个8位的二进制数，每8位之间用点隔开，每个8位整数可以转换成一个0～255的十进制整数。因此，我们看到的IPv4地址类似于这种形式：192.168.10.12。

IPv6的使用，不仅能解决IPv4网络地址资源数量的问题，而且也解决了多种接入设备连入互联网的障碍。IPv6地址的格式为X:X:X:X:X:X:X:X，其中每个X以十六进制表示，例如：ABED:EF02:2346:6289:ABDD:EF11:2315:6781。

IP地址是一种低级协议，Java提供了InetAddress类对IP地址进行封装，并提供了解析IP地址的主机名称、获取本机IP地址等方法。

6) InetAddress类

Java中的InetAddress是一个代表IP地址的对象。IP地址可以由字节数组和字符串来分别表示，而InetAddress将IP地址以对象的形式进行封装，可以更方便地操作和获取其属性。

InetAddress的构造函数不是公开的(public)，因此需要通过它提供的静态方法来获取。其常用方法如表11-2所示。

表11-2 InetAddress类的常用方法

方 法	描 述
InetAddress getLocalHost()	获取本地计算机的InetAddress对象
String getbyname(String host)	凭主机名获取IP地址
getAllByName(String host)	根据主机名返回其所有可能的InetAddress对象

获取本地和远程主机的InetAddress地址，代码如下。

```
1.   public class GetInetAddress {
2.       public static void main(String[] args) {
3.           try {
4.               // 获取本地主机的InetAddress对象
5.               InetAddress inetAddress = InetAddress.getLocalHost();
6.               System.out.println("ip地址=" + inetAddress.getHostAddress());
7.               String hostName = inetAddress.getHostName();
8.               System.out.println("主机名=" + hostName);
9.               System.out.println("************");
10.
11.              //通过域名获取，只获取一个
12.              InetAddress inetAddressBaidu = InetAddress.getByName
                     ("www.baidu.com");
13.              System.out.println(inetAddressBaidu);
14.
15.              System.out.println("************");
16.              //通过域名获取所有的集群服务器IP
17.              InetAddress[] inetAddressBaidus = InetAddress.getAllByName
                     ("www.baidu.com");
18.              for (int i = 0; i < inetAddressBaidus.length; i++) {
19.                  System.out.println(inetAddressBaidus[i].getHostAddress());
20.              }
21.          } catch (UnknownHostException e) {
22.              e.printStackTrace();
23.          }
24.
25.      }
```

```
26.  }
```

运行结果如图 11-4 所示。

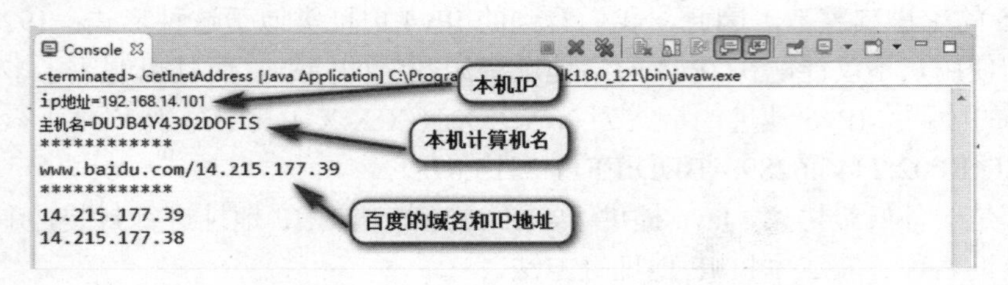

图 11-4　获取百度的 IP 地址和域名

7)　DNS

DNS(Domain Name System，域名系统)可以把 IP 地址映射成一个字符串，这样可以方便用户记忆。通过域名，最终得到该域名对应的 IP 地址的过程叫作域名解析。

8)　端口

端口是一个整数，当我们要访问一台计算机上的某个服务时，用端口号才能区别这个服务与同一台计算机上的其他服务程序。同一台计算机上运行的不同的应用程序从不同的端口取数据。同一台机器上不能有两个程序使用同一个端口，端口号可以从 0～65535，其中周知端口号为 1024 以下，所以我们自定义的程序端口号要大于 1024。表 11-3 中列举了一些常用的协议对应的端口。

表 11-3　常用的协议和对应的端口

方　法	对应的端口号
Telnet 协议	23
简单邮件传输协议 SMTP	25
文件传输协议 FTP	21
超文本传输协议 HTTP	80

2. URL 与 URLConnection

URL(Uniformed Resource Location，统一资源定位)可以理解成一个指向互联网资源的“指针”，通过 URL 可以获得互联网资源的相关信息。例如，通过 URL 可以获得一个 InputStream 对象，从而获取网络资源的信息，以及一个到 URL 所应用的对象连接 URLConnection。

URLConnection 对象可以向所代表的 URL 发送请求和读取 URL 的资源。

通常，创建一个和 URL 的连接，需要如下几个步骤。

第一步，创建 URL 对象，并通过调用 openconnection 方法获得 URLConnection 对象。

第二步，设置 URLConnection 参数和普通请求属性。

第三步，向远程资源发送请求。

第四步，远程资源变为可用，程序可以访问远程资源的头字段和通过输入流来读取远程资源返回的信息。

如果只是发送 GET 方式的请求，建立连接即可；如果需要发送 POST 方式的请求，则需要获取 URLConnection 对象所对应的输出流来发送请求。GET 请求传递参数的方式是直接将参数放入 URL 的资源后面。

获取百度 Logo，代码如下。

```
1.  public class MyDownload {
2.      public static void downPic(String path,DownListener downListener){
3.          ByteArrayOutputStream out=null;
4.          try {
5.              URL url = new URL(path);
6.              HttpURLConnection  conn=(HttpURLConnection)
    url.openConnection();
7.              conn.setReadTimeout(5000);
8.              conn.setConnectTimeout(5000);//设置连接超时
9.              conn.setRequestMethod("GET");//用 GET 的方式发请求
10.             conn.setRequestProperty("accept", "*/*");
11.             conn.setRequestProperty("connection", "Keep-Alive");
12.
13.             InputStream input=conn.getInputStream();
14.
15.             out =new ByteArrayOutputStream();
16.             byte[] b=new byte[1024];
17.
18.             int n=0;
19.             while((n=input.read(b))>0)
20.             {
21.                 out.write(b, 0, n);
22.             }
23.             downListener.success(out.toByteArray());
24.
25.         } catch (Exception e) {
26.            downListener.error("图片下载失败");
27.             e.printStackTrace();
28.         }
29.     }
30.
31.     interface DownListener{
32.         public void success(byte[] b);
33.         public void error(String error);
34.     }
35. }
```

在 MyDownload 类中有一个接口 DownListener，在接口中声明两个方法：public void success(byte[] b)用于将成功下载的远程资源回调给调用者，public void error(String error)用于提示出错信息。在 DownListener 类的 downPic()方法中处理远程资源的连接和下载，并通过接口的方法传给调用者。

在类 GetBaiduPic 中有一个按钮，单击后进行百度 Logo 的下载，代码如下。

```
1.  public class GetBaiduPic extends javax.swing.JFrame {
2.      private JLabel jLabel1;
3.      private JButton jButton1;
4.      String url="https://www.baidu.com/img/bd_logo1.png";
5.      public static void main(String[] args) {
6.          SwingUtilities.invokeLater(new Runnable() {
7.              public void run() {
8.                  GetBaiduPic inst = new GetBaiduPic();
```

```
9.              inst.setLocationRelativeTo(null);
10.             inst.setVisible(true);
11.         }
12.     });
13. }
14.
15. public GetBaiduPic() {
16.     super();
17.     initGUI();
18. }
19.
20. private void initGUI() {
21.     try {
22.         setDefaultCloseOperation(WindowConstants.DISPOSE_ON_CLOSE);
23.         getContentPane().setLayout(null);
24.         {
25.             jLabel1 = new JLabel();
26.             getContentPane().add(jLabel1);
27.             jLabel1.setText("jLabel1");
28.             jLabel1.setBounds(0, 0, 480, 280);
29.         }
30.         {
31.             jButton1 = new JButton();
32.             getContentPane().add(jButton1);
33.             jButton1.setText("点击开始下载图片");
34.
35.             jButton1.setBounds(10, 10, 140, 30);
36.         }
37.         pack();
38.         setSize(520, 300);
39.         jButton1.addMouseListener(new MouseAdapter() {
40.
41.             @Override
42.             public void mouseClicked(MouseEvent e) {
43.                 MyDownload.downPic(url, new DownListener() {
44.
45.                     @Override
46.                     public void success(byte[] b) {
47.                         jLabel1.setIcon(new ImageIcon(b));
48.                     }
49.
50.                     @Override
51.                     public void error(String error) {
52.                         jLabel1.setText("图片下载失败");
53.                     }
54.                 });
55.             }
56.
57.         });
58.     } catch (Exception e) {
59.         //add your error handling code here
60.         e.printStackTrace();
61.     }
62. }
63. }
```

运行程序，结果如图 11-5 所示。

图 11-5 下载百度 Logo

3. 百度地图 Web 服务 API

百度地图 Web 服务 API 为开发者提供 HTTP/HTTPS 接口，即开发者通过 HTTP/HTTPS 形式发起检索请求，获取返回 JSON 或 XML 格式的检索数据。用户可以基于此开发 JavaScript、C#、C++、Java 等语言的地图应用。

全球逆地理编码服务(又名 Geocoder)是一类 Web API 接口服务，用户可以通过该功能，将位置坐标解析成对应的行政区划数据以及周边高权重地标地点分布情况，整体描述坐标所在的位置。

注册成为百度地图开发者后，创建应用并获取密钥，将经纬度通过 HTTP 请求发送到百度地图开放平台。通过全球逆地理编码服务功能，用户即可获取服务器返回的地理位置信息。具体的请求参数、返回结果参数、服务状态码等，可以登录百度地图开发平台查看开发文档。

GET 请求地址如下所示。

```
1.    //GET 请求
2.    http://api.map.baidu.com/reverse_geocoding/v3/?ak=您的ak
3.    &output=json&coordtype=wgs84ll&location=31.225696563611,121.49884033194
```

【任务实施】

1. 任务分析

(1) 连接设备并使用串口调试工具进行调试。

(2) 通过串口获取北斗定位设备采集的数据并解析经纬度数据。

(3) 通过百度地图 API 获取地理位置信息。

2. 任务实施

1) 连接设备并使用串口调试工具进行调试

参考物联网工程实训系统中的相关操作，将北斗定位模块按图 11-3 进行连接， 将 RS485 转 RS232 转接头连接到电脑串口，记下识别出来的 COM 口号。打开随书资源“./项目 11”目录下的串口调试工具“XCOM”，相关参数设置如图 11-6 所示，在发送窗口输入采集指令“01 03 00 05 00 23 14 12”，如果接收区能收到图 11-6 所示的数据，则说明北斗定位模块工作正常。

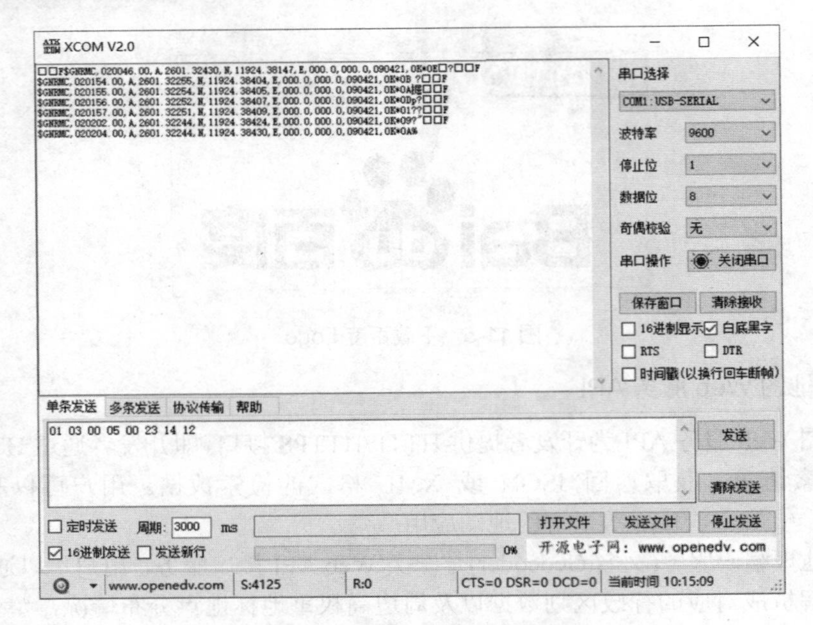

图 11-6　串口调试工具请求数据

2)　通过串口获取北斗定位设备采集的数据并解析经纬度数据

新建工程 project11_task1，把随书提供的资源："./ 项目 11"文件夹的 lib 下的 3 个 jar 文件添加到工程的 lib 文件夹下，并添加到 Build Path 中，如图 11-7 所示。

图 11-7　添加 jar 包到工程中

新建包 com.newland.device，在包下新建类 BeiDouAuto.java，在类中定义采集北斗定位模块信息的定时器、采集时间间隔、定位信息采集指令等属性，北斗定位模块的定位信息采集指令可在随书资源"./项目 11"文件夹中提供的北斗模块开发文档中查询，如图 11-8 所示。

读取定位数据（RMC）

命令帧：01 03 00 05 00 23 14 12

地址	功能码	寄存器起始地址	寄存器个数	CRC 校验
0x01	0x03	0x00 0x05	0x00 0x23	0x14 0x12

响应帧：

地址	功能码	数据长度	数据	CRC 校验
0x01	0x03	0x46	70 字节数据	两字节校验

图 11-8　北斗定位模块读取定位数据协议

接下来，在构造函数中初始化定义的属性，构造函数需要添加 SerialPort 参数，在初始化 BeiDouAuto 类对象时，将北斗模块连接的串口对象传递进来，代码如下。

```
1.   public class BeiDouAuto {
2.    //采集北斗定位信息定时器
3.    Timer timer;
4.    //定时查询状态时间(毫秒)
5.    private long period;
6.    //获取定位信息指令
7.    byte[] order;
8.    //北斗模块串口
9.    SerialPort port;
10.   //存储定位信息Map
11.   public static Map<String,Double> map;
12.
13.   public BeiDouAuto(SerialPort port) {
14.    this.port = port;
15.    period = 2000;
16.    map = new HashMap<String, Double>();
17.    order = new byte[] {0x01,0x03,0x00,0x05,0x00,0x23,0x14,0x12};
18.    addListener();
19.   }
20.  }
```

在 BeiDouAuto 类中分别封装单次采集定位信息的方法、持续采集定位信息的方法，以及停止持续采集定位信息的方法。持续采集定位信息的方法通过定时器每隔 2s 发送请求指令，代码如下。

```
1.    /**
2.     * 获取单次定位信息
3.     */
4.    public void getLocationOneTime() {
5.     try {
6.      SerialPortManager.sendToPort(port, order);
7.     } catch (SendDataToSerialPortFailure e) {
8.      // TODO Auto-generated catch block
9.      e.printStackTrace();
10.    } catch (SerialPortOutputStreamCloseFailure e) {
11.     // TODO Auto-generated catch block
12.     e.printStackTrace();
13.    }
14.   }
15.
16.   /**
17.    * 持续获取定位信息
18.    */
19.   public void start() {
20.    if(timer != null) return;
21.
22.    timer = new Timer("BeiDouAuto_Timer", true);// 定时器(true: 使用守护线程/
   后台线程)
23.    timer.schedule(new TimerTask() {
24.      @Override
25.     public void run() {
26.      try {
27.       SerialPortManager.sendToPort(port, order);
28.      } catch (Exception e) {
```

```
29.       e.printStackTrace();
30.       timer.cancel();// 停止定时器(查询设备状态)
31.     }
32.    }
33.  }, 0, period);
34. }
35.
36. /**
37.  * 停止持续获取定位信息
38.  */
39. public void stop() {
40.  timer.cancel();
41. }
```

在 BeiDouAuto 类中定义 addListener 方法，用来为串口添加事件监听，并在监听器中解析按照北斗设备协议解析定位数据，设备协议如表 11-4 所示。在开发过程中，可以根据实际需求解析所需数据，这里只对经纬度进行解析。

表 11-4 GNRMC 解析

字段	符 号	含 义	取值范围	举 例	备 注
1	$	语句起始符			
2	GNRMC	RMC 协议头			RMC 协议头，GNRMC 表示联合定位
3	hhmmss.ss	UTC 时间	时时分分秒秒.秒秒	072905.00	北京东八区需要时+8
4	A	定位状态	A/V		A-有效，V-无效
5	ddmm.mmmmm	纬度	度度分分.分分分分分	3640.46260	计算要转为度：36 度+40.46260 分，40.46260/60=0.67438 度，所以为 36.67438 度
6	a	纬度方向	N/S		N-北纬，S-南纬
7	ddmm.mmmmm	经度	度度分分.分分分分分	11707.54950	计算要转为度：117 度+07.54950 分，07.54950/60=0.12583 度，所以为 117.12583 度
8	a	经度方向	E/W		E-东经，W-西经
9	x.xxx-xxx.x	对地速度	节	123.2	地速率 节单位 地面速率 000.0～999.9 节，Knot
10	x.xxx-xxx.x	对地航向	度	000.0～359.9	地面航向(000.0～359.9 度，以正北为参考基准)
11	xxxxxx	日期	日月年	050119	2019 年 1 月 5 日
12	aa	天线状态	OK/OP/OR		OK 代表天线正常(OK)；OP 代表开路(OPEN)；OR 代表天线短路(SHORT)

续表

字段	符 号	含 义	取值范围	举 例	备 注
13	*	语句结束符			
14	24	校验和	对‘$’和‘*’之间的数据(不包括这两个字符)按字节进行异或运算，用十六进制数值表示		

解析得到的经纬度信息保存到静态 Map 容器中，代码如下。

```
1.    /**
2.     * 添加串口监听，解析定位信息
3.     */
4.   private void addListener() {
5.    try {
6.     SerialPortManager.addListener(port, new SerialPortEventListener() {
7.
8.      @Override
9.      public void serialEvent(SerialPortEvent arg0) {
10.
11.      try {
12.       switch(arg0.getEventType())
13.       {
14.       case SerialPortEvent.DATA_AVAILABLE:
15.        if(port!=null)
16.        {
17.
18.         byte[] bytes = SerialPortManager.readFromPort(port);
19.
20.         String res = new String(bytes);
21.         System.out.println(res);
22.
23.
24.         if(res.length()==75) {
25.          String[] ress= res.split(",");
26.          //解析经度
27.          String longStr = ress[3];
28.          int headLong = Integer.parseInt(longStr.substring(0, 2));
29.          double tailLong = Double.parseDouble(longStr.substring(2))/60;
30.          double longitude = headLong+tailLong;
31.          BigDecimal bdLong = new BigDecimal(longitude);
32.          double longitude2 = bdLong.setScale(6,BigDecimal.ROUND_HALF_UP).
               doubleValue();
33.          System.out.println("经度: "+longitude2);
34.          map.put("longitude",longitude2);
35.          //解析纬度
36.          String latStr = ress[5];
37.          int headLat = Integer.parseInt(latStr.substring(0,3));
38.          double tailLat = Double.parseDouble(latStr.substring(3))/60;
39.          double latitude = headLat+tailLat;
40.          BigDecimal bdLat = new BigDecimal(latitude);
41.          double latitude2 = bdLat.setScale(6,BigDecimal.ROUND_HALF_UP).
               doubleValue();
42.          System.out.println("纬度: "+latitude2);
43.          map.put("latitude", latitude2);
```

```
44.         }
45.       }
46.     }
47.   } catch (Exception e) {
48.     // TODO Auto-generated catch block
49.     e.printStackTrace();
50.   }
51.  }
52. });
53.
54. } catch (TooManyListeners e) {
55.   // TODO Auto-generated catch block
56.   e.printStackTrace();
57. }
58. }
```

3) 通过百度地图 API 获取地理位置信息

打开百度地图开放平台 http://lbsyun.baidu.com/，如图 11-9 所示。

图 11-9 百度地图开放平台界面

单击图 11-9 中的“登录”按钮，如果已有百度账号可以直接登录，未拥有百度账号的，必须先注册百度账号后再完成登录。

登录成功后，进入控制台，在应用管理模块中创建一个应用，如图 11-10 所示。

创建应用时，由于应用还在开发阶段，暂时不对 IP 做任何限制，如图 11-11 所示。应用创建成功后可以在“我的应用”中获取该应用的密钥备用。

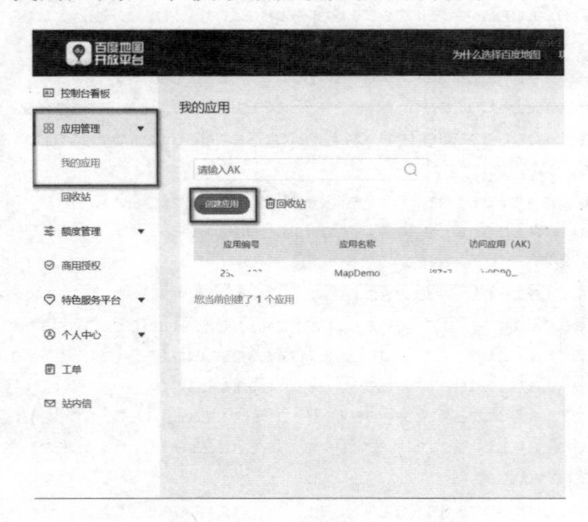

图 11-10 创建百度地图应用

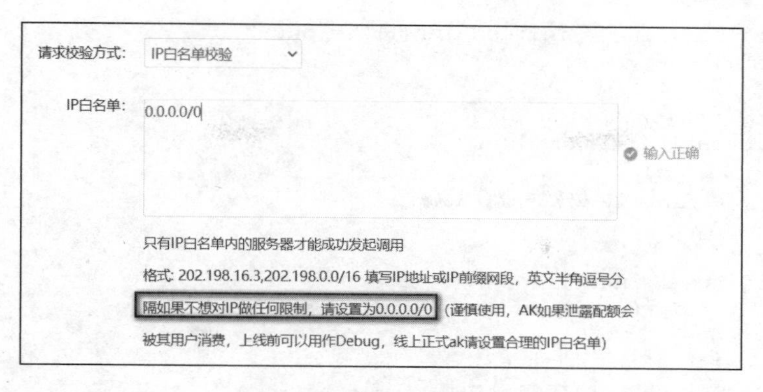

图 11-11　填写 IP 白名单

新建包 com.newland.main，在包下新建类 HttpHelper.java，在类中定义发送 HTTP 请求必须用到的常量，并通过知识解析中提到的 URL 与 URLConnection，分别封装发送 GET 请求、POST 请求等方法，代码如下。

```
1.  public class HttpHelper {
2.   public static final int BUFFSIZE = 2048; //缓冲区大小 2M
3.   public static final int TIMEOUT = 6000; //超时时间设置为 6 秒
4.   /**
5.    * 向指定 URL 发送 GET 方法的请求
6.    *
7.    * @param url
8.    *           发送请求的 URL
9.    * @param param
10.   *           请求参数，请求参数应该是 name1=value1&name2=value2 的形式。
11.   * @return URL 所代表远程资源的响应结果
12.   */
13.  public static String sendGet(String url) {
14.   String result = "";
15.   BufferedReader in = null;
16.   try {
17.    String urlNameString = url;
18.    URL realUrl = new URL(urlNameString);
19.    // 打开和 URL 之间的连接
20.    URLConnection connection = realUrl.openConnection();
21.    // 设置通用的请求属性
22.    connection.setRequestProperty("accept", "*/*");
23.    connection.setRequestProperty("connection", "Keep-Alive");
24.    connection.setRequestProperty("user-agent",
25.      "Mozilla/4.0 (compatible; MSIE 6.0; Windows NT 5.1;SV1)");
26.    //设置超时时间
27.    connection.setConnectTimeout(TIMEOUT);
28.    connection.setReadTimeout(TIMEOUT);
29.    //建立实际的连接
30.    connection.connect();
31.    // 获取所有响应头字段
32.    Map<String, List<String>> map = connection.getHeaderFields();
33.
34.    // 定义 BufferedReader 输入流来读取 URL 的响应，注意返回的是 UTF-8 的格式
35.    in = new BufferedReader(new InputStreamReader(
36.      connection.getInputStream(),"UTF-8"));
37.    String line;
```

```
38.     while ((line = in.readLine()) != null) {
39.      result += line;
40.     }
41.    } catch (Exception e) {
42.     e.printStackTrace();
43.    }
44.    // 使用finally块来关闭输入流
45.    finally {
46.     try {
47.      if (in != null) {
48.       in.close();
49.      }
50.     } catch (Exception e2) {
51.      e2.printStackTrace();
52.     }
53.    }
54.    return result;
55.   }
56.
57.
58.   /**
59.    * 向指定URL发送GET方法的请求
60.    *
61.    * @param url
62.    *            发送请求的URL
63.    * @param param
64.    *            请求参数，请求参数应该是 name1=value1&name2=value2 的形式。
65.    * @return URL所代表远程资源的响应结果
66.    */
67.   public static String sendGet(String url, String param) {
68.    String urlNameString = url + "?" + param;
69.    return sendGet(urlNameString);
70.   }
71.
72.   /**
73.    *  向指定 URL 发送POST方法的请求
74.    *
75.    * @param url
76.    *         发送请求的 URL
77.    * @param param
78.    *            请求参数，请求参数应该是 name1=value1&name2=value2 的形式。
79.    * @return 所代表远程资源的响应结果
80.    */
81.   public static String sendPost(String url, String param) {
82.    ...
83.   }
84.
85.   /* 从输入流中获取byte流
86.    * @param inStream
87.    * @return byte[]
88.    * @throws Exception
89.    */
90.   public static byte[] readInputStream(InputStream inStream) throws
      Exception {
91.    ByteArrayOutputStream outStream = new ByteArrayOutputStream();
92.    byte[] buffer = new byte[BUFFSIZE];
93.    int len = 0;
94.    while ((len = inStream.read(buffer)) != -1) {
```

```
95.     outStream.write(buffer, 0, len);
96.    }
97.   return outStream.toByteArray();
98.  }
99. }
```

在 com.newland.main 包下新建类 BaiDuMapUtil.java，在类中定义 getLocationInfo 方法，该方法接收经纬度作为参数，发送 HTTP 请求后返回 JSON 格式的经纬度信息。注意！需要将百度地图密钥填入访问路径指定位置，代码如下。

```
1.   public class BaiDuMapUtil {
2.    HttpHelper httpHelper;
3.    public BaiDuMapUtil() {
4.     httpHelper = new HttpHelper();
5.    }
6.    public String getLocationInfo(Double longitude,Double latitude) {
7.
8.     BigDecimal bdLong = new BigDecimal(longitude);
9.     double longitude2 =
     bdLong.setScale(6,BigDecimal.ROUND_HALF_UP).doubleValue();
10.
11.    BigDecimal bdlat = new BigDecimal(latitude);
12.    double latitude2 =
     bdlat.setScale(6,BigDecimal.ROUND_HALF_UP).doubleValue();
13.    String urlParameters = "ak=您的
     ak=json&coordtype=wgs84ll&location="+longitude2+","+latitude2;
14.    String path =
     "http://api.map.baidu.com/reverse_geocoding/v3/?"+urlParameters;
15.    String location = httpHelper.sendGet(path);
16.    return location;
17.   }
18.  }
```

在 com.newland.main 目录下新建 App 类作为程序入口，在类中添加 main 函数，在 main 函数中先打开北斗设备对应的串口，创建 BeiDouAuto 类对象，调用 getLocationOneTime 方法采集一次经纬度，再创建 BaiDuMapUtil 对象，调用 getLocationInfo 方法获取地理位置信息，代码如下。

```
1.   public class App {
2.
3.    public static void main(String[] args) {
4.     HttpHelper httpHelper = new HttpHelper();
5.     try {
6.      SerialPort port = SerialPortManager.openPort("COM2", 9600);
7.      BeiDouAuto bdAuto = new BeiDouAuto(port);
8.      bdAuto.getLocationOneTime();
9.      Thread.sleep(5000);
10.     BaiDuMapUtil bdUtil = new BaiDuMapUtil();
11.     String location = bdUtil.getLocationInfo(26.023463,119.415923);
12.     System.out.println(location);
13.     JSONObject object = JSON.parseObject(location);
14.     JSONObject result = object.getJSONObject("result");
15.     System.out.println(result.getString("formatted_address"));
16.     JSONObject addressComponent = result.getJSONObject("addressComponent");
17.     System.out.println(addressComponent.getString("country"));
18.     System.out.println(addressComponent.getString("country_code_iso"));
19.     System.out.println(addressComponent.getString("province"));
20.     System.out.println(addressComponent.getString("city"));
```

```
21.     System.out.println(addressComponent.getString("district"));
22.     System.out.println(addressComponent.getString("adcode"));
23.     System.out.println(addressComponent.getString("street"));
24.     System.out.println(addressComponent.getString("direction"));
25.
26.
27.    } catch (SerialPortParameterFailure e) {
28.     // TODO Auto-generated catch block
29.     e.printStackTrace();
30.    } catch (NotASerialPort e) {
31.     // TODO Auto-generated catch block
32.     e.printStackTrace();
33.    } catch (NoSuchPort e) {
34.     // TODO Auto-generated catch block
35.     e.printStackTrace();
36.    } catch (PortInUse e) {
37.     // TODO Auto-generated catch block
38.     e.printStackTrace();
39.    } catch (InterruptedException e) {
40.     // TODO Auto-generated catch block
41.     e.printStackTrace();
42.    }
43.   }
44.  }
```

3. 运行结果

运行程序，可以从控制台上看到当前的地理位置信息，如图 11-12 所示。

```
Problems  Javadoc  Declaration  Search  Console
&lt;terminated&gt; App (12) [Java Application] D:\java\jre\bin\javaw.exe
{"status":0,"result":{"location":{"lng":119.42730460795389,"lat":26.02670501462449},"formatted_address":"福建省
福建省福州市马尾区马江路15
中国
CHN
福建省
福州市
马尾区
350105
马江路
南
```

图 11-12 当前的地理位置信息

任务 2 将经纬度数据上报到云平台

【任务描述】

本任务要求编程实现将采集的经纬度上报到云平台。基于任务 1 的封装获取北斗定位模块采集的经纬度信息后，通过云平台提供的 TCP，先与云平台建立连接，然后将经纬度数据上报云平台。任务清单如表 11-5 所示。

表 11-5 任务清单

任务课时	2 课时	任务组员数量	建议 3 人
任务组采用设备	1 个北斗定位模块 1 个 RS485 转 RS232 转接头 1 台 PC 电源、USB 转串口、连接线若干		

【拓扑图】

本任务的拓扑图如图 11-13 所示。

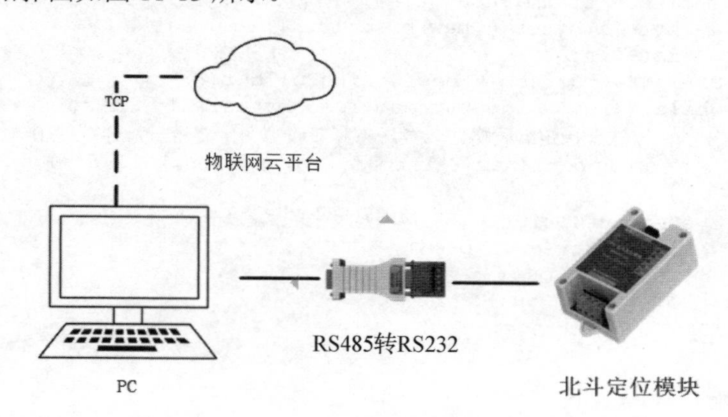

图 11-13 拓扑图

【知识解析】

1. 基于 TCP 的 Socket 套接字

1) Socket 和 ServerSocket

Socket 通常也称为“套接字”，用于描述 IP 地址和端口，是一个通信链的句柄。应用程序通常通过“套接字”向网络发出请求或者应答网络请求。

Socket 类库在 java.net 包中。Socket 是建立网络连接时使用的。在连接成功时，应用程序两端都会产生一个 Socket 实例对象。可以通过这个实例对象获取输入流和输出流，从而可以通过输入流获取网络对端传递过来的数据，通过输出流可以把数据传递到网络对端。

ServerSocket 用于服务器端，在 java.net 包中，通常两个通信端进行通信，必须有一端负责监听，另外一端负责呼叫。ServerSocket 对象使用 accept()方法监听客户端的 Socket 连接请求，如果收到一个客户端 Socket 的连接请求，该方法将返回一个新的 Socket 对象与客户端进行通信。如果没有连接，它将一直处于等待状态。

也就是说，ServerSocket 只用于服务器端，Socket 在服务端与客户端都存在。

2) 使用 Socket 进行通信

当客户端和服务端产生了各自的 Socket 后，程序就可以通过各自的 Socket 进行通信。Socket 提供如下两个方法来获取输入流和输出流。

getInputStream()方法获得网络连接输入，同时返回一个 InputStream 对象实例。

getOutputStream()方法连接的另一端将得到输入，同时返回一个 OutputStream 对象实例。

TCP 的服务器端，代码如下。

```
1.   public class MyService {
2.       public static void main(String[] args) throws Exception {
3.            // 监听指定的端口 9999
4.            ServerSocket server = new ServerSocket(9999);
5.
6.            // server 将一直等待连接的到来
7.            System.out.println("等待客户端的连接...");
8.            Socket socket = server.accept();
9.            // 建立好连接后，从 socket 中获取输入流，并建立缓冲区进行读取
10.           InputStream inputStream = socket.getInputStream();
11.           byte[] bytes = new byte[1024];
12.           int len;
13.        StringBuilder sb = new StringBuilder();
14.        while ((len = inputStream.read(bytes)) != -1) {
15.             //注意指定编码格式，发送方和接收方一定要统一，建议使用 UTF-8
16.             sb.append(new String(bytes, 0, len,"UTF-8"));
17.           }
18.        System.out.println("客户端发过来的信息是: " + sb);
19.            inputStream.close();
20.           socket.close();
21.           server.close();
22.         }
23.  }
```

TCP 的客户端，代码如下。

```
1.   public class MyClient {
2.      public static void main(String args[]) throws Exception {
3.         // 与服务端建立连接 参数是服务端的 IP 和端口
4.         Socket socket = new Socket("127.0.0.1", 9999);
5.         // 建立连接后获得输出流
6.         OutputStream outputStream = socket.getOutputStream();
7.         String message="hi 我又来了";
8.         outputStream.write(message.getBytes("UTF-8"));
9.         outputStream.close();
10.        socket.close();
11.      }
12.  }
```

服务器端通过 new ServerSocket(int port)绑定了一个端口，并在该端口监听客户端的连接请求，用 accept()接收客户端的连接请求。客户端通过 new Socket(host，port)获取 Socket 对象。该构造方法中两个参数的作用如下：第一个是要连接的服务器地址；第二个是连接服务器的端口号。然后可以获得该 Socket 对象中的输入流和输出流发送和读取的数据。该客户端向服务器端发送一个字符串，服务器端收到客户端发来的字符串，把收到的字符串输出到控制台。

运行程序，结果如图 11-14 所示。

通常情况下，服务器不应只接收一个客户端的请求，而应该通过循环调用 accept()方法不断接收来自客户端的所有请求。

2. 基于 UDP 的数据包传送

UDP 的特点是无连接、不可靠，基于数据包的通信，传递的是数据包，它用 DatagramPacket 类对数据进行封装。

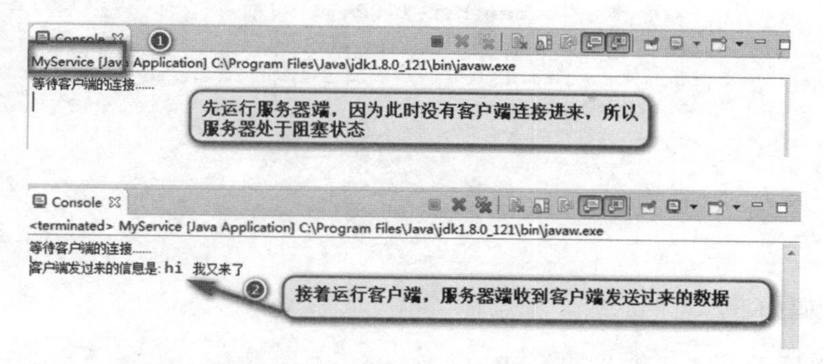

图 11-14 服务器端与客户端通信

1) DatagramPacket 数据包

该类用来实现无连接的包投递服务。每个数据包仅根据包中包含的地址信息把数据送往目的地，传送的多个数据包可能走不同的路由，DatagramPacket 类的常用构造方法如表 11-6 所示，也可能按不同的顺序到达目的地。

表 11-6 DatagramPacket 类的常用构造方法

方 法	描 述
DatagramPacket(byte[] buf, int length, InetAddress address, int port)	创建数据包实例，用来将存放在 buf 中的长度为 length 的包数据发送到 address 参数指定的地址和 port 参数指定的端口号的主机
DatagramPacket(byte[] buf, int length)	创建数据包实例，用来接收长度为 length 的包数据并存放到 buf 中

2) DatagramSocket 数据报套接字

该类是用于发送和接收数据的数据报套接字，可通过构造方法绑定端口号进行初始化 DatagramSocket 对象，初始化后通过 DatagramSocket 对象的 send(DatagramPacket dp)进行数据包的发送、receive(DatagramPacket dp)进行数据包的接收。

服务器端代码如下。

```
1.   public class UdpServer {
2.      public static void main(String[] args) throws Exception {
3.         // 1.定义一个 UDP socket，通常会侦听一个端口
4.         DatagramSocket ds = new DatagramSocket(6789);
5.
6.         // 2.定义一个数据报包，用于存放接收来的数据
7.         byte[] buf = new byte[1024];
8.         DatagramPacket dp = new DatagramPacket(buf, buf.length);
9.         while (true) {
10.           System.out.println("在等待客户端发送数据...");
11.           // 3.通过 socket 服务的方法
```

```
12.             ds.receive(dp);
13.
14.             // 4.通过数据报包提供的方法取出对端发送过来的数据
15.             String ip = dp.getAddress().getHostAddress();
16.             System.out.println("ip=" + ip);
17.             byte[] data = new byte[1024];
18.             data = dp.getData();// 真实数据
19.             System.out.println("客户端发送过来的数据: " + new String(data, 0,
                  dp.getLength()));
20.         }
21.         // 5.关闭 socket
22.         // ds.close();
23.     }
24. }
```

客户端代码如下。

```
1.  public class UdpClient {
2.
3.      /**
4.       * 客户端负责往服务器发送数据
5.       * @throws SocketException
6.       */
7.      public static void main(String[] args) throws Exception {
8.          //1.socket
9.          DatagramSocket ds = new DatagramSocket();
10.
11.         //2.数据报包 提供数据,
12.         String data="hello udp";
13.         byte[] buf=data.getBytes();
14.         DatagramPacket dp = new DatagramPacket(buf, buf.length,
15.                 InetAddress.getByName("127.0.0.1"), 6789);
16.
17.         //3.通过 socket 服务的发送功能将数据发送出去
18.         ds.send(dp);
19.
20.         //4.关闭 socket
21.         ds.close();
22.     }
23. }
```

程序运行，结果如图 11-15 所示。

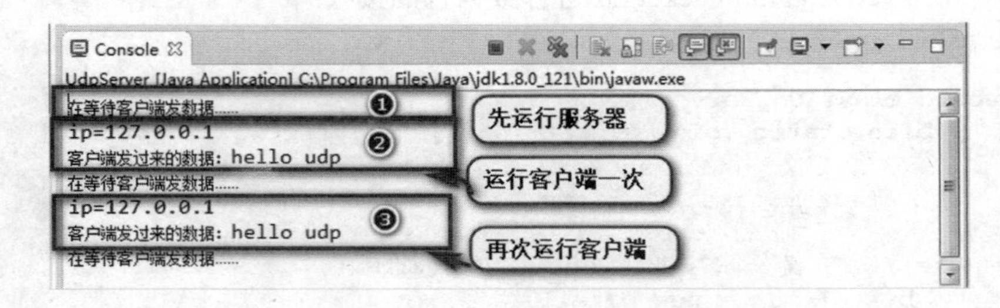

图 11-15　UDP 的服务器端与客户端通信

可以看到，服务器一直在等待客户端发送数据过来，一有数据就接收并显示出来。

【任务实施】

1. 任务分析

(1) 连接设备并使用串口调试工具进行调试。
(2) 在云平台上创建项目并添加传感器。
(3) 物联网云平台的接入协议分析。
(4) 按物联网云平台的接入协议封装对应的包。
(5) 编写向物联网云平台发送数据包的线程。
(6) 整合实现数据上报功能。

2. 任务实施

1) 连接设备并使用串口调试工具进行调试

参考任务 1 中的任务实施进行调试，确保北斗定位模块能正常采集定位数据。

2) 云平台创建项目并添加传感器

打开物联网云平台：http://www.nlecloud.com/，注册账号并登录云平台，切换到“开发者中心”，单击“新增项目”选项，在弹出来的“添加项目”界面中填写：项目名称为经纬度数据上报(可以自定义)；行业类别为智慧城市；联网方案为 WIFI，然后单击“下一步”按钮，如图 11-16 所示。

图 11-16 添加项目

在弹出来的“添加设备”界面中填写：设备名称为 BEIDOU(可以自定义)；通信协议为 TCP；设备标识为 PKR123456(可以自定义)，单击“确定添加设备”按钮，如图 11-17 所示。

设备添加完成后，即创建好 “经纬度数据上报”项目，如图 11-18 所示。

图 11-17　添加设备

单击图 11-18 中的“1 个设备”，可以看到项目信息，如图 11-19 所示。记下设备标识、传输密钥以及设备 ID，在以后的程序编写中需要这些信息。单击图 11-19 中的设备名称“BEIDOU”处，可以进入“添加传感器”界面，按图 11-20、图 11-21 中的信息添加经度和纬度传感器，传感器类型为“自定义”，标识名必须是 longitude 和 latitude。

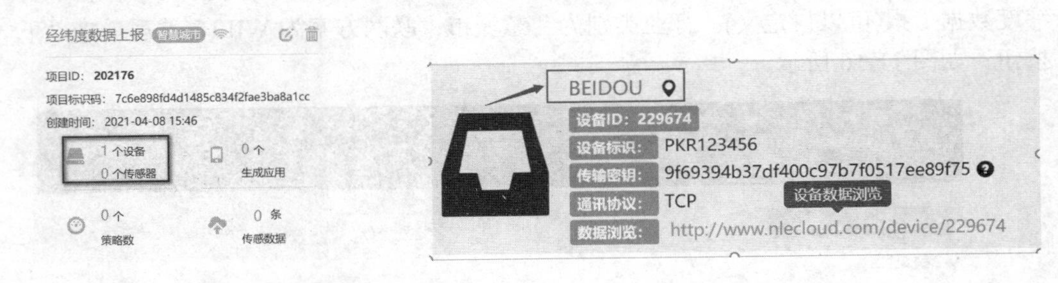

图 11-18　经纬度数据上报项目

图 11-19　项目信息

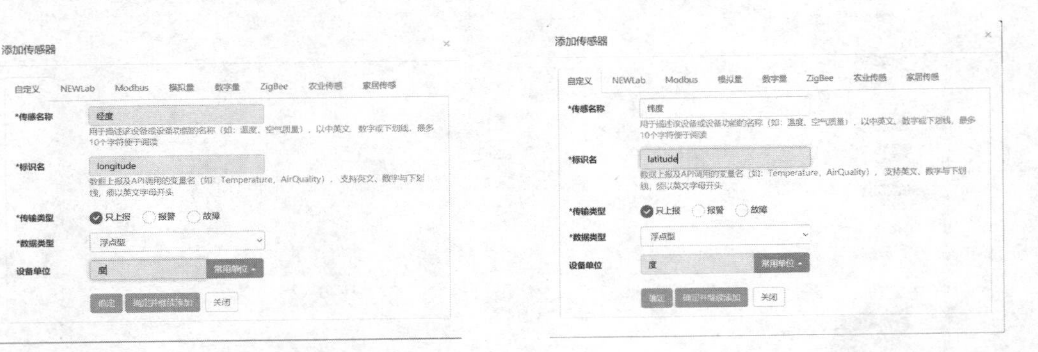

图 11-20　添加经度传感器

图 11-21　添加纬度传感器

3)　物联网云平台的接入协议分析

若要将采集的传感器数据上传到物联网云平台，得先与云平台建立连接，云平台有一套规定的接入流程：一般是先向云平台 ndp.nlecloud.com 发送 TCP 连接请求，端口是 8600。物联网云平台收到请求后进行响应，连接成功后，上报传感器数据和响应，要控制设备就上传命令请求和响应，因为是 TCP 连接，所以要与云平台一直维持 TCP 心跳连接和响应，这些请求和响应的格式都需要符合云平台规定的接入协议，具体如下。

①协议基于 TCP，协议全部以 JSON 格式定义。

② 接入流程。

第一步：访问平台 http://www.nlecloud.com，注册账号。

第二步：平台以项目为单位，一个项目下可以有多个设备，每个设备可以包含多个传感器，所以依次添加项目、设备、传感器，添加设备时在“通信协议”栏中可以选择 TCP\MQTT\HTTP 等方式。

第三步：操作以上功能时顺便记录下设备标识(device)、设备传输密钥(key)、传感器和执行器的 API 标识等信息。

第四步：硬件设备发送 TCP 连接请求到平台，平台接入服务器地址为 ndp.nlecloud.com 或 IP 地址为 120.77.58.34，TCP 端口号为 8600、8700、8800。连接成功后发送以下各项数据报文与平台进行交互。

③ 请求类型。

请求类型说明如表 11-7 所示。

表 11-7 请求类型说明

类型值	含义	方向
1	CONN_REQ：连接请求	C(client)→S(server)
2	CONN_RESP：连接响应	S→C
3	PUSH_DATA：上报数据	C→S
4	PUSH _ACK：上报数据确认	S→C
5	CMD_REQ：命令请求	S→C
6	CMD_RESP：命令响应	C→S
7	PING_REQ：心跳请求	S→C
8	PING_RESP：心跳响应	C↔S
其他值	保留	

④ 连接请求 C(client)→S(server)。

当设备建立 TCP 连接到指定端口后，需要发送连接请求报文，请求报文格式如下。

```
{
    "t": 1,
    "device": "P123456789",
    "key":"9861d43a0733415ab5424ee7d0f1c685",
    "ver":"v1.1"
}
```

连接请求说明如表 11-8 所示。

⑤ 连接响应 S(server)→C(client)。

硬件设备客户端发送连接请求后，服务器端会发送响应消息，响应报文格式如下。

```
{
    "t": 2,
    "status": 0
}
```

连接响应说明如表 11-9 所示。

表 11-8　连接请求说明

JSON 键	JSON 值	说　明	消息示例
t	1	固体数字 1，代表连接请求	
device	设备标识	在平台上添加不同设备时设备标识如下。 (1) 新大陆网关：进入网关设置→【参数设置】→【系统参数】中的序列号 (2) 新大陆农业网关：浏览器登录农业网关设置页面→【设备状态】中的设备编号 (3) 新大陆家居网关：家居网关主界面左上角的一行序列号 (4) 其他 MCU/SOC/网关/手机等设备：可自行输入一个唯一的标识用于与平台连接	PF12345678
Key	传输密钥	在平台上添加设备时自动生成的一串字符串(32 位长度的字符串)，该值在全局内具备唯一性	9861d43a0733415ab5424ee7d0f1c685
ver	客户端代码版本号	可以是自己拟定的一组客户端代码版本号值	V1.1

表 11-9　连接响应说明

JSON 键	JSON 值	说　明	消息示例
t	2	固体数字 2，代表连接响应	2
status	状态结果	用一个字节表示，根据验证情况，枚举值如下。 0：握手连接成功。 1：握手连接失败-协议错误。 2：握手连接失败-未添加设备。 3：握手连接失败-设备鉴权失败。 4：握手连接失败-未授权。 5-255：保留值	0

当 status 为非 0 失败时：服务器端不主动断开设备的连接，将保留 35s 后重新发起连接请求。

⑥　数据上报 C(client)→S(server)。

设备与服务器建立连接后，便可以进行传感数据的上报上传。上报报文格式如下。

```
{
    "t": 3,
    "datatype":1,
    "datas": { 见下表说明 } 或 [ 见下表说明 ],
    "msgid": 123
}
```

数据上报说明如表 11-10 所示。

表 11-10 数据上报说明

<table>
<tr><th>JSON 键</th><th>JSON 值</th><th>说　明</th><th>报文示例</th></tr>
<tr><td>t</td><td>3</td><td>固体数字 3，代表数据上报</td><td>3</td></tr>
<tr><td>datatype</td><td>数据上报格式类型</td><td>具体为 datas 属性内的传感数据格式类型如下。
= 1：JSON 格式 1 字符串。
= 2：JSON 格式 2 字符串。
= 3：JSON 格式 3 字符串</td><td>1</td></tr>
<tr><td rowspan="3">datas</td><td rowspan="3">要上报的传感数据数组</td><td>该属性根据 datatype 类型的不同，可以上报多个传感器数据，也可以上报同一传感器的多条数据，其中 apitag1 为传感的标识名，value 为传感值，可以是数字、浮点、字符串、二进制数(最大 48 字节大小)
数据类型为 1(JSON 格式 1 字符串)：
"datas":
{
"apitag1": "value1",
"apitag2": value2,
…
}</td><td>"datas":
{
"temperature":
23.5,
"rgb-r": "#999",
…
}</td></tr>
<tr><td>数据类型为 2(JSON 格式 2 字符串)：
apitag1 与 value 数据格式同上，datetime1 须是 yyyy-mm-dd hh:mm:ss 格式
"datas":
{
"apitag1":{"datetime1":"value1"},
"apitag2": {"datetime2":"value2"},
…
}</td><td>"datas":
{
"temperature":
{"2015-03-22
22:31:12":22.5},
…
}</td></tr>
<tr><td>数据类型为 3(JSON 格式 3 字符串)示例：
value 数据格式同上。
dt 须是 yyyy-mm-dd hh:mm:ss 格式
"datas":
[
{
"apitag":"temperature",
"datapoints":
[
{
"dt":"2018-01-22 22:22:22", //可选
"value": 36.5 //数字"浮点"字符串
}
]
},
{
"apitag": "location",
"datapoints":
[…]
},
{ … }
]</td><td></td></tr>
<tr><td>msgid</td><td>消息编号</td><td>由客户端生成的一个用于表示该条报文的编号，用于服务器下发“上报响应”时原样带回</td><td>123</td></tr>
</table>

⑦　数据上报响应 S(server)→C(client)。

设备进行任何一次传感数据上报后，服务器端会下发确认信息，报文格式如下。

```
{
    "t": 4,
    "msgid": 123,
    "status":0
}
```

数据上报响应说明如表 11-11 所示。

表 11-11　数据上报响应说明

JSON 键	JSON 值	说　明	消息示例
t	4	固体数字 4，代表数据上报响应	4
msgid	消息编号	由服务器端原样返回客户端上一次上报数据的消息 ID 值	123
status	状态结果	用一个字节表示如下。 0：上报成功。 1：上报失败。 其他：保留值	0

4)　按物联网云平台的接入协议封装对应的包

新建工程 project11_task2，把随书资源“./项目 11“文件夹下的“dll”目录下的两个文件复制到 jre 安装目录下，把 lib 下的 3 个 jar 文件添加到工程下的“lib”文件夹中，并添加到 Build Path 中，如图 11-22 所示。

图 11-22　添加 jar 包到工程中

新建包 com.nle.gateway，在包下新建类 Package.java，在类中定义物联网云平台规定的连接请求和响应、上报传感器数据和响应、命令请求和响应、TCP 心跳连接和响应的类型值，以后这些类型值在程序中就用请求和响应类型 t 来识别，同时还定义了连接请求的回应状态值。云平台接收上报数据的 JSON 格式有三种，在这里我们先用第一种，所以 JSON_TYPE_1 值为 1，代码如下。

```
1.   public abstract class Package {
2.    //应答状态码
3.    public static final int CONN_REQ = 1;
4.    public static final int CONN_RESP = 2;
5.    public static final int PUSH_DATA = 3;
6.    public static final int PUSH_ACK = 4;
7.    public static final int CMD_REQ = 5;
8.    public static final int CMD_RESP = 6;
9.    public static final String VER = "v3.1";
10.
```

```
11.  public static final int STATUS_OK = 0;
12.  public static final int STATUS_FAIL = 1;
13.  public static final int STATUS_FAIL_NO_DEVICE = 2;
14.  public static final int STATUS_FAIL_DEVICE_UNAUTHORIZED = 3;
15.  public static final int STATUS_FAIL_UNAUTHORIZED = 4;
16.
17.  public static final int JSON_TYPE_1 = 1;
18.  public static final int JSON_TYPE_2 = 2;
19.  public static final int JSON_TYPE_3 = 3;
20.
21.  public static final String  PING_RESP = "$#AT#";//"$#AT#\r"
22.  public static final String  PING_REQ = "$OK##\\n";//"$OK##\r"
23.
24.  protected int t;
25.
26.  public abstract boolean checkResp(RespPackage resp);
27.
28. }
```

当程序向物联网云平台发起连接或者上报数据或者接受从云平台下发的控制指令后，云平台会发送响应消息。

新建类RespPackage.java代表上述请求的各种响应，isActive()用于测试返回值是否为1，代码如下。

```
1.   //云平台响应连接的报文格式
2.   public class RespPackage{
3.    /**
4.     * 发送包中t值的含义如下。
5.     * 1: 表示连接握手包
6.     * 3: 表示上传信息包
7.     *
8.     * 返回包中t值的含义:
9.     * 2: 表示连接握手包返回的确认信息包
10.    * 4: 表示上传信息后返回的确认信息包
11.    * 5: 表示继电器控制命令包
12.    */
13.   private Integer t;
14.   private Integer status;
15.   private Integer msgid;
16.   private String data;
17.   private String apitag;
18.
19.         //get 和 set 方法省略
20.   public boolean isActive() {
21.    return "1".equals(data);
22.   }
23.  }
24.
```

新建类ConnPackage.java代表连接，继承自Package，在构造方法中指明包的类型是连接请求，在重写的checkResp()方法中判断是不是请求连接的响应，代码如下。

```
1.   //连接云平台报文格式
2.   public class ConnPackage  extends Package{
3.    private String device;
4.    private String key;
5.    private String ver;
```

```
6.
7.   public ConnPackage(String device, String key) {
8.    this.device = device;
9.    this.key = key;
10.
11.   t = Package.CONN_REQ;
12.   ver = Package.VER;
13.  }
14.
15.  public boolean checkResp(RespPackage resp) {//监测报文
16.   int t = resp.getT();
17.   int status = resp.getStatus();
18.   if(Package.CONN_RESP == t && Package.STATUS_OK == status) {
19.    return true;
20.   }
21.   return false;
22.  }
23.
24.  @Override
25.  public String toString() {//组包代码
26.   return "{'t':"+t+",'device':'" + this.device + "','key':'" + this.key +
    "','ver':'" + this.ver + "'}";
27.  }
28. }
```

新建类 DataPackage.java 代表要上传的数据包，继承自 Package，在构造方法中指明包的类型是上传数据请求，上传的数据格式为 JSON_TYPE_1，在重写的 checkResp()方法中判断是不是上传数据的响应，代码如下。

```
1.  //向云平台发送数据报文格式
2.  public class DataPackage  extends Package{
3.   private int datatype;
4.   private String apitag;
5.   private String value;
6.   private int msgid;
7.   public DataPackage(String apitag, String value) {
8.    this.apitag = apitag;
9.    this.value = value;
10.
11.   this.msgid = (int)(Math.random() * 1000);
12.   t = Package.PUSH_DATA;
13.   datatype = Package.JSON_TYPE_1;
14.  }
15.
16.
17.  @Override
18.  public String toString() {
19.   return "{'t':"+t+",'datatype':"+datatype+",'datas':{'"+apitag+"':
    "+value+"},'msgid':'"+msgid+"'}";
20.  }
21.
22.
23.  @Override
24.  public boolean checkResp(RespPackage resp) {
25.   int t = resp.getT();
26.   int status = resp.getStatus();
27.   int msgid = resp.getMsgid();
28.   if(Package.PUSH_ACK == t && Package.STATUS_OK == status && this.msgid ==
      msgid) {
```

```
29.     return true;
30.    }
31.   return false;
32.  }
33.
34. }
```

新建类 DataBeiDou.java 代表要上传的经纬度数据包，在构造方法中指明包的类型是上传数据请求，上传的数据格式为 JSON_TYPE_1，代码如下。

```
1.  public class DataBeiDou {
2.   private String t;
3.   private Map<String,Double> datas;
4.   private String msgid;
5.   private String datatype;
6.   public DataBeiDou(String t, Map<String, Double> datas, String msgid,
    String datatype) {
7.    super();
8.    this.t = t;
9.    this.datas = datas;
10.   this.msgid = msgid;
11.   this.datatype = datatype;
12.  }
13.  public String getT() {
14.   return t;
15.  }
16.  public void setT(String t) {
17.   this.t = t;
18.  }
19.  public Map<String, Double> getDatas() {
20.   return datas;
21.  }
22.  public void setDatas(Map<String, Double> datas) {
23.   this.datas = datas;
24.  }
25.  public String getMsgid() {
26.   return msgid;
27.  }
28.  public void setMsgid(String msgid) {
29.   this.msgid = msgid;
30.  }
31.  public String getDatatype() {
32.   return datatype;
33.  }
34.  public void setDatatype(String datatype) {
35.   this.datatype = datatype;
36.  }
37. }
```

5) 编写向物联网云平台发送数据包的线程

软网关程序运行起来后，首先需要使用 Socket 与物联网云平台建立 TCP 连接。Socket 通道建好后，可以得到输入流和输出流，当向云平台传送数据时，就可以利用输出流进行写操作。用 GatewaySendThread 线程做发送的事情，云平台的响应暂时不做处理。为了操作数据包，GatewaySendThread 也提供了一些具体的方法，代码如下。

```
1.  public class GatewaySendThread extends Thread {
2.   //传输的字符编码
3.   private static final String CHARSET = "GB2312";
```

```
4.    //初始化日志
5.    private Logger log = Logger.getLogger("GatewaySendThread");
6.
7.    private boolean isContinue;//是否继续发送标记
8.    private Socket socket;//Tcp 连接对象
9.    private BufferedReader bReader;//读取输入流数据
10.   private BufferedWriter bwriter;//数据写如输出流
11.
12.   private String ip;
13.   private int port;
14.   private String device;
15.   private String key;
16.
17.   //构造函数
18.   public GatewaySendThread(String ip, int port, String device, String key) {
19.    this.setName("GatewaySendThread");
20.    this.setDaemon(true);
21.
22.    this.ip = ip;
23.    this.port = port;
24.    this.device = device;
25.    this.key = key;
26.   }
27.
28.   public void run() {
29.    try {
30.     //实例化 Socket
31.     socket = new Socket(ip, port);
32.     socket.setSoTimeout(0);//超时时间设置成无限长
33.     //初始化缓冲区
34.     bwriter = new BufferedWriter(new OutputStreamWriter(socket.
       getOutputStream()));
35.     bReader = new BufferedReader(new InputStreamReader(socket.
       getInputStream(), CHARSET));
36.
37.     //1. 向服务器发送连接请求
38.     if(isContinue = sendConnReq()) {
39.      //2. 发送信息包
40.      while(isContinue) {
41.       System.out.println(String.valueOf(BeiDouAuto.map.get("longitude")));
42.       System.out.println(String.valueOf(BeiDouAuto.map.get("latitude")));
43.       DataBeiDou dd = new DataBeiDou("3",BeiDouAuto.map,"123","1");
44.       String jsonStr = JSON.toJSONString(dd);
45.       System.out.println(jsonStr);
46.       sendDatagram(jsonStr);
47.       Thread.sleep(2000);
48.      }
49.     }else {
50.      log.warning("请求连接失败");
51.     }
52.
53.    } catch (IOException | InterruptedException e) {
54.     e.printStackTrace();
55.    }finally {
56.     if(socket != null) {
57.      try {
58.       socket.close();
59.      } catch (IOException e) {
```

```
60.        e.printStackTrace();
61.      }
62.    }
63.
64.    }
65.   }
66.
67.
68.  /**
69.   * 发送连接请求
70.   * @return
71.   * @throws IOException
72.   */
73.  private boolean sendConnReq() throws IOException {
74.   ConnPackage connPkg = new ConnPackage(device, key);
75.   sendDatagram(connPkg);
76.
77.   String resp = bReader.readLine();
78.   log.info("resp:" + resp);
79.
80.
81.   RespPackage respPkg = JSONObject.parseObject(resp, RespPackage.class);
82.   return connPkg.checkResp(respPkg);
83.  }
84.
85.  /**
86.   * 发送报文
87.   * @param content 报文
88.   * @throws IOException
89.   */
90.  private void sendDatagram(String content) throws IOException {
91.   log.info("send:" + content);
92.   bwriter.write(content);
93.   bwriter.flush();
94.  }
95.
96.  /**
97.   * 发送报文
98.   * @param content 报文
99.   * @throws IOException
100.  */
101. private void sendDatagram(Package pkg) throws IOException {
102.  sendDatagram(pkg.toString());
103. }
104.
105. /**
106.  * 网关停止运行
107.  */
108. public void stopGateway() {
109.  isContinue = false;
110.  this.interrupt();
111.
112. }
113.
114. public BufferedReader getBufferedReader() {
115.  return bReader;
116. }
117. }
```

6） 整合实现数据上报功能

新建包 com.newland.device，将任务 1 封装的北斗设备采集工具类 BeiDouAuto 复制进来。在 com.newland.main 包中新建 Frame 作为程序入口，新建 main 函数，定义访问云平台必需的参数变量。

再一次检查设备连接，确认 COM 口，以下代码中假设用的是 COM2，初始化 BeiDouAuto 对象，调用 start 方法开始持续采集经纬度数据。初始化 GatewaySendThread 对象，调用 start 方法开始上报经纬度数据到云平台，代码如下。

```
1.  public class Frame {
2.
3.   public static void main(String[] args) {
4.    //云平台ip地址
5.    String ip = "ndp.nlecloud.com";
6.    //云平台通信端口
7.    int port = 8600;
8.    //网关表示
9.    String device = "您的云平台项目中设备标识";
10.   //通信密钥
11.   String key = "您的设备传输密钥";
12.   try {
13.    SerialPort portBd = SerialPortManager.openPort("COM1", 9600);
14.    BeiDouAuto bdAuto = new BeiDouAuto(portBd);
15.    bdAuto.start();
16.   } catch (SerialPortParameterFailure | NotASerialPort | NoSuchPort |
      PortInUse e) {
17.    // TODO Auto-generated catch block
18.    e.printStackTrace();
19.   }
20.   GatewaySendThread gatewatSendThread = new GatewaySendThread(ip, port,
      device, key);
21.   gatewatSendThread.start();
22.  }
23.
24. }
```

3. 运行结果

运行程序，可以从控制台上看到首先发送请求连接包并在云平台上收到程序上报的经纬度数据，如图 11-23、图 11-24 所示。

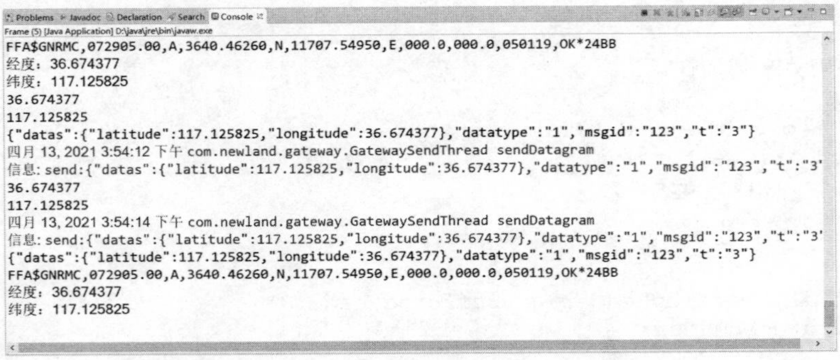

```
Problems  Javadoc  Declaration  Search  Console
Frame (5) [Java Application] D:\java\jre\bin\javaw.exe
FFA$GNRMC,072905.00,A,3640.46260,N,11707.54950,E,000.0,000.0,050119,OK*24BB
经度：36.674377
纬度：117.125825
36.674377
117.125825
{"datas":{"latitude":117.125825,"longitude":36.674377},"datatype":"1","msgid":"123","t":"3"}
四月 13, 2021 3:54:12 下午 com.newland.gateway.GatewaySendThread sendDatagram
信息: send:{"datas":{"latitude":117.125825,"longitude":36.674377},"datatype":"1","msgid":"123","t":"3"
36.674377
117.125825
四月 13, 2021 3:54:14 下午 com.newland.gateway.GatewaySendThread sendDatagram
信息: send:{"datas":{"latitude":117.125825,"longitude":36.674377},"datatype":"1","msgid":"123","t":"3"
{"datas":{"latitude":117.125825,"longitude":36.674377},"datatype":"1","msgid":"123","t":"3"}
FFA$GNRMC,072905.00,A,3640.46260,N,11707.54950,E,000.0,000.0,050119,OK*24BB
经度：36.674377
纬度：117.125825
```

图 11-23 控制台显示经纬度数据和上报的数据

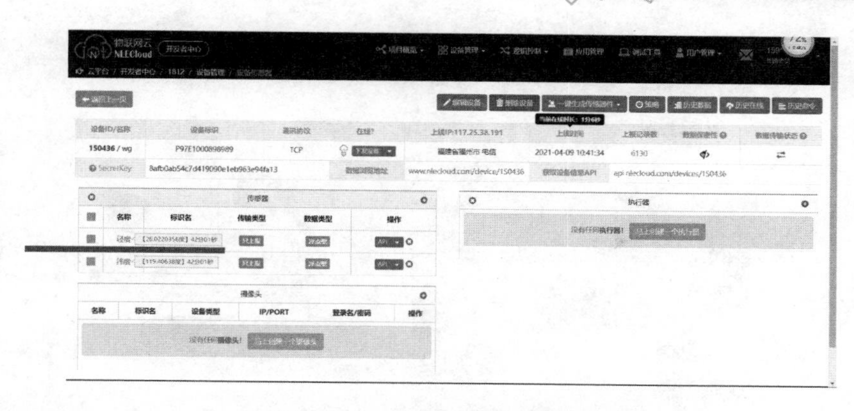

图 11-24 云平台收到上报的信息

至此，北斗设备的经纬度数据上报云平台功能已实现完毕。完整的项目源码请参阅随书资料中的 project11_task2。

思考与练习

1. 简述 Java 发起 HTTP 请求的过程。
2. 简述北斗设备的请求和响应帧结构。
3. 简述百度地图 SDK 集成与调用 API 的过程。
4. 简述 TCP 与 UDP 的区别。
5. 简述 ServerSocket 和 Socket 的区别。
6. 简述物联网云平台接入流程。

项目12 智慧园区系统综合实现（串口篇）

【项目描述】

园区通常是一个比较大的区域，传统的管理方法主要依靠人工巡逻、人为判断这样比较主观的方式，人力成本高、出错率高。智慧园区项目通过物联网和云计算的应用，提高了运营效率，降低了管理成本。本项目为综合开发串口篇，整合了前面所讲的知识点和案例，以完成智慧园区的注册登录、设备监控界面、设备参数配置、设备数据采集等功能模块，通过模拟实战让读者更深入地学习和了解物联网开发。任务清单如表 12-1 所示。

表 12-1　任务清单

任务课时	建议实训	任务组员数量	建议 3 人
任务组采用设备	1 个 ADAM-4150+ 1 个照明灯 1 个风扇 1 个红外对射 1 个二氧化碳传感器 1 个噪声传感器 1 个微动开关 1 个行程开关 1 个 ZigBee 四输入模块 1 个 ZigBee 节点盒 1 个 ZigBee 温湿度传感器		

续表

任务组采用设备	1 个 ZigBee 协调器 1 个电动推杆 7 个继电器 1 个三色灯 1 个 RS485 转 RS232 转接头 电源、USB 转串口、连接线若干

界面设计如图 12-1 所示。

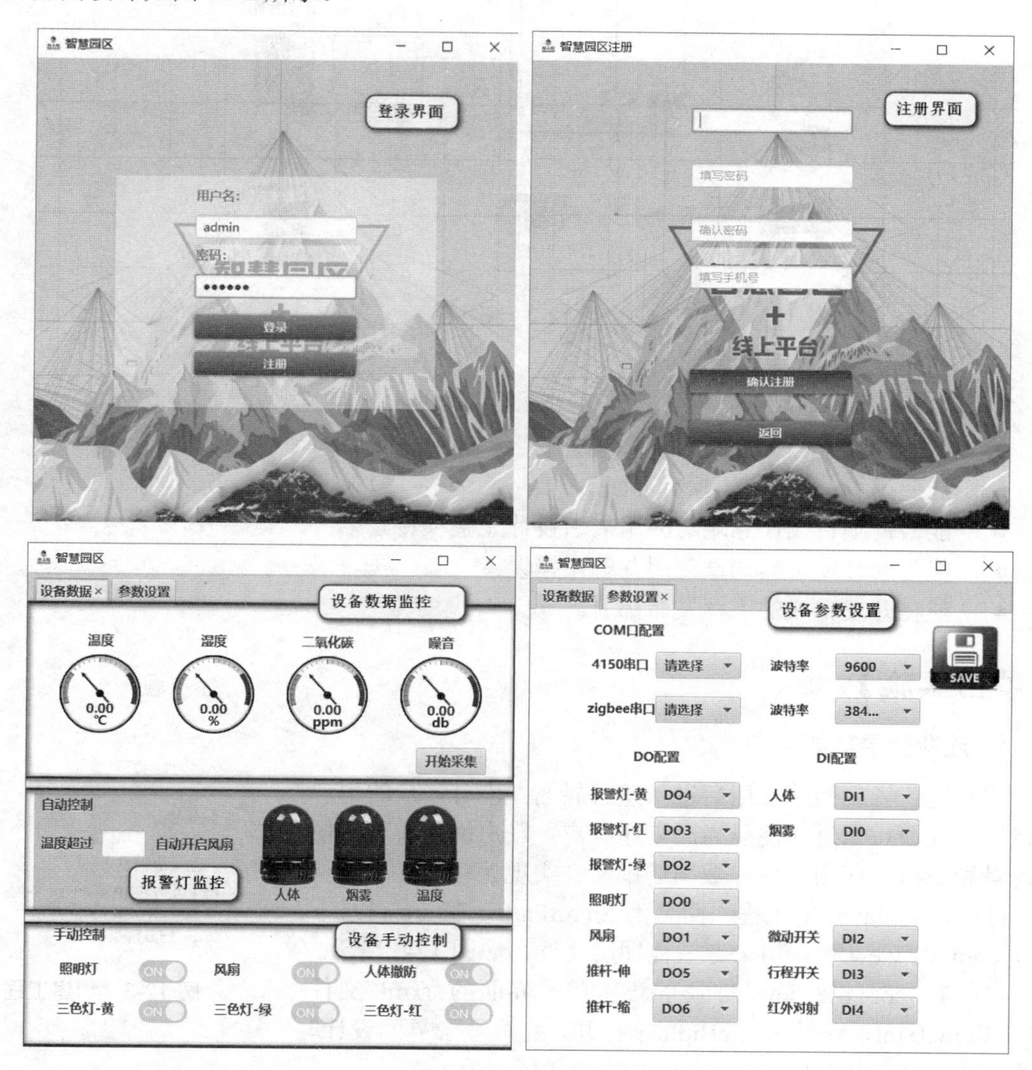

图 12-1 界面设计

【拓扑图】

本项目设备安装整合了门禁安防与室内环境模块，拓扑图如图 12-2 所示。

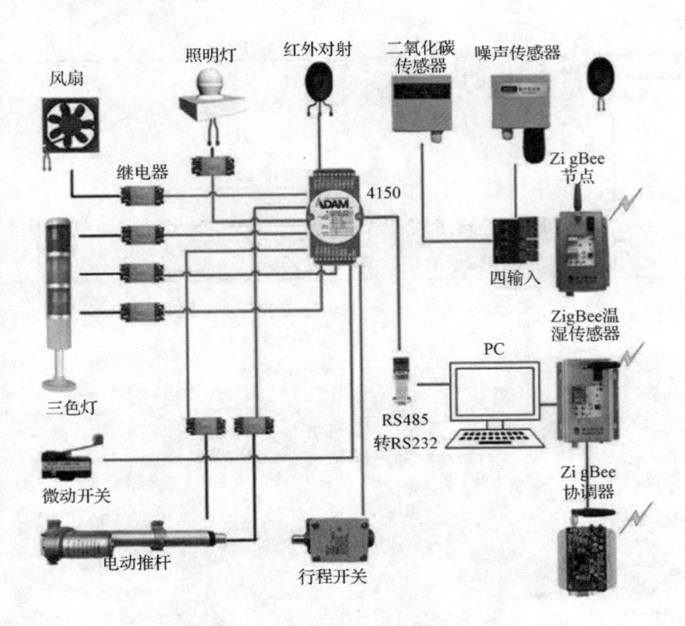

图 12-2　拓扑图

【技能目标】

- 能对项目进行 MVC 分层结构设计。
- 能使用 JavaFX 制作项目界面。
- 能根据项目设计思路与应用需求设计完整项目。
- 能借助开发工具调试应用并解决错误。
- 能将开发完成的工程发布部署。

【项目实施】

1. 注册登录功能

用户运行智慧园区应用后进入登录界面，没有账户的用户可以通过注册按钮进入注册界面创建账户。账户创建成功后便可登录智慧园区应用。用户数据通过文件实现保存。

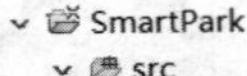

图 12-3　创建工程

(1) 新建 JavaFX 工程，命名为 SmartPark，在 src 目录下创建 com.nle.view 包，用来存放视图层文件，如图 12-3 所示。

(2) 在 com.nle.view 目录下新建登录界面的 fxml 文件 LoginView.fxml，通过 SceneBuilder 打开，完成登录界面设计，为界面上的控件添加 id 和监听，如图 12-4 所示。

(3) 在 src 目录下创建 com.nle.ctrl 包，用来存放控制层文件，在该目录下创建 LoginViewController 类，并将 SceneBuilder 自动生成的属性与方法添加至该类中，如图 12-5、图 12-6 所示。

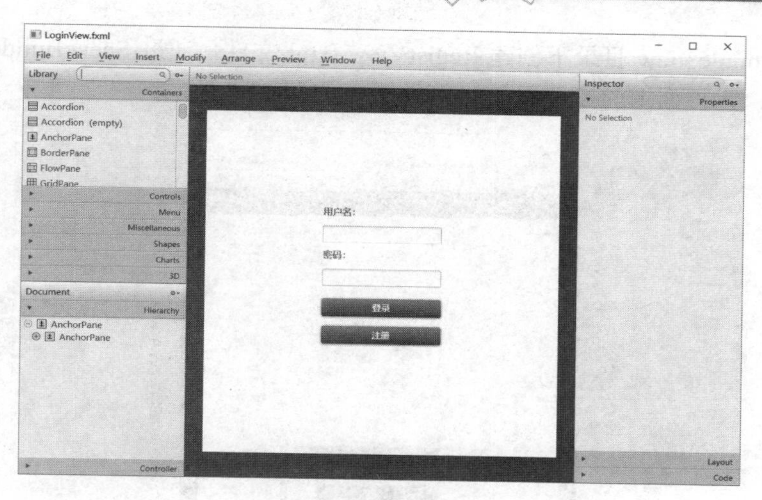

图 12-4　登录界面设计

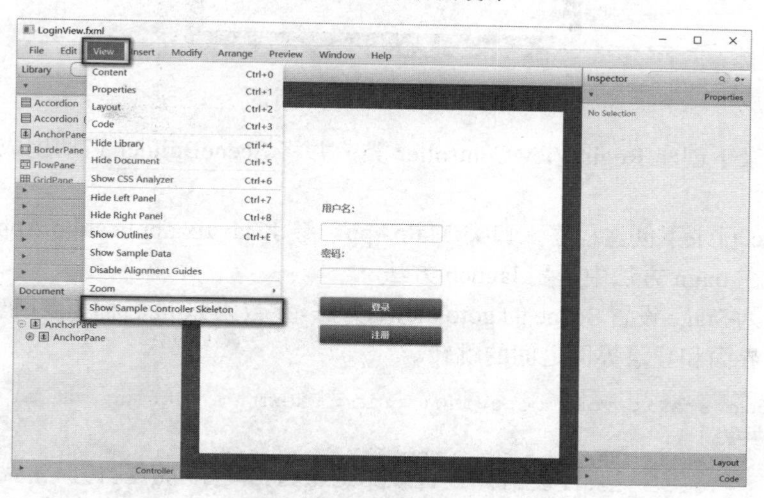

图 12-5　自动生成属性和方法

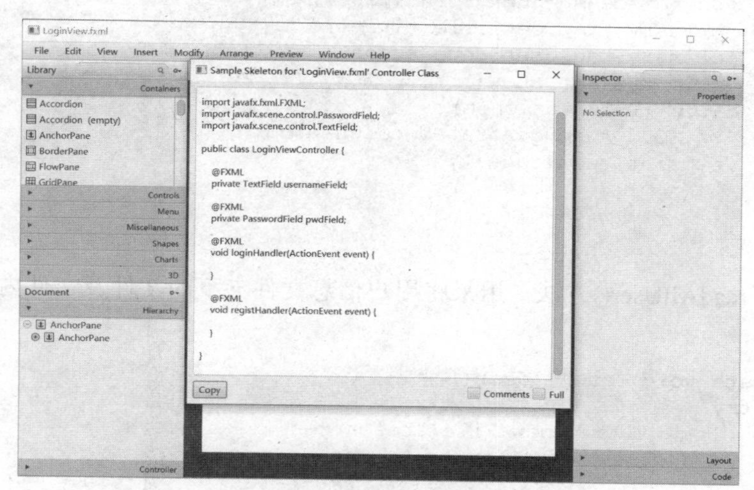

图 12-6　自动生成属性和方法

(4) 在 com.nle.view 目录下创建 RegistView.fxml 文件，通过 SceneBuilder 打开，完成注册界面设计，为界面上的控件添加 id 和监听，如图 12-7 所示。

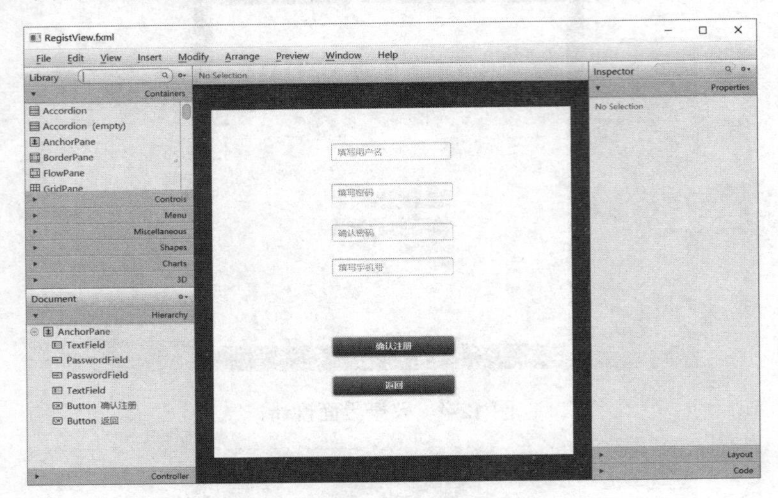

图 12-7　注册界面设计

(5) 在目录下创建 RegistViewController 类，并将 SceneBuilder 自动生成的属性与方法添加至该类中。

(6) 在 src 目录下创建程序入口类 MainApp，继承 javafx.application.Application，重写 start 方法，并在 main 方法中调用 launch 方法。

(7) 定义为 Stage 设置 Scene 的 gotoView()方法，参数为 fxml 名称和要跳转界面的标题。用来实现注册界面和登录界面之间的跳转。

```
1.  public static void gotoView(String viewName, String title) {
2.      try {
3.          FXMLLoader loader = new FXMLLoader();
4.          loader.setLocation(MainApp.class.getResource(viewName+".fxml"));
5.          AnchorPane pane = loader.load();
6.          Scene scene = new Scene(pane);
7.          primaryStage.setTitle(title);
8.          primaryStage.setScene(scene);
9.          primaryStage.show();
10.     } catch (IOException e) {
11.         // TODO Auto-generated catch block
12.         e.printStackTrace();
13.     }
14.
15. }
```

(8) 定义 readAllUser()方法，用来将用户信息文件夹下的文件及文件内容保存在一个 HashMap 中。

```
1.  private void readAllUser() {
2.      try {
3.          File file = new File("resources/files/users");
4.          if(!file.exists()) {
5.
6.              file.mkdirs();
7.
```

```
8.           }
9.
10.          String[] files = file.list();
11.
12.          for (int i = 0; i < files.length; i++) {
13.
14.              FileInputStream fis = new FileInputStream(new File("resources/
                   files/users/"+files[i]));
15.
16.              BufferedReader bufr = new BufferedReader(new InputStreamReader
                   (fis));
17.
18.              String pwd = bufr.readLine();
19.
20.              users.put(files[i].substring(0, files[i].lastIndexOf(".")), pwd);
21.
22.              bufr.close();
23.          }
24.      } catch (FileNotFoundException e) {
25.          // TODO Auto-generated catch block
26.          e.printStackTrace();
27.      } catch (IOException e) {
28.          // TODO Auto-generated catch block
29.          e.printStackTrace();
30.      }
31.
32.  }
```

(9) 为登录界面的注册按钮添加监听，用户单击按钮跳转到注册界面进行注册。直接在监听方法中调用 MainApp 的 gotoView()方法即可。

```
1.   @FXML
2.   void regHandler(ActionEvent event) {
3.       MainApp.gotoView("RegView", "智慧园区注册界面");
4.   }
```

(10) 在根目录下创建 resource/files 文件夹用来保存用户信息文件。

(11) 为注册界面的注册按钮添加监听。用户按要求输入基本信息后，单击注册按钮，获取输入框中的用户名和密码，判断用户名不存在后，为用户创建用户信息文件，并将密码存入该文件。

```
1.   /**
2.    * 注册按钮
3.    * @param event
4.    * @throws IOException
5.    */
6.   @FXML
7.   void registHandler(ActionEvent event) throws IOException {
8.
9.       Alert alert = new Alert(AlertType.WARNING);
10.      alert.setTitle("温馨提示");
11.      String username = usernameField.getText();
12.      String pwd = pwdField.getText();
13.      String confirmPwd = confirmPwdField.getText();
14.      if(StringUtil.checkNotNull(username)) {
15.          File file = new File("resources/files/users/"+username+".txt");
16.          file.createNewFile();
17.          if(StringUtil.checkStrLength(pwd, 6)) {
```

```
18.             if(pwd.equals(confirmPwd)) {
19.                PrintWriter out = new PrintWriter(file);
20.                out.println(pwd);
21.                out.flush();
22.                out.close();
23.                //注册信息回填登录界面
24.                loginController.getUsernameField().setText(username);
25.                loginController.getPwdField().setText(pwd);
26.                loginController.getUsers().put(username, pwd);
27.                alert.setContentText("注册成功");
28.                alert.show();
29.                MainApp.gotoView("LoginView", "智慧园区登录界面");
30.             }else {
31.                alert.setContentText("两次密码不一致");
32.                alert.show();
33.             }
34.          }else {
35.             alert.setContentText("密码必须是 6 位以上");
36.             alert.show();
37.          }
38.       }else {
39.          alert.setContentText("请正确填写用户名");
40.          alert.show();
41.       }
42.    }
```

(12) 为注册界面的返回按钮添加监听，用户单击按钮跳转到登录界面。直接在监听方法中调用 MainApp 的 gotoView()方法即可。

```
1.    @FXML
2.    void backHandler(ActionEvent event) {
3.       MainApp.gotoView("LoginView", "智慧园区登录界面");
4.    }
```

(13) 为登录界面的登录按钮添加监听，用户单击登录按钮后对用户信息进行判断，登录成功后跳转至功能页面 FunctionView(在下一个步骤中创建)。

```
1.    /*
2.     * 登录按钮监听
3.     *
4.     */
5.    @FXML
6.    public void loginHandler() {
7.       String username = usernameField.getText();
8.       String pwd = pwdField.getText();
9.       if(StringUtil.checkNotNull(username)) {
10.         if(users.containsKey(username)) {
11.            if(users.get(username).equals(pwd)) {
12.               MainApp.gotoView("FunctionView", "智慧园区功能界面");
13.            }else {
14.               alert.setContentText("密码错误! ");
15.               alert.show();
16.            }
17.         }else {
18.            alert.setContentText("用户名不存在! ");
19.            alert.show();
20.         }
21.      }else {
```

```
22.         alert.setContentText("用户名不能为空");
23.         alert.show();
24.     }
25. }
```

2. 设备监控界面

用户登录成功后跳转到设备监控界面。监控界面分为三部分：设备数据监控，报警灯信息，手动控制开关。用户可以通过应用查看设备实时数据、报警灯报警情况，也可以手动控制设备。设备数据采集部分的代码需要在设备参数配置后完成，本小节先完成界面和部分功能。

(1) 在 com.nle.view 目录下新建功能界面的 fxml 文件 FunctionView.fxml，通过 SceneBuilder 打开，完成功能界面设计，设备数据和参数配置使用 TabPane 控件实现界面切换。为界面上的控件添加 id 和监听，如图 12-8 所示。

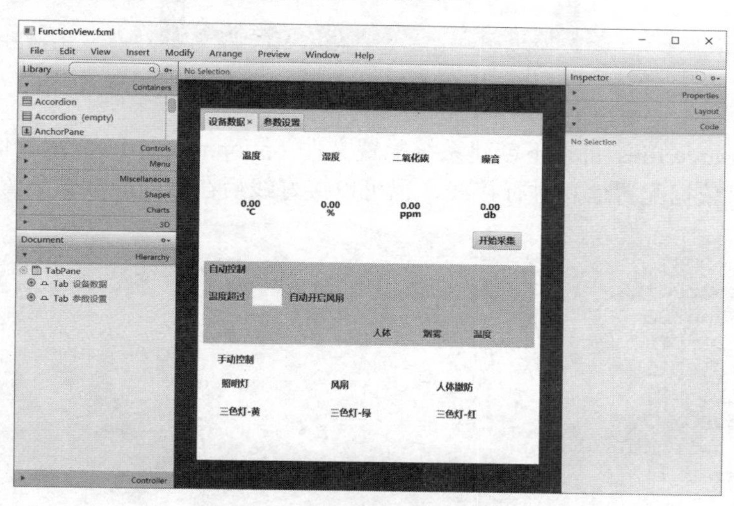

图 12-8 设备监控界面设计

(2) 在 com.nle.ctrl 目录下创建 FunctionViewController 类，并将 SceneBuilder 自动生成的属性与方法添加至该类中。

(3) 在根目录 resource 文件夹下新增 images 文件夹，保存界面样式需要用到的图片，如图 12-9 所示。

(4) 本项目界面的样式采用可复用的 css 文件来实现，将资源包中的 css 文件添加到工程与 fxml 文件相同的目录下，并在 SceneBuilder 中设置引用就可以实现界面样式的美化。有兴趣的读者可以自行学习，如图 12-10、图 12-11 所示。

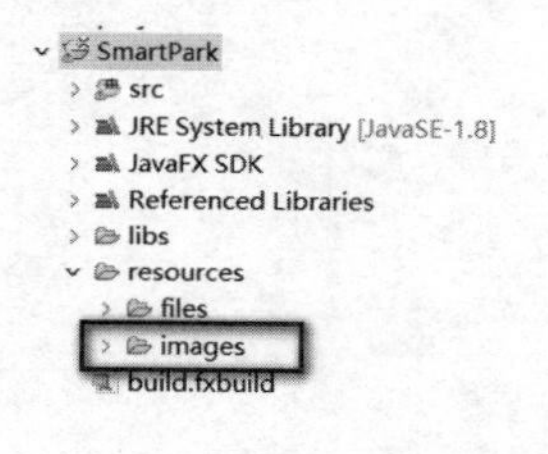

图 12-9 添加图片文件

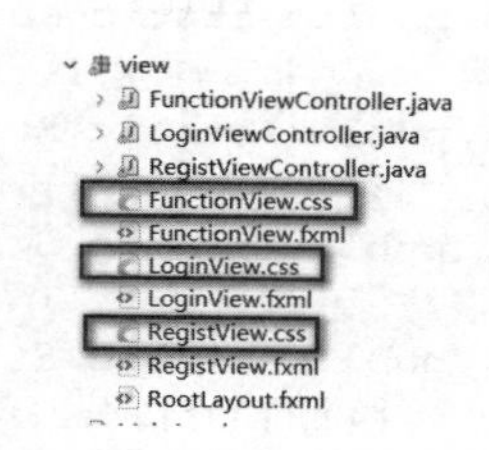

图 12-10 添加 css 文件

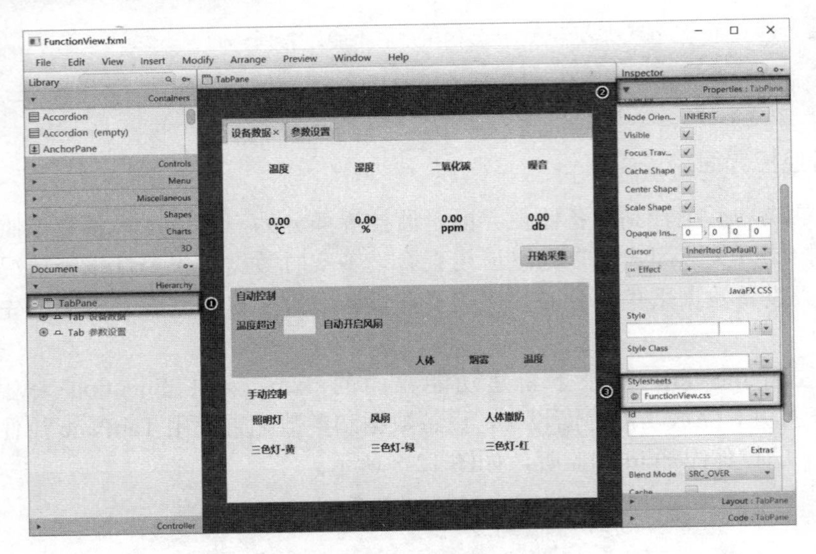

图 12-11　关联 css 文件

(5) 在 resource/files 目录下创建系统配置文件 option.properties。配置信息如下。用户在安装设备时可依据配置信息进行接线，也可以接好线后在后面的参数配置功能中配置。

```
1.   microswitchDI=DI2
2.   redDO=DO3
3.   opendoorDO=DO5
4.   closedoorDO=DO6
5.   smokeDI=DI0
6.   humanDI=DI1
7.   greenDO=DO2
8.   yellowDO=DO4
9.   limitswitchDI=DI3
10.  infraredDI=DI4
11.  lightDO=DO0
12.  blowerDO=DO1
```

(6) 在 src 下创建 com.nle.util 包，在该目录下创建 OptionUtils 类，用来存储系统配置信息和 DO 口的指令。

```
1.   public class OptionUtils {
2.       //配置文件
3.       public static Properties properties = new Properties();
4.       //可用串口及比特率
5.       public static String adamCom;
6.       public static int adamBt;
7.       public static String zigbeeCom;
8.       public static int zigbeeBt;
9.       //黄色报警灯 DO 口
10.      public static String yellowDO;
11.      //红色报警灯 DO 口
12.      public static String redDO;
13.      //绿色报警灯 DO 口
14.      public static String greenDO;
15.      //照明灯 DO 口
16.      public static String lightDO;
17.      //风扇 DO 口
```

```
18.     public static String blowerDO;
19.     //烟雾采集 DI 口
20.     public static String smokeDI;
21.     //人体采集 DI 口
22.     public static String humanDI;
23.     //行程开关 DI 口
24.     public static String limitswitchDI;
25.     //微动开关 DI 口
26.     public static String microswitchDI;
27.     //开门 DO 口
28.     public static String opendoorDO;
29.     //关门 DO 口
30.     public static String closedoorDO;
31.     //红外对射
32.     public static String infraredDI;
33.     //控制指令
34.     public static Map<String,byte[][]> controlCmd = new HashMap<>();
35.
36.     static {
37.         controlCmd.put("DO0", new byte[][] {{0x01, 0x05,0x00, 0x10, (byte)
              0xFF, 0x00, (byte) 0x8D, (byte) 0xFF}
38.             ,{0x01, 0x05,0x00, 0x10, 0x00, 0x00, (byte) 0xCC, (byte) 0x0F}});
39.         controlCmd.put("DO1", new byte[][] {{0x01, 0x05,0x00, 0x11, (byte)
              0xFF, 0x00, (byte) 0xDC, (byte) 0x3F}
40.             ,{0x01, 0x05,0x00, 0x11, 0x00, 0x00, (byte) 0x9D, (byte) 0xCF}});
41.         controlCmd.put("DO2", new byte[][] {{0x01, 0x05,0x00, 0x12, (byte)
              0xFF, 0x00, (byte) 0x2C, (byte) 0x3F}
42.             ,{0x01, 0x05,0x00, 0x12, 0x00, 0x00, (byte) 0x6D, (byte) 0xCF}});
43.         controlCmd.put("DO3", new byte[][] {{0x01, 0x05,0x00, 0x13, (byte)
              0xFF, 0x00, (byte) 0x7D, (byte) 0xFF}
44.             ,{0x01, 0x05,0x00, 0x13, 0x00, 0x00, (byte) 0x3C, (byte) 0x0F}});
45.         controlCmd.put("DO4", new byte[][] {{0x01, 0x05,0x00, 0x14, (byte)
              0xFF, 0x00, (byte) 0xCC, (byte) 0x3E}
46.             ,{0x01, 0x05,0x00, 0x14, 0x00, 0x00, (byte) 0x8D, (byte) 0xCE}});
47.         controlCmd.put("DO5", new byte[][] {{0x01, 0x05,0x00, 0x15, (byte)
              0xFF, 0x00, (byte) 0x9D, (byte) 0xFE}
48.             ,{0x01, 0x05,0x00, 0x15, 0x00, 0x00, (byte) 0xDC, (byte) 0x0E}});
49.         controlCmd.put("DO6", new byte[][] {{0x01, 0x05,0x00, 0x16, (byte)
              0xFF, 0x00, (byte) 0x6D, (byte) 0xFE}
50.             ,{0x01, 0x05,0x00, 0x16, 0x00, 0x00, (byte) 0x2C, (byte) 0x0E}});
51.         controlCmd.put("DO7", new byte[][] {{0x01, 0x05,0x00, 0x17, (byte)
              0xFF, 0x00, (byte) 0x3C, (byte) 0x3E}
52.             ,{0x01, 0x05,0x00, 0x17, 0x00, 0x00, (byte) 0x7D, (byte) 0xCE}});
53.     }
54. }
```

(7) 在 FunctionViewController 类中添加读取配置信息的方法 initOptions()。并在 initialize()方法中调用。

```
1.   /**
2.    * 读取配置文件信息
3.    * @throws IOException
4.    */private void initOptions() throws IOException {
5.      File file = new File("resources/files/options.properties");
6.      if(!file.exists()) {
7.         file.getParentFile().mkdirs();
8.         file.createNewFile();
9.      }
```

```
10.     FileInputStream fis = new FileInputStream(file);
11.     OptionUtils.properties.load(fis);
12.     OptionUtils.yellowDO = OptionUtils.properties.getProperty("yellowDO");
13.     OptionUtils.redDO = OptionUtils.properties.getProperty("redDO");
14.     OptionUtils.greenDO = OptionUtils.properties.getProperty("greenDO");
15.     OptionUtils.lightDO = OptionUtils.properties.getProperty("lightDO");
16.     OptionUtils.blowerDO = OptionUtils.properties.getProperty("blowerDO");
17.     OptionUtils.co2DI = OptionUtils.properties.getProperty("co2DI");
18.     OptionUtils.noiseDI = OptionUtils.properties.getProperty("noiseDI");
19.     OptionUtils.humanDI = OptionUtils.properties.getProperty("humanDI");
20. }
```

3. 设备参数配置

串口名称，波特率，4150 采集器，连接设备的 DI、DO 口都需要根据设备的安装来配置。在配置文件中已经配置了各个设备 IO 口的默认值，串口需要根据运行的环境来确定，故不需要文件保存。用户登录成功后，可以根据实际情况进行参数配置。

(1) 在 FunctionView.fxml 中添加参数配置界面，完成控件添加，设置控件的 id 和监听，并将属性和监听方法添加到 FunctionViewController 中，如图 12-12 所示。

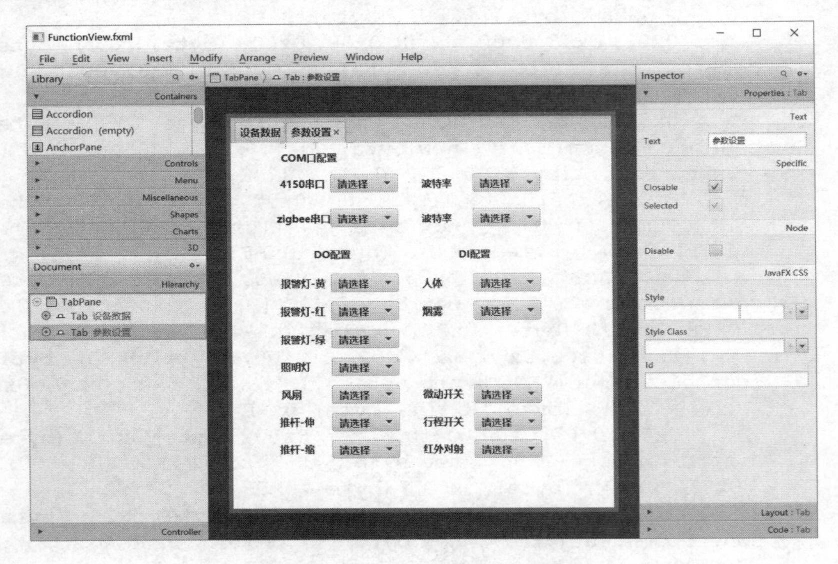

图 12-12　设备参数配置界面

(2) 在 FunctionViewController 中添加 initComboBox 方法，初始化各下拉列表框数据，并在 initialize 方法中调用。

```
1.  private void initComboBox() {
2.
3.      adamBt.setValue("9600");
4.
5.      zigbeeBt.setValue("38400");
6.
7.      adamBt.getItems().addAll("9600", "38400", "115200");
8.
9.      zigbeeBt.getItems().addAll("9600", "38400", "115200");
10.
11.     yellowDO.getItems().addAll("DO0", "DO1", "DO2", "DO3", "DO4", "DO5",
          "DO6", "DO7");
```

```
12.
13.     lightDO.getItems().addAll("DO0", "DO1", "DO2", "DO3", "DO4", "DO5",
          "DO6", "DO7");
14.
15.     blowerDO.getItems().addAll("DO0", "DO1", "DO2", "DO3", "DO4", "DO5",
          "DO6", "DO7");
16.
17.     greenDO.getItems().addAll("DO0", "DO1", "DO2", "DO3", "DO4", "DO5",
          "DO6", "DO7");
18.
19.   redDO.getItems().addAll("DO0", "DO1", "DO2", "DO3", "DO4", "DO5", "DO6",
          "DO7");
20.
21.     opendoorDO.getItems().addAll("DO0", "DO1", "DO2", "DO3", "DO4", "DO5",
          "DO6", "DO7");
22.
23.   closedoorDO.getItems().addAll("DO0", "DO1", "DO2", "DO3", "DO4", "DO5",
          "DO6", "DO7");
24.
25.     humanDI.getItems().addAll("DI0", "DI1", "DI2", "DI3", "DI4", "DI5",
          "DI6");
26.
27.     smokeDI.getItems().addAll("DI0", "DI1", "DI2", "DI3", "DI4", "DI5",
          "DI6");
28.
29.     microswitchDI.getItems().addAll("DI0", "DI1", "DI2", "DI3", "DI4",
          "DI5", "DI6");
30.
31.     limitswitchDI.getItems().addAll("DI0", "DI1", "DI2", "DI3", "DI4",
          "DI5", "DI6");
32.
33.     infraredDI.getItems().addAll("DI0", "DI1", "DI2", "DI3", "DI4", "DI5",
          "DI6");
34. }
```

(3) 在SceneBuilder中打开FunctionView.fxml，为TabPane控件添加onSelectionChanged监听，并在FunctionViewController中完成初始化可用串口的代码，以实现用户在监控界面和配置界面之间切换时更新可用串口信息，如图12-13所示。

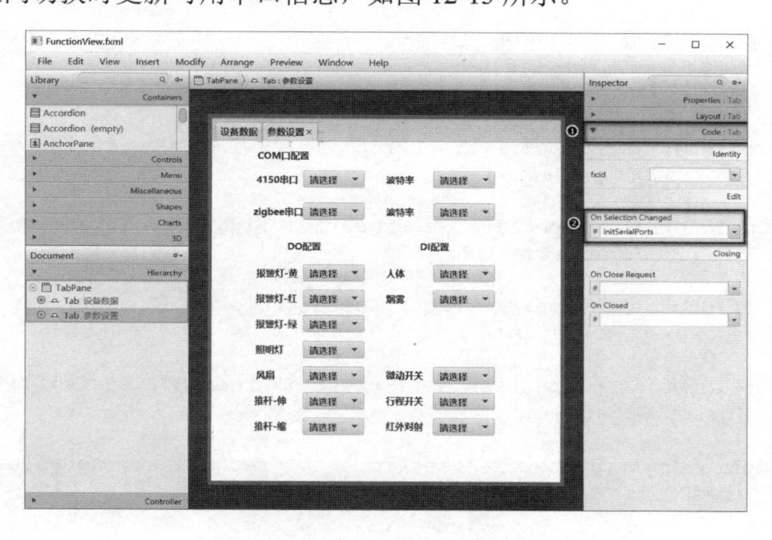

图12-13 为TabPane添加监听

```
1.   /**
2.    * 初始化可用串口
3.    */
4.   @FXML
5.   private void initSerialPorts() {
6.
7.       List serialPorts = SerialPortManager.findPort();
8.
9.       adamCom.getItems().clear();
10.
11.      adamCom.getItems().addAll(serialPorts);
12.
13.      zigbeeCom.getItems().clear();
14.
15.      zigbeeCom.getItems().addAll(serialPorts);
16.  }
```

(4) 用户在参数配置界面将串口和 IO 口配置设置好后，单击“保存”按钮，将配置信息保存到 Option.properties 中，然后调用 initOptions 方法，将配置更新到 OptionUtils 中。

```
1.   /**
2.    *  保存设置信息
3.    * @param event
4.    * @throws IOException
5.    */
6.   @FXML
7.   private void saveOption(ActionEvent event) throws IOException {
8.
9.       // 设置串口和波特率
10.      // 4150 串口配置
11.      OptionUtils.adamCom = String.valueOf(adamCom.getValue());
12.
13.      OptionUtils.adamBt =
     Integer.parseInt(String.valueOf(adamBt.getValue()));
14.
15.      //ZigBee 串口配置
16.      OptionUtils.zigbeeCom = String.valueOf(zigbeeCom.getValue());
17.
18.      OptionUtils.zigbeeBt = Integer.parseInt(String.valueOf(zigbeeBt.
            getValue()));
19.
20.      // 保存传输口配置
21.      OptionUtils.properties.setProperty("yellowDO", String.valueOf
           (yellowDO.getValue()));
22.
23.      OptionUtils.properties.setProperty("lightDO", String.valueOf
           (lightDO.getValue()));
24.
25.      OptionUtils.properties.setProperty("blowerDO", String.valueOf
          (blowerDO.getValue()));
26.
27.      OptionUtils.properties.setProperty("greenDO", String.valueOf
           (greenDO.getValue()));
28.
29.      OptionUtils.properties.setProperty("redDO", String.valueOf
           (redDO.getValue()));
30.
```

```
31.     OptionUtils.properties.setProperty("humanDI", String.valueOf
          (humanDI.getValue()));
32.
33.     OptionUtils.properties.setProperty("smokeDI", String.valueOf
          (smokeDI.getValue()));
34.
35.     OptionUtils.properties.setProperty("microswitchDI", String.valueOf
          (microswitchDI.getValue()));
36.
37.     OptionUtils.properties.setProperty("limitswitchDI", String.valueOf
          (limitswitchDI.getValue()));
38.
39.     OptionUtils.properties.setProperty("infraredDI", String.valueOf
          (infraredDI.getValue()));
40.
41.     OptionUtils.properties.setProperty("opendoorDO", String.valueOf
          (opendoorDO.getValue()));
42.
43.     OptionUtils.properties.setProperty("closedoorDO", String.valueOf
          (closedoorDO.getValue()));
44.
45.     OptionUtils.properties.store(new FileOutputStream(new File("resources/
          files/options.properties")), null);
46.
47.     initOptions();
48.
49.     Alert tip = new Alert(AlertType.INFORMATION);
50.
51.     tip.setTitle("提示");
52.
53.     tip.setContentText("参数配置成功");
54.
55.     tip.show();
56.
57. }
```

4. 控制报警灯

当设备监控到发生紧急情况的时候，应用自动触发对应报警灯，并通知监控界面。触发条件的判断就在采集线程中进行，每次采集到最新数据都判断是否需要报警。

在 com.nle.thread 目录下创建一条专门控制报警灯闪烁的线程 FlickTask，当需要触发报警灯时，只要修改一个标记即可让报警灯闪烁。

```
1.  /**
2.   * 报警灯闪烁线程
3.   * @author admin
4.   *
5.   */
6.  public class FlickTask implements Runnable {
7.      //报警总开关
8.      private boolean alarm;
9.      //红色是否报警
10.     private boolean redAlarm;
11.     //绿色是否报警
12.     private boolean greenAlarm;
13.     //黄色是否报警
14.     private boolean yellowAlarm;
```

```
15.     //顺序闪烁是否报警
16.     private boolean mixAlarm;
17.
18.     private SerialPort portAdam;
19.
20.     private AlarmLampManager lampManager;
21.
22.     private boolean flickStop;
23.
24.     public FlickTask(SerialPort portAdam) {
25.        this.portAdam = portAdam;
26.        alarm = true;
27.        lampManager = new AlarmLampManager(portAdam);
28.     }
29.
30.     @Override
31.     public void run() {
32.        while(!flickStop) {
33.           while(alarm) {
34.              try {
35.                 if(redAlarm) {
36.                    lampManager.lampOn('r');
37.                    Thread.sleep(500);
38.                    lampManager.lampOff('r');
39.                    Thread.sleep(500);
40.                 }
41.                 if(greenAlarm) {
42.                    lampManager.lampOn('g');
43.                    Thread.sleep(500);
44.                    lampManager.lampOff('g');
45.                    Thread.sleep(500);
46.                 }
47.                 if(yellowAlarm) {
48.                    lampManager.lampOn('y');
49.                    Thread.sleep(500);
50.                    lampManager.lampOff('y');
51.                    Thread.sleep(500);
52.                 }
53.                 if(mixAlarm) {
54.                    lampManager.lampOn('r');
55.                    Thread.sleep(200);
56.                    lampManager.lampOn('g');
57.                    Thread.sleep(200);
58.                    lampManager.lampOn('y');
59.                    Thread.sleep(1000);
60.                    lampManager.lampOff('r');
61.                    Thread.sleep(200);
62.                    lampManager.lampOff('g');
63.                    Thread.sleep(200);
64.                    lampManager.lampOff('y');
65.                    Thread.sleep(1000);
66.                 }
67.              } catch (InterruptedException e) {
68.                 // TODO Auto-generated catch block
69.                 e.printStackTrace();
70.              }
71.              try {
72.                 Thread.sleep(100);
73.              } catch (InterruptedException e) {
74.                 // TODO Auto-generated catch block
```

```
75.                     e.printStackTrace();
76.                 }
77.             }
78.             try {
79.                 Thread.sleep(100);
80.             } catch (InterruptedException e) {
81.                 // TODO Auto-generated catch block
82.                 e.printStackTrace();
83.             }
84.         }
85.     }
86.
87.     public boolean isRedAlarm() {
88.         return redAlarm;
89.     }
90.     public void setRedAlarm(boolean redAlarm) {
91.         this.redAlarm = redAlarm;
92.     }
93.     public boolean isGreenAlarm() {
94.         return greenAlarm;
95.     }
96.     public void setGreenAlarm(boolean greenAlarm) {
97.         this.greenAlarm = greenAlarm;
98.     }
99.     public boolean isYellowAlarm() {
100.        return yellowAlarm;
101.    }
102.    public void setYellowAlarm(boolean yellowAlarm) {
103.        this.yellowAlarm = yellowAlarm;
104.    }
105.    public boolean isMixAlarm() {
106.        return mixAlarm;
107.    }
108.    public void setMixAlarm(boolean mixAlarm) {
109.        this.mixAlarm = mixAlarm;
110.    }
111.    public SerialPort getPortAdam() {
112.        return portAdam;
113.    }
114.    public void setPortAdam(SerialPort portAdam) {
115.        this.portAdam = portAdam;
116.    }
117.    public boolean isAlarm() {
118.        return alarm;
119.    }
120.    public void setAlarm(boolean alarm) {
121.        this.alarm = alarm;
122.    }
123.    public boolean isFlickStop() {
124.        return flickStop;
125.    }
126.    public void setFlickStop(boolean flickStop) {
127.        this.flickStop = flickStop;
128.    }
129. }
```

5. 设备数据采集和控制

设备的各项参数配置好后，用户在设备数据界面单击“开始采集”按钮，即可开始通过线程定时采集设备数据，并将数据更新到监控界面的控件上，实现设备监控功能。采集过程中检测到红外对射有人需要入园时，发送指令让推杆模拟开门。当设备数据超过设定标准时，例如温度超过用户设定温度或有火焰时，触发相应报警灯，并自动打开风扇。同时，用户也可通过界面下方的不同按钮控制不同的设备，也可以控制安防检测的设防和撤防。

(1) 手动控制按钮必须在用户单击“开始采集”按钮之后才可使用。在 FunctionViewController 中添加 btnHandler 方法，控制开关按钮的可用状态。在 initialize 方法中先将按钮设置为不可用。

```
1.   /**
2.    * 手动控制按钮是否可用
3.    * @param clickable
4.    */
5.   private void btnHandler(boolean clickable) {
6.       lampBtn.setDisable(clickable);
7.       blowerBtn.setDisable(clickable);
8.       redlampBtn.setDisable(clickable);
9.       yellowlampBtn.setDisable(clickable);
10.      greenlampBtn.setDisable(clickable);
11.      humanAlarmBtn.setDisable(clickable);
12.  }
```

(2) 在 src 目录下创建 com.nle.listener 包，创建 4150 采集器的监听类 AdamListener，负责人体、烟雾传感器数据，当检测到人体或烟雾时，触发对应报警灯。

```
1.   /**
2.    * 4150 采集器监听
3.    * @author admin
4.    *
5.    */
6.   public class AdamListener implements SerialPortEventListener{
7.
8.       private SerialPort port;
9.
10.      private FunctionViewController functionController;
11.
12.      private FlickTask ft;
13.
14.      private Label redAlarmPic;
15.
16.      private Label greenAlarmPic;
17.
18.      private Label yellowAlarmPic;
19.
20.      public AdamListener(SerialPort port, FunctionViewController
           functionController) {
21.
22.          this.port = port;
23.
24.          this.functionController = functionController;
25.
26.          ft = functionController.getFlickTask();
27.
28.          redAlarmPic = functionController.getAlarmPic1();
29.
30.          greenAlarmPic = functionController.getAlarmPic2();
31.
```

```
32.         yellowAlarmPic = functionController.getAlarmPic3();
33.     }
34.
35.     @Override
36.     public void serialEvent(SerialPortEvent arg0) {
37.
38.         try {
39.             switch(arg0.getEventType())
40.             {
41.             case SerialPortEvent.DATA_AVAILABLE:
42.                 if(port!=null)
43.                 {
44.
45.                     byte[] res=SerialPortManager.readFromPort(port);
46.                     String s= ByteUtils.byteArrayToHexString(res);
47.                     //System.out.println("从串口收到的数据是="+s);
48.                     //如果是采集数据指令响应数据
49.                     if(s.substring(2, 4).equals("01")) {
50.
51.                         String msg = s.substring(6, 8);
52.
53.                         //获取每个通道数值
54.                         int result = Integer.parseInt(msg, 16);
55.
56.                         String str = ByteUtils.toBinary7(result);
57.
58.                         String index = "6543210";
59.                         //判断微动开关
60.                         int indexMicroswitch = index.indexOf
                               (OptionUtils.microswitchDI.charAt(2));
61.                         if(str.charAt(indexMicroswitch)=='1') {
62.                             if(CloseDoorTask.openAble) {
63.                                 SerialPortManager.sendToPort(port, OptionUtils.
                                      controlCmd.get(OptionUtils.opendoorDO)[0]);
64.                             }
65.                         }
66.
67.                         //判断行程开关，推杆顶到行程开关时，关闭开门继电器，停止开门
68.                         int indexLimitswitch = index.indexOf(OptionUtils.
                              limitswitchDI.charAt(2));
69.                         if(str.charAt(indexLimitswitch)=='1') {
70.                             SerialPortManager.sendToPort(port, OptionUtils.
                                  controlCmd.get(OptionUtils.opendoorDO)[1]);
71.                             if(CloseDoorTask.closeAble) {
72.                                 //开启关门线程，15 秒后自动关门
73.                                 Thread closeDoor = new Thread(new
                                      CloseDoorTask(port));
74.                                 closeDoor.start();
75.                             }
76.
77.                         }
78.
79.                         //判断红外对射
80.                         int indexInfrared = index.indexOf(OptionUtils.
                              infraredDI.charAt(2));
81.                         if(str.charAt(indexInfrared)=='0') {
82.                             if(CloseDoorTask.openAble) {
83.                                 SerialPortManager.sendToPort(port, OptionUtils.
                                      controlCmd.get(OptionUtils.opendoorDO)[0]);
```

```
84.                     }
85.                 }
86.
87.                 //判断人体入侵
88.                 if(!functionController.getHumanAlarmBtn().isSelected()) {
89.                     int indexHuman = index.indexOf(OptionUtils.
                         humanDI.charAt(2));
90.                     if(str.charAt(indexHuman)=='0') {
91.                         openAlarmLamp('r');
92.
93.                     }else {
94.                         closeAlarmLamp('r');
95.                     }
96.                 }
97.
98.                 //判断烟雾 DI1 口
99.                 if(str.charAt(5)=='1') {
100.
101.                    openAlarmLamp('g');
102.
103.                }else {
104.
105.                    closeAlarmLamp('g');
106.
107.                }
108.
109.            }
110.
111.        }
112.    }
113.    } catch (NumberFormatException e) {
114.        // TODO Auto-generated catch block
115.        e.printStackTrace();
116.    } catch (FileNotFoundException e) {
117.        // TODO Auto-generated catch block
118.        e.printStackTrace();
119.    } catch (SendDataToSerialPortFailure e) {
120.        // TODO Auto-generated catch block
121.        e.printStackTrace();
122.    } catch (SerialPortOutputStreamCloseFailure e) {
123.        // TODO Auto-generated catch block
124.        e.printStackTrace();
125.    }
126.
127. }
128.
129. /**
130.  *
131.  * 开启报警灯，并通知界面
132.  *
133.  * @param lampColor 报警颜色
134.  * @throws FileNotFoundException
135.  */
136. public void openAlarmLamp(char lampColor) throws FileNotFoundException {
137.
138.     switch (lampColor) {
139.     case 'r':
140.         ft.setRedAlarm(true);
141.         break;
```

```
142.         case 'g':
143.             ft.setGreenAlarm(true);
144.             break;
145.         case 'y':
146.             ft.setYellowAlarm(true);
147.             break;
148.         }
149.         //更新报警灯图片
150.         Image image = new Image(new FileInputStream(new File("resources/
               images/bjd_on.gif")), 72, 105, false, false);
151.         //子线程更新 JavaFX 界面时不能直接更新
152.         Platform.runLater(new Runnable() {
153. 
154.             @Override
155.             public void run() {
156. 
157.                 switch (lampColor) {
158.                 case 'r':
159.                     redAlarmPic.setGraphic(new ImageView(image));
160.                     break;
161.                 case 'g':
162.                     greenAlarmPic.setGraphic(new ImageView(image));
163.                     break;
164.                 case 'y':
165.                     yellowAlarmPic.setGraphic(new ImageView(image));
166.                     break;
167.                 }
168. 
169.             }
170.         });
171. 
172.     }
173. 
174.     /**
175.      *
176.      * 关闭报警灯，并通知界面
177.      *
178.      * @param lampColor 报警颜色
179.      * @throws FileNotFoundException
180.      */
181.     public void closeAlarmLamp(char lampColor) throws
       FileNotFoundException {
182. 
183.         switch (lampColor) {
184.         case 'r':
185.             ft.setRedAlarm(false);
186.             break;
187.         case 'g':
188.             ft.setGreenAlarm(false);
189.             break;
190.         case 'y':
191.             ft.setYellowAlarm(false);
192.             break;
193.         }
194.         //更新报警灯图片
195.         Image image = new Image(new FileInputStream(new File("resources/
               images/bjd_off.png")), 72, 105, false, false);
196.         //子线程更新 JavaFX 界面时不能直接更新
197.         Platform.runLater(new Runnable() {
```

```
198.
199.            @Override
200.            public void run() {
201.                switch (lampColor) {
202.                case 'r':
203.                    redAlarmPic.setGraphic(new ImageView(image));
204.                    break;
205.                case 'g':
206.                    greenAlarmPic.setGraphic(new ImageView(image));
207.                    break;
208.                case 'y':
209.                    yellowAlarmPic.setGraphic(new ImageView(image));
210.                    break;
211.                }
212.            }
213.        });
214.
215.    }
216. }
```

(3) 在 com.nle.listener 包中创建 ZigBee 协调器数据采集监听类 ZigBeeListener。源码在配套资源/SmartPark/com/nle/listener 目录下。

(4) 在 src 目录下创建 com.nle.thread 包，创建 4150 采集器数据采集线程 CheckTask，定时发送采集指令。

```
1.   /**
2.    * 4150 采集器采集指令发送线程
3.    * @author admin
4.    */
5.   public class CheckTask implements Runnable{
6.       private SerialPort portAdam;
7.       private boolean checkStop;
8.       public CheckTask(SerialPort portAdam) {
9.           this.portAdam = portAdam;
10.      }
11.
12.      @Override
13.      public void run() {
14.          while(!checkStop) {
15.              try {
16.                  byte[] order = new byte[] {0x01, 0x01, 0x00, 0x00, 0x00, 0x07,
                     0x7D, (byte) 0xC8};
17.                  SerialPortManager.sendToPort(portAdam, order);
18.                  Thread.sleep(1000);
19.              } catch (InterruptedException e) {
20.                  // TODO Auto-generated catch block
21.                  e.printStackTrace();
22.              } catch (SendDataToSerialPortFailure e) {
23.                  // TODO Auto-generated catch block
24.                  e.printStackTrace();
25.              } catch (SerialPortOutputStreamCloseFailure e) {
26.                  // TODO Auto-generated catch block
27.                  e.printStackTrace();
28.              }
29.          }
30.      }
31.      public boolean isCheckStop() {
32.          return checkStop;
```

```
33.     }
34.     public void setCheckStop(boolean checkStop) {
35.         this.checkStop = checkStop;
36.     }
37. }
```

(5) 在FunctionViewController中定义两个全局的SerialPort属性，分别为portAdam和portZigBee。在“开始采集”按钮的监听方法中根据配置信息初始化两个串口对象，并添加串口监听器。

```
1.  /**
2.  * 采集控制按钮
3.  * @throws FileNotFoundException
4.   */
5.  @FXML
6.  private void handleCollect() {
7.      if(collectBtn.getText().equals("开始采集")) {
8.          if(StringUtil.checkNotNull(OptionUtils.adamCom)&&!"null".
               equals(OptionUtils.adamCom)) {
9.              collectBtn.setText("停止采集");
10.             try {
11.                 portAdam = SerialPortManager.openPort(OptionUtils.adamCom,
                       OptionUtils.adamBt);
12.                 portZigBee = SerialPortManager.openPort
                       (OptionUtils.zigbeeCom, OptionUtils.zigbeeBt);
13.                 //开启报警灯闪烁线程
14.                 flickTask = new FlickTask(portAdam);
15.                 Thread threadFlick = new Thread(flickTask);
16.                 threadFlick.start();
17.                 //开启检测任务
18.                 checkTask = new CheckTask(portAdam);
19.                 Thread threadCheckHuman = new Thread(checkTask);
20.                 threadCheckHuman.start();
21.                 //添加串口监听
22.                 SerialPortManager.addListener(portAdam, new AdamListener
                       (portAdam, this));
23.                 //添加ZigBee监听
24.                 SerialPortManager.addListener(portZigBee, new ZigBeeListener
                       (portZigBee, portAdam, this));
25.                 btnHandler(false);
26.             } catch (SerialPortParameterFailure e) {
27.                 // TODO Auto-generated catch block
28.                 e.printStackTrace();
29.             } catch (NotASerialPort e) {
30.                 // TODO Auto-generated catch block
31.                 e.printStackTrace();
32.             } catch (NoSuchPort e) {
33.                 // TODO Auto-generated catch block
34.                 e.printStackTrace();
35.             } catch (PortInUse e) {
36.                 // TODO Auto-generated catch block
37.                 e.printStackTrace();
38.             } catch (TooManyListeners e) {
39.                 // TODO Auto-generated catch block
40.                 e.printStackTrace();
41.             }
42.         }else {
43.             alert.setContentText("串口未设置");
```

```
44.             alert.show();
45.         }
46.     }else {
47.         collectBtn.setText("开始采集");
48.         SerialPortManager.closePort(portAdam);
49.         SerialPortManager.closePort(portZigBee);
50.         //关闭报警灯闪烁线程
51.         flickTask.setFlickStop(true);
52.         //关闭检测任务
53.         checkTask.setCheckStop(true);
54.         //添加串口监听
55.         SerialPortManager.removeListener(portAdam);
56.         //添加 ZigBee 监听
57.         SerialPortManager.removeListener(portZigBee);
58.         btnHandler(true);
59.     }
60. }
```

6. 手动控制设备

在功能界面的控制器 FunctionViewController 中添加手动控制按钮的监听。用户单击按钮时切换报警灯、照明灯、风扇的开关状态。

```
1.  /**
2.   * 单击开灯按钮调用
3.   */
4.  @FXML
5.  private void handleLamp() {
6.      try {
7.          byte[] order = null;
8.          if(lampBtn.isSelected()) {
9.              order = OptionUtils.controlCmd.get(OptionUtils.lightDO)[0];
10.         }else {
11.             order = OptionUtils.controlCmd.get(OptionUtils.lightDO)[1];
12.         }
13.         SerialPortManager.sendToPort(portAdam, order);
14.     } catch (Exception e) {
15.         e.printStackTrace();
16.     }
17.
18. }
19.
20. /**
21.  * 单击风扇按钮调用
22.  */
23. @FXML
24. private void handleBlower() {
25.     try {
26.         byte[] order = null;
27.         if(blowerBtn.isSelected()) {
28.             order = OptionUtils.controlCmd.get(OptionUtils.blowerDO)[0];
29.         }else {
30.             order = OptionUtils.controlCmd.get(OptionUtils.blowerDO)[1];
31.         }
32.         SerialPortManager.sendToPort(portAdam, order);
33.     } catch (Exception e) {
34.         e.printStackTrace();
35.     }
36. }
```

```
37.
38.  /**
39.   * 控制红色报警灯
40.   */
41.  @FXML
42.  private void handleRedLamp() {
43.      try {
44.          byte[] order = null;
45.          if(redlampBtn.isSelected()) {
46.              order = OptionUtils.controlCmd.get(OptionUtils.redDO)[0];
47.          }else {
48.              order = OptionUtils.controlCmd.get(OptionUtils.redDO)[1];
49.          }
50.          SerialPortManager.sendToPort(portAdam, order);
51.      } catch (Exception e) {
52.          e.printStackTrace();
53.      }
54.  }
55.  /**
56.   * 控制黄色报警灯
57.   */
58.  @FXML
59.  private void handleYellowLamp() {
60.      try {
61.          byte[] order = null;
62.          if(yellowlampBtn.isSelected()) {
63.              order = OptionUtils.controlCmd.get(OptionUtils.yellowDO)[0];
64.          }else {
65.              order = OptionUtils.controlCmd.get(OptionUtils.yellowDO)[1];
66.          }
67.          SerialPortManager.sendToPort(portAdam, order);
68.      } catch (Exception e) {
69.          e.printStackTrace();
70.      }
71.  }
72.  /**
73.   * 控制绿色报警灯
74.   */
75.  @FXML
76.  private void handleGreenLamp() {
77.      try {
78.          byte[] order = null;
79.          if(greenlampBtn.isSelected()) {
80.              order = OptionUtils.controlCmd.get(OptionUtils.greenDO)[0];
81.          }else {
82.              order = OptionUtils.controlCmd.get(OptionUtils.greenDO)[1];
83.          }
84.          SerialPortManager.sendToPort(portAdam, order);
85.      } catch (Exception e) {
86.          e.printStackTrace();
87.      }
88.  }
```

7. 运行结果

检查设备连接无误后，运行 MainApp.java 程序，输入管理员账户(用户名：admin，密码：123456)，单击“登录”按钮后，登录成功自动跳转到功能页面与参数配置页面，如图 12-14 所示。

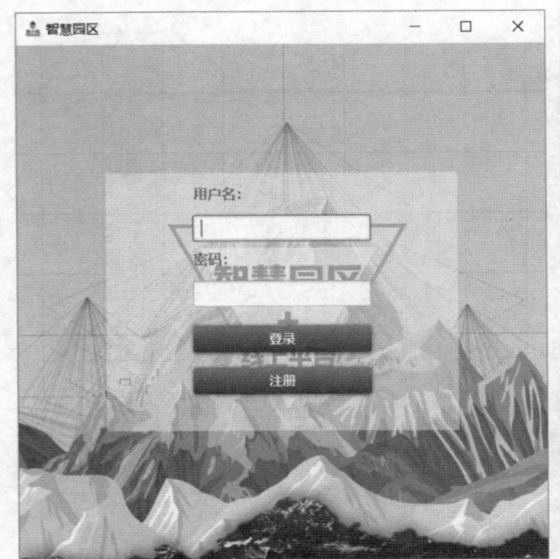
智慧园区
用户名:
密码:
登录
注册

智慧园区注册
填写密码
确认密码
填写手机号
+
线上平台
确认注册
返回

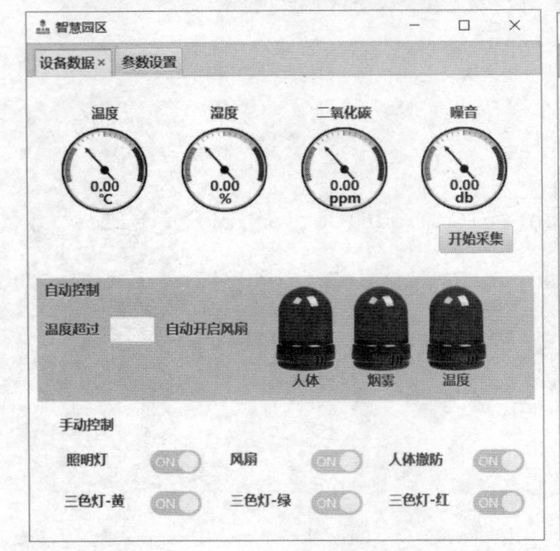
智慧园区
设备数据 ×
参数设置
温度
湿度
二氧化碳
噪音
0.00 ℃
0.00 %
0.00 ppm
0.00 db
开始采集
自动控制
温度超过
自动开启风扇
人体
烟雾
温度
手动控制
照明灯
风扇
人体撤防
三色灯-黄
三色灯-绿
三色灯-红
ON

智慧园区
设备数据
参数设置 ×
COM口配置
4150串口
请选择
波特率
9600
SAVE
zigbee串口
请选择
波特率
384...
DO配置
DI配置
报警灯-黄
DO4
人体
DI1
报警灯-红
DO3
烟雾
DI0
报警灯-绿
DO2
照明灯
DO0
风扇
DO1
微动开关
DI2
推杆-伸
DO5
行程开关
DI3
推杆-缩
DO6
红外对射
DI4

图 12-14　运行结果

智慧园区环境实时监测（云平台篇）

【项目描述】

本项目实现园区环境的实时监测。监测的数据来源有两种，第一种是采集的传感数据经 LoRa 网关上传到物联网云平台，第二种是采集到的传感数据经 NB-IoT 上传到物联网云平台。

第一种数据流向：利用物联网工程实训系统 2.0 中的 PC 端智能环境云软件生成模拟的传感器数据，通过串口下发到 6 个模拟传感器 NewSensor(简称 NS)的设备上。NS 通过 LoRa 通道 LoRa 节点把数据传递给 LoRa 网关。LoRa 网关通过轮询方式采集 NS 传感器数据并将数据上传至物联网云服务平台，并由云平台长期保存。Java 客户端程序从云平台取数据进行展示。

第二种数据流向：NB-IoT 节点上接传感器，经过采集后通过 NB-IoT 网络上传至物联网云平台，并由云平台长期保存。Java 客户端程序从云平台取数据进行展示。

本项目任务清单如表 13-1 所示。

表 13-1 任务清单

任务课时	建议实训	任务组员数量	建议 3 人
任务组采用设备	4 个 New Sensor 通用版 2 个 NewSensor LoRa 版 1 个 NB-IoT 节点 1-空气质量传感器模块 1 个 NB-IoT 节点 2-可燃气体传感器模块 2 个 LoRa 节点 1 个路由器 1 个 LoRa 网关 1 个 RS232- RS485 转接头 电源、USB 转串口、连接线若干		

【拓扑图】

本项目拓扑图如图 13-1 所示。

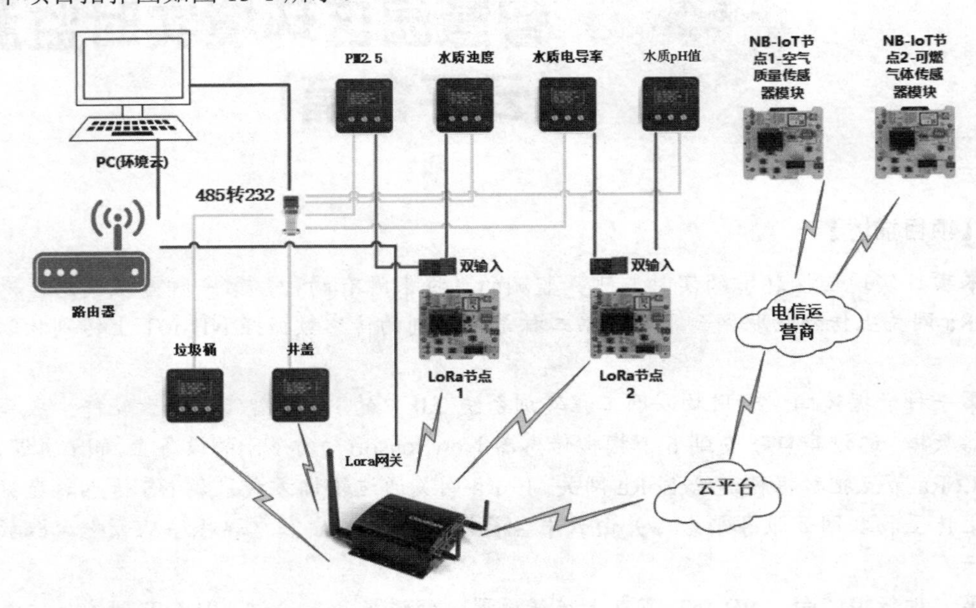

图 13-1 拓扑图

【技能目标】

- 能熟练使用 JavaFX 设计制作 UI。
- 能综合应用 Java 的网络通信技术。
- 能综合应用 Java 的多线程技术。
- 能综合应用 Java 的串口通信技术。
- 能编写与物联网云服务平台交互的 Java 程序。

任务清单如表 13-2 所示。

表 13-2 任务清单

任务课时	12 课时	任务组员数量	建议 3 人
任务组采用设备	PC 安装好智能环境云 6 个 NS 模块 2 个 LoRa 节点 1 个 LoRa 网关 2 个 NB-IoT 节点		

【项目实施】

1. 设备配置注意事项

本项目用到比较多的设备，所有的设备连接、烧写、配置需要有物联网工程相关知识，这里不展开叙述。为确保项目成功实施，需注意以下事项。

(1) LoRa 网关配置注意事项。

LoRa 网关参数配置如图 13-2 所示。

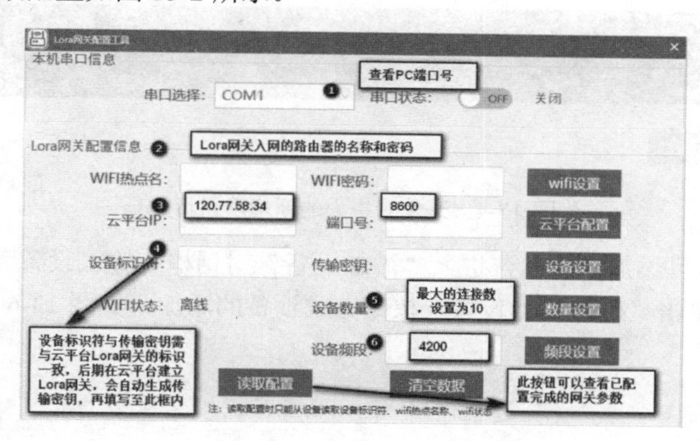

图 13-2 LoRa 网关参数配置

(2) NS 配置注意事项。

NS 传感器的数据来源于环境云，请参考物联网工程实训系统中关于智能环境云的设置，用 NS 配置工具配置 NS 时具体的传感器名、标识名、地址如图 13-3 所示。

2. 搭建物联网云服务平台项目

(1) 注册与登录物联网云平台。

(请参考物联网工程实训系统中的操作步骤，此处略)。

(2) 新增项目和添加网关设备。

登录成功后，进入开发者中心的新增项目页面，如图 13-4 所示。

步骤 1：在开发者中心的项目列表页面上，单击左上角的“新增项目”按钮，在弹出的“添加项目”对话框中设置项目名称和行业类别，“行业类别”设置为“智慧城市”，单击“下一步”按钮完成新增项目的操作，如图 13-5 所示。

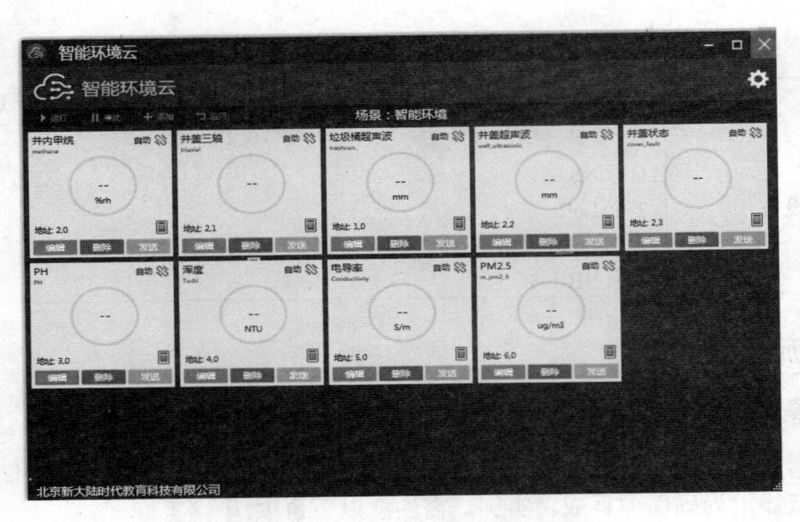

图 13-3　智能环境云配置参数界面

图 13-4　开发者中心的新增项目页面

步骤 2：新项目添加完成后，进入“添加设备”对话框，输入设备名称和设备标识(标识名必须唯一)，单击“确定添加设备”按钮完成设备的添加，如图 13-6 所示。

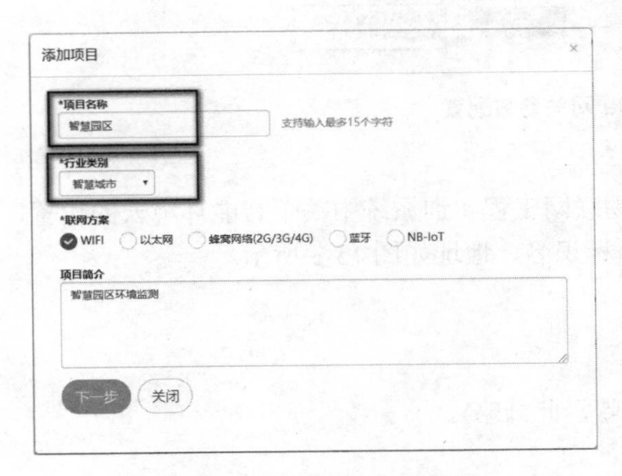

图 13-5　“添加项目”对话框

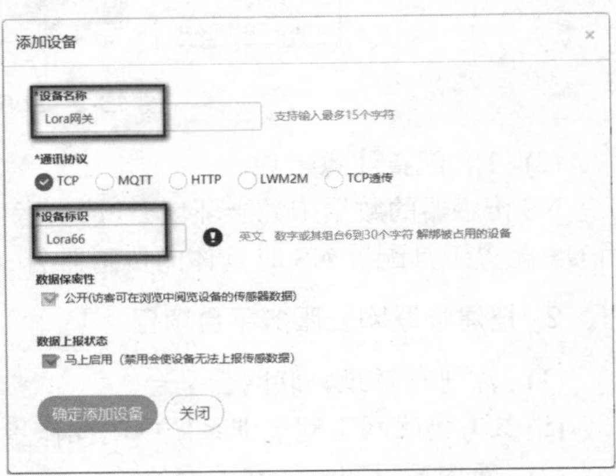

图 13-6　“添加设备”对话框

步骤 3：添加设备完成后，自动返回到开发中心的项目列表页面。在此页面上可以看到新增的项目，此项目中有一个新增设备，如图 13-7 所示。

步骤 4：单击 1个设备 0个传感器 按钮，进入设备列表页面，可查看新添加的设备，如图 13-8 所示。

图 13-7 开发者中心的项目列表页面

图 13-8 项目的设备列表页面

步骤 5： 在项目的设备列表页面上，单击“添加设备”按钮，再添加其他设备“一氧化碳(NB)”和 “可燃气(NB)”，如图 13-9 和图 13-10 所示。

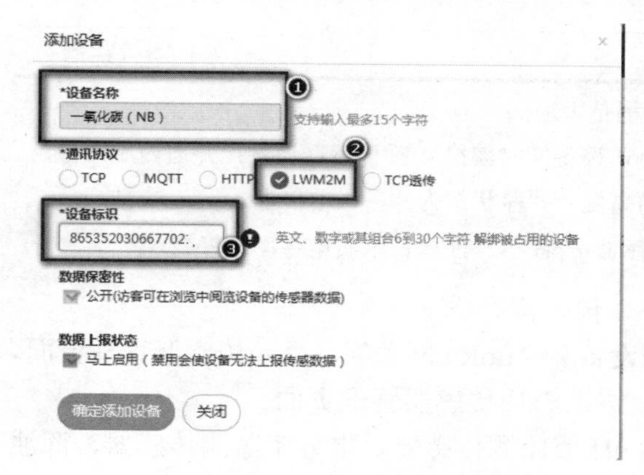

图 13-9 添加一氧化碳(NB)设备

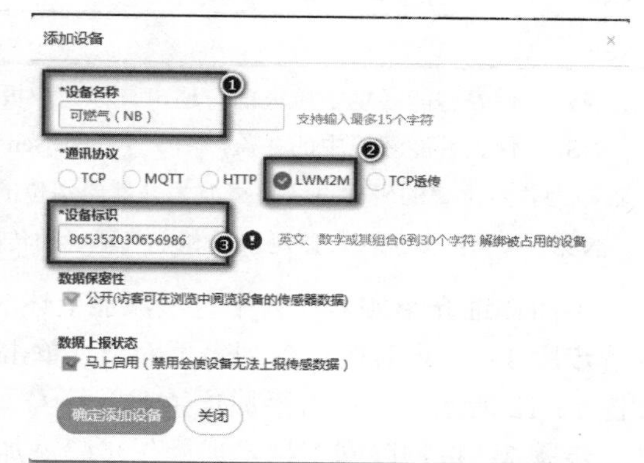

图 13-10 添加可燃气(NB)设备

步骤 6： 设备添加完成后，再查看项目的设备列表页面，如图 13-11 所示。

图 13-11 项目的设备列表页面

(3) 给 LoRa 网关设备添加传感器。

本项目使用 6 个 NewSensor(简称 NS)设备用于模拟真实传感器，其中 NS1、NS2 为 LoRa 版，NS3～NS6 为 AO 版。需要注意的是：采用 LoRa 模式的 NS1 和 NS2(内置 LoRa 节点)中数据直接被 LoRa 网关采集，采用 AO 模式的 NS3～NS6 输出 4～20mA 电流(需外加 LoRa 节点)，才能被 LoRa 网关采集。

6 个 NewSensor 设备的说明如表 13-3 所示。

表 13-3　NS 设备说明

<table>
<tr><th>设备名称</th><th colspan="4">传感器名称</th><th>传输方式</th><th>监测场合</th></tr>
<tr><td>NS1</td><td colspan="4">超声波</td><td rowspan="2">LoRa 传输</td><td>垃圾桶</td></tr>
<tr><td>NS2</td><td>甲烷</td><td>三轴</td><td>超声波</td><td>井盖状态</td><td>井盖</td></tr>
<tr><td>NS3</td><td colspan="4">PH</td><td rowspan="4">AO 传输</td><td rowspan="3">水质环境</td></tr>
<tr><td>NS4</td><td colspan="4">浑度</td></tr>
<tr><td>NS5</td><td colspan="4">水质电导率</td></tr>
<tr><td>NS6</td><td colspan="4">PM2.5</td><td>空气环境</td></tr>
</table>

注：NS1：代表智能环境中垃圾桶，超声波标识垃圾桶是否溢满。

NS2：代表智能环境中的井盖，用一个 NewSensor 设备同时模拟 4 种传感器设备：井盖内甲烷值，三轴即为井盖位置的坐标点，超声波为井盖内水位的高度，井盖状态为井盖当前信息。

NS3～NS4：分别代表监测水质和空气环境的传感器：pH 值、浊度、水质电导率、PM2.5。

下面详细介绍如何在云平台上添加上述 NS 传感器。

步骤 1：在项目的设备列表页面上，单击设备名“LoRa 网关”，进入传感器管理页面，如图 13-12 所示，再单击添加传感器图标 ⊕，进入添加传感器管理页面。

步骤 2：按物联网工程实训操作步骤添加 pH 值监测传感器、电导率监测传感器、浑浊度监测传感器、PM2.5 值监测传感器、井盖状态监测传感器、井盖超声波监测传感器、井盖三轴监测传感器、井内甲烷监测传感器、智能垃圾桶监测传感器等，添加结果如图 13-13 所示。

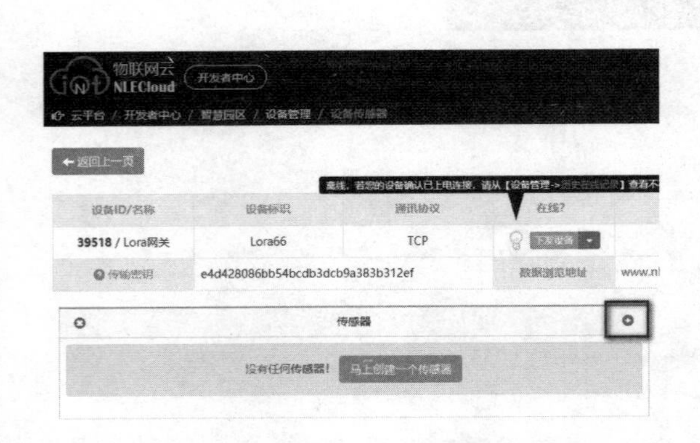

图 13-12　LoRa 网关设备的传感器管理页面

图 13-13　添加传感器

(4) 给 NB-IoT 设备添加传感器(一氧化碳)。

步骤 1：在项目的设备列表页面上，单击设备名“一氧化碳(NB)”，进入传感器管理页面，当新增设备时采用 LWM2M 通信协议设备，系统默认会添加两个传感器，如图 13-14 所示。

步骤 2：本设备只监测一氧化碳值(标识名 CarbonMonoxidesStr)，所以可以删除标识名为 FlammableGas 的传感器，单击此传感器右边的 ✪ 图标删除，删除后如图 13-15 所示。

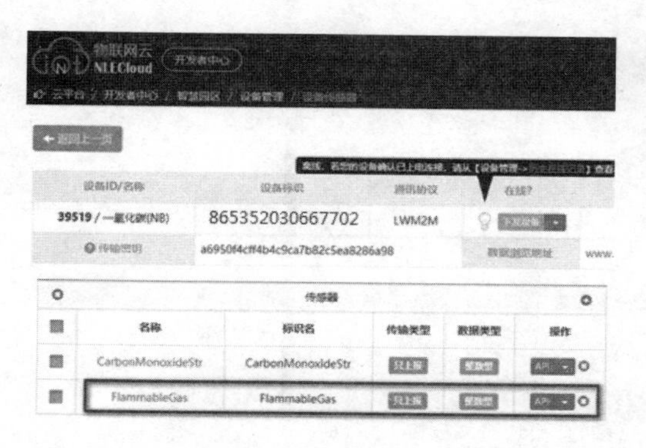

图 13-14 一氧化碳(NB)设备的传感器管理页面(1)

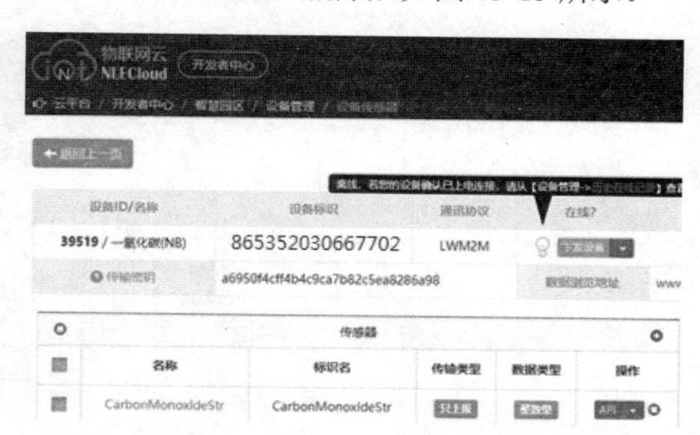

图 13-15 一氧化碳(NB)设备的传感器管理页面(2)

(5) 给 NB-IoT 设备添加传感器(可燃气)。

步骤 1：在项目的设备列表页面上，单击设备名“可燃气(NB)”，进入传感器管理页面，当新增设备时采用 LWM2M 通信协议，系统默认会添加两个传感器，如图 13-16 所示。

步骤 2：本设备只监测可燃气值(标识名 FlammableGas)，所以可以删除标识名为 CarbonMonoxidesStr 的传感器，单击此传感器右边的 ✪ 图标删除，删除后如图 13-17 所示。

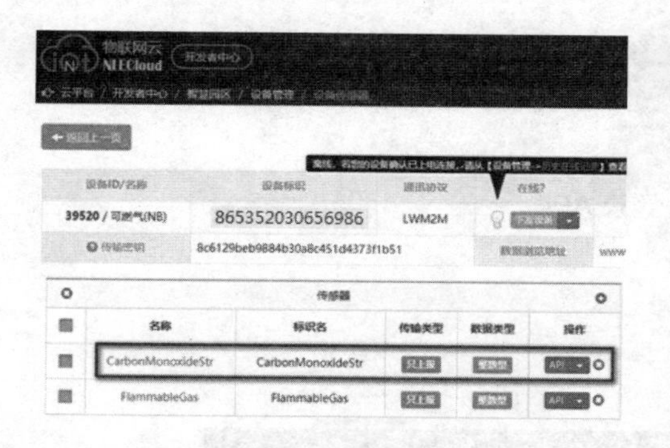

图 13-16 可燃气(NB)设备的传感器管理页面(1)

图 13-17 可燃气(NB)设备的传感器管理页面(2)

3. 云平台 API 在线调试工具的使用

要登录并获取云平台数据，需要调用云平台提供的一套 API。接口概览网址为 http://api.nlecloud.com/doc/api，如图 13-18 所示。

单击左边列表框中选项，在右边可以得到请求方式和地址以及参数说明等，比如，单击“登录账号 API”，结果如图 13-19 所示。

Java 程序从云平台获取数据，可以参照相关格式进行数据请求并从响应中获取对应的请求结果。为了让读者明白请求和响应结果的用法，先利用云平台提供的 API 接口进行数据测试，以验证传感器数据确实上传到了云平台，接下来详细介绍 API 在线调试工具的使用。

图 13-18　云平台接口概览页面

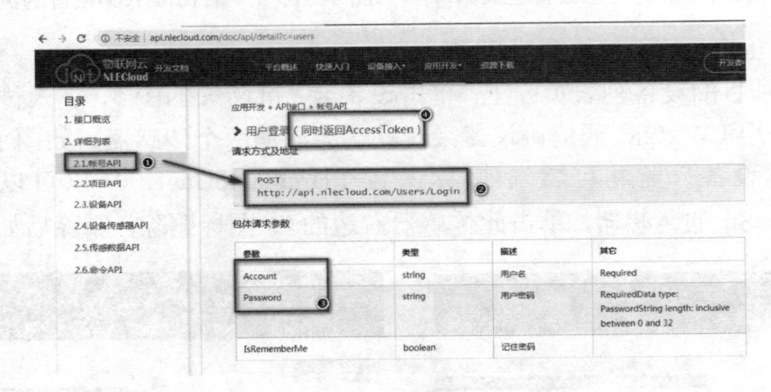

图 13-19　登录 API

(1)　测试登录云平台是否成功。

打开 API 在线调试工具，如图 13-20 所示。

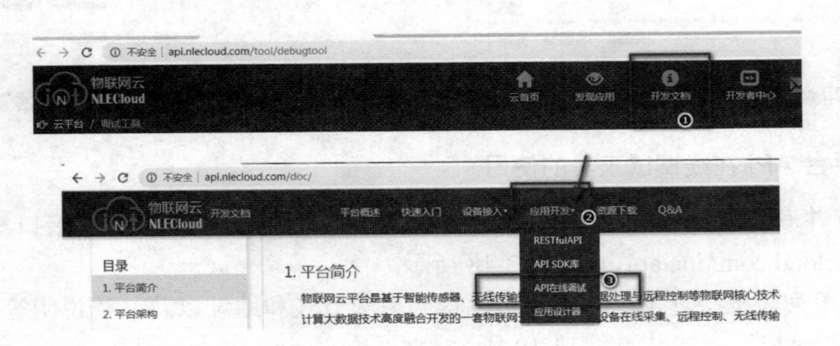

图 13-20　进入 API 在线调试

从“登录账号 API”中可以看出，用户名和密码发送到云平台，走的是 POST 方式，如果成功则返回 AccessToken 值，之后的每个请求都必须携带 AccessToken 值，所以需要保存这个 AccessToken 值，如图 13-21 所示。

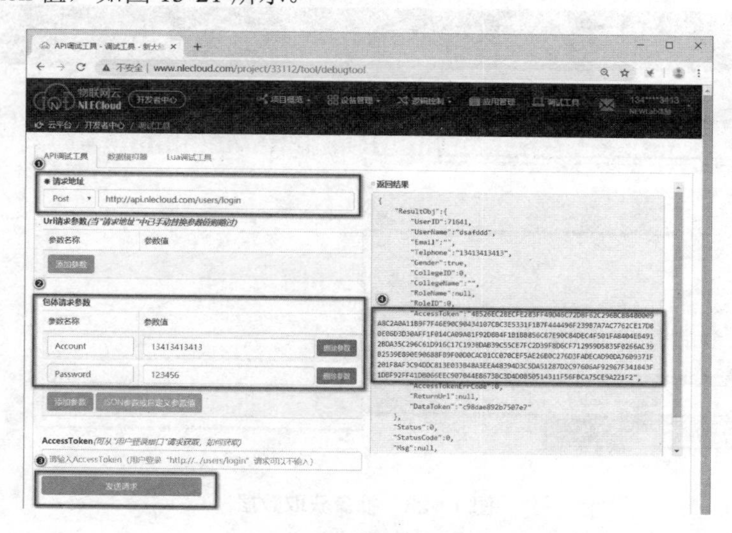

图 13-21　测试登录

图 13-21 中①选择 Post，②输入登录云平台的账号和密码(注意要输入自己的账号和密码，此处是模拟的账号和密码)，单击发送请求，④是返回结果，如果登录成功，把返回结果中的 AccessToken 值复制并粘贴到③处。

(2)　测试批量查询设备最新数据。

有了 AccessToken 值后，如果要批量获取设备最新数据，可以选择“设备 API”，查阅请求方式和地址，如图 13-22 所示。

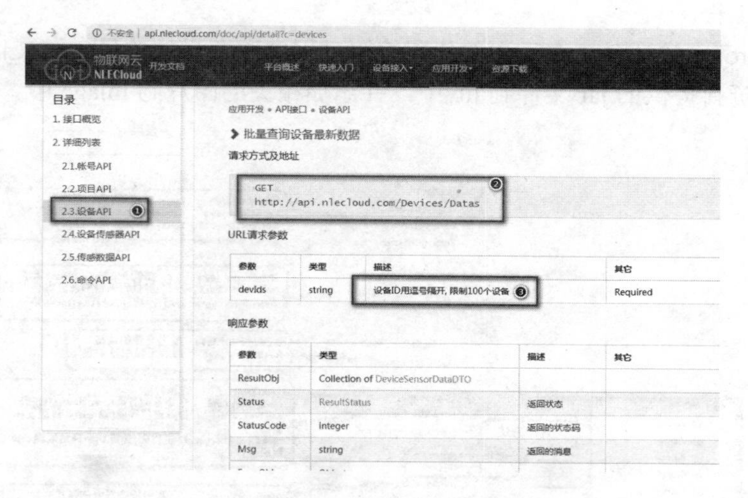

图 13-22　设备 API 说明

接着回到 API 测试工具页面，按请求方式和地址及 URL 请求参数填写图 13-23，其中②的参数值分别从图 13-21 中获取，参数填完后单击“发送请求“，在右边可以看到返回结果列出了每个设备中包含的传感器的 ApiTag 和对应的值 Value，如果 Java 程序想获取这些值，只需要按格式解析就行了。

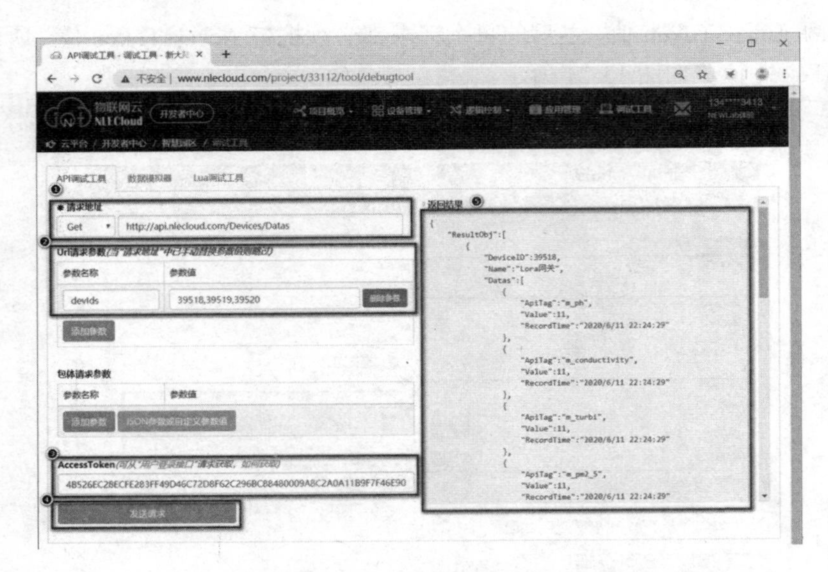

图 13-23　批量获取数据

其他 API 接口有兴趣的读者可继续测试，返回结果分析可参考 API 接口说明文档。

4. 实时监测园区环境的功能实现

经过上述操作，初步验证了空气环境、水质环境、井盖状态、垃圾桶状态的实时监测数据已上传至物联网云服务平台，现需要编写 PC 端应用程序实时显示园区环境监测值。首先需要使用 JavaFX 技术制作 UI 界面，再通过启动一个定时器，定时从云平台获取实时数据，并同时更新 UI 的显示值。

(1)　创建工程。

创建工程 project13_task，在后面的代码编写过程中逐渐添加的目录和文件如图 13-24 所示，添加随书资料提供的 jar 文件到 lib 中，并添加相关的图片到 image 中。

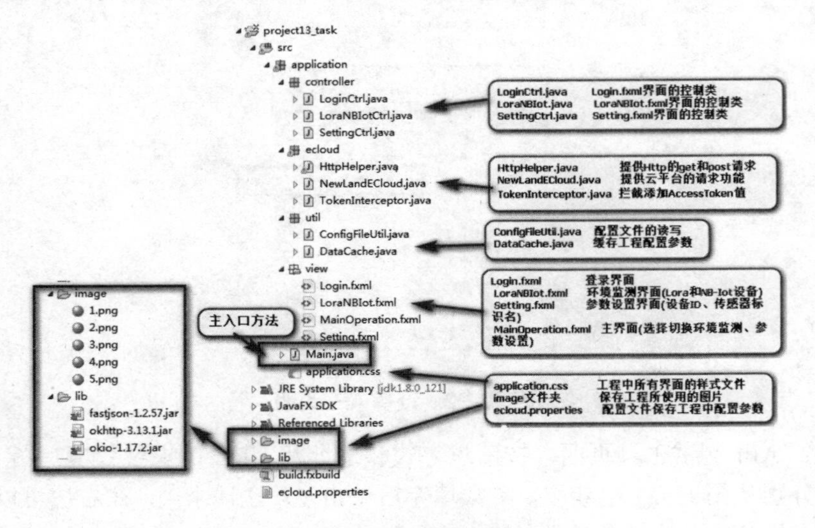

图 13-24　工程目录结构

(2) 设计 UI。

步骤 1： 通过拖曳方式制作登录界面，如图 13-25 所示。

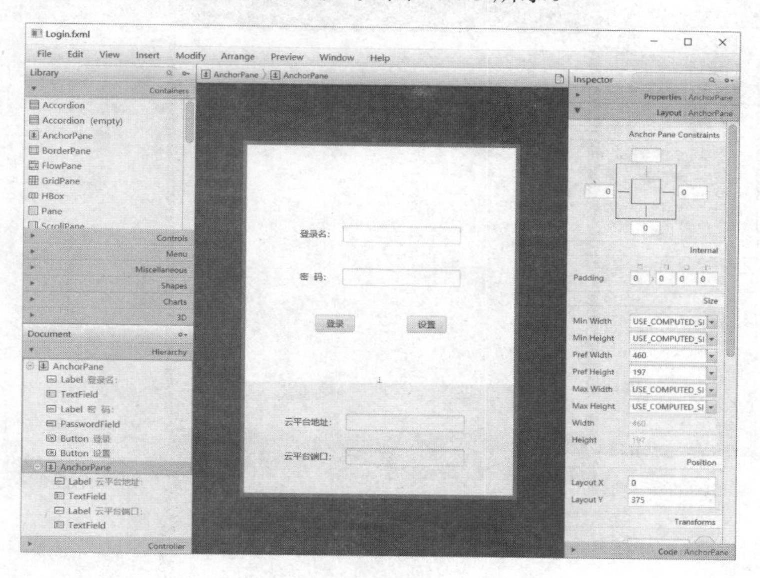

图 13-25 登录界面

步骤 2： 通过拖曳方式制作环境实时监测界面，如图 13-26 所示。

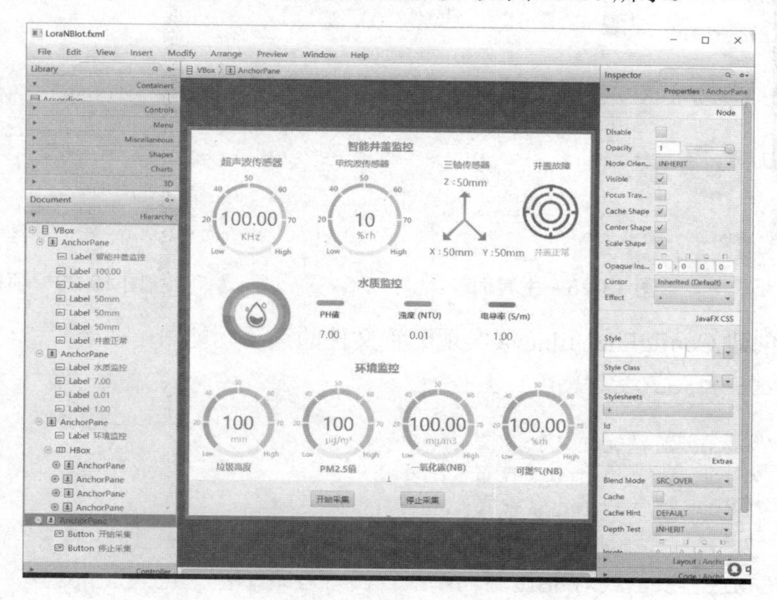

图 13-26 环境实时监测界面

步骤 3： 通过拖曳方式制作参数设置界面，如图 13-27 所示。

步骤 4： 通过拖曳方式制作主界面，如图 13-28 所示。

(3) 配置文件读写操作。

步骤 1： 在工程目录下添加 ecloud.properties 文件，如图 13-29 所示。

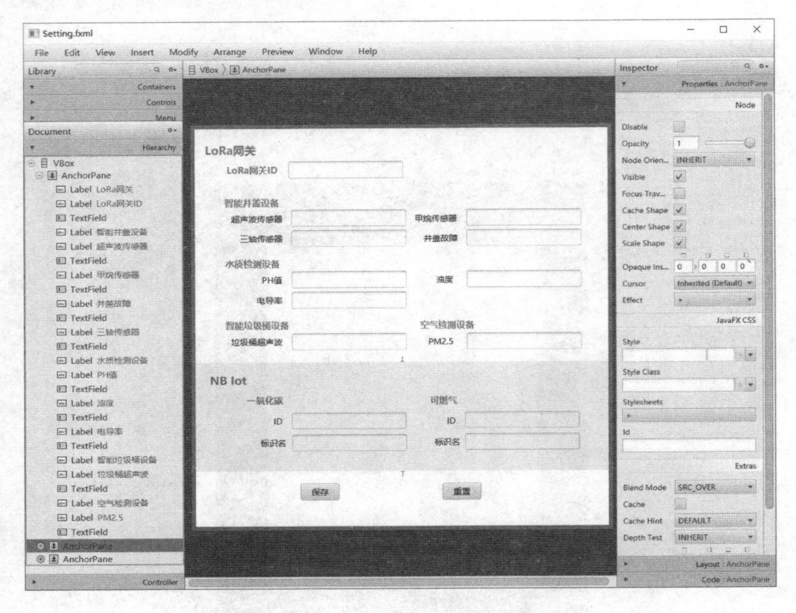

图 13-27　参数设置界面

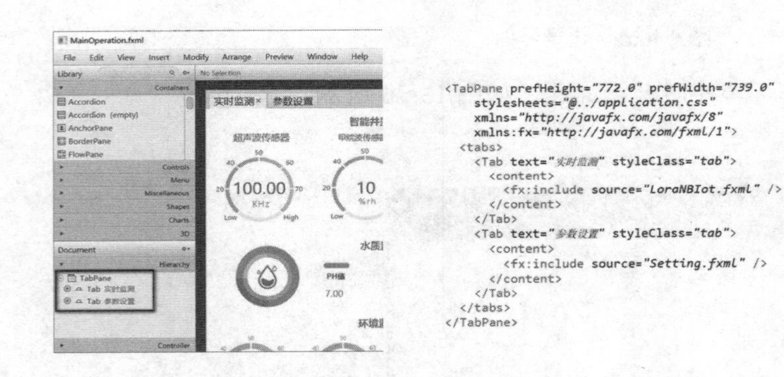

图 13-28　主界面

Package Explorer
LoraNBIot
src
JRE System Library [jre1.8.0_191]
JavaFX SDK
Referenced Libraries
lib
build.fxbuild
ecloud.properties

图 13-29　项目配置文件

步骤 2：创建 ConfigFileUtil.java 实现配置文件的读写功能。

```
1.   public class ConfigFileUtil {
2.       /**
3.        * 保存或修改配置文件
4.        * @param filepath 配置文件路径
5.        * @param values 参数值
6.        * @return
7.        */
8.       public static boolean saveOrUpdateData(String filepath, Map<String,
           String> values) {return true;}
9.
10.      /**
11.       * 获取配置文件参数值
12.       * @param filepath
13.       * @return
14.       */
15.      public static Map loadData(String filepath) {return null;}
16.  }
```

步骤 3： 完成配置文件的保存或修改功能。

```
1.   /**
2.    * 保存或修改配置文件
3.    * @param filepath 配置文件路径
4.    * @param values 参数值
5.    * @return
6.    */
7.  public static boolean saveOrUpdateData(String filepath, Map<String,
      String> values) {
8.
9.      Properties prop = (Properties)loadData(filepath);
10.     if(prop == null) prop = new Properties();
11.     prop.putAll(values);
12.
13.     OutputStream out = null;
14.     try {
15.         out = new FileOutputStream(filepath);
16.         prop.store(out, "");
17.
18.         return true;
19.     } catch (FileNotFoundException e) {
20.         e.printStackTrace();
21.     } catch (IOException e) {
22.         e.printStackTrace();
23.     }finally {
24.         if(out != null) {
25.             try {
26.                 out.close();
27.             } catch (IOException e) {
28.                 e.printStackTrace();
29.             }
30.         }
31.     }
32.
33.     return false;
34. }
```

步骤 4： 完成读取配置文件中的参数值。

```
1.   /**
2.    * 获取配置文件参数值
3.    * @param filepath
4.    * @return
5.    */
6.   @SuppressWarnings("rawtypes")
7.   public static Map loadData(String filepath) {
8.      File file = new File(filepath);
9.      if(!file.exists()) {
10.         return null;
11.     }
12.
13.     Properties prop = new Properties();
14.     InputStream in = null;
15.     try {
16.         in = new FileInputStream(filepath);
17.         prop.load(in);
18.
19.         return (Map) prop;
```

```
20.     } catch (FileNotFoundException e) {
21.         e.printStackTrace();
22.     } catch (IOException e) {
23.         e.printStackTrace();
24.     }finally {
25.         if(in != null) {
26.             try {
27.                 in.close();
28.             } catch (IOException e) {
29.                 e.printStackTrace();
30.             }
31.         }
32.     }
33.     return null;
34. }
```

(4) 缓存配置参数。

步骤 1：创建 DataCache.java 实现配置参数的缓存，并添加以下方法。

```
1.  package application.util;
2.
3.  import java.lang.reflect.Field;
4.  import java.util.HashMap;
5.  import java.util.Map;
6.
7.  import application.ecloud.NewLandECloud;
8.
9.  public class DataCache {
10.     //项目标题
11.     public static final String TITLE = "";
12.     //配置文件
13.     private static final String FILE_PATH = "ecloud.properties";
14.     //云平台操作对象
15.     private static NewLandECloud nlecloud;
16.
17.     private static String address;                  //物联网云服务平台网址
18.     private static String port;                     //物联网云服务平台端口号
19.
20.     private static String gatewayDeviceIdLoRa;      //LoRa 网关 ID
21.     private static String well_ultrasonic;          //井盖超声波传感器标识名
22.     private static String methane;                  //井内甲烷传感器标识名
23.     private static String triaxial;                 //井盖三轴传感器标识名
24.     private static String cover_fault;              //井盖状态传感器标识名
25.     private static String m_ph;                     //pH 值传感器标识名
26.     private static String m_turbi;                  //浑度传感器标识名
27.     private static String m_conductivity;           //电导率传感器标识名
28.     private static String m_pm2_5;                  //PM2.5 传感器标识名
29.     private static String trashcan_ultrasonic;      //垃圾桶超声波传感器标识名
30.
31.     public static String carbonMonoxideDeviceIdNB;    //一氧化碳设备 ID(NB-Iot 设备)
32.     public static String carbonMonoxide; //一氧化碳传感器标识名(NB-Iot 设备)
33.     public static String flammableGasDeviceIdNB; //可燃气设备 ID(NB-Iot 设备)
34.     public static String flammableGas;    //可燃气传感器标识名(NB-Iot 设备)
35.     static {
36.         //取配置文件参数，并初始化类变量(缓存变量)
37.         initConfig();
```

```
38.     }
39.
40.     /**
41.      * 缓存并保存物联网云服务平台地址和端口号
42.      * @param address 物联网云服务平台地址
43.      * @param port 物联网云服务平台端口号
44.      * @return
45.      */
46.     public static boolean saveECloudConnParam(String address, String port)
          {return false;}
47.     /**
48.      * 缓存并保存传感器标识名
49.      * @param gatewayDeviceIdLoRa LoRa 网关 ID
50.      * @param well_ultrasonic 井盖超声波传感器标识名
51.      * @param methane 井内甲烷传感器标识名
52.      * @param triaxial 井盖三轴传感器标识名
53.      * @param cover_fault 井盖状态传感器标识名
54.      * @param m_ph pH 值传感器标识名
55.      * @param m_turbi 浑度传感器标识名
56.      * @param m_conductivity 电导率传感器标识名
57.      * @param trashcan_ultrasonic 垃圾桶超声波传感器标识名
58.      * @param m_pm2_5 PM2.5 传感器标识名
59.      * @param carbonMonoxideDeviceIdNB 一氧化碳设备 ID(NB-Iot 设备)
60.      * @param carbonMonoxide 一氧化碳传感器标识名(NB-Iot 设备)
61.      * @param flammableGasDeviceIdNB 可燃气设备 ID(NB-Iot 设备)
62.      * @param flammableGas 可燃气传感器标识名(NB-Iot 设备)
63.      * @return 是否保存成功
64.      */
65.     public static boolean saveDeviceIDAndSensorName(String
                gatewayDeviceIdLoRa,
66.             String well_ultrasonic, String methane,
67.             String triaxial, String cover_fault, String m_ph, String m_turbi,
68.             String m_conductivity, String trashcan_ultrasonic, String
                  m_pm2_5,
69.             String carbonMonoxideDeviceIdNB, String carbonMonoxide,
70.             String flammableGasDeviceIdNB, String flammableGas) {return
                  false;}
71.
72.     /**
73.      * 读取配置文件参数，并初始化类变量(缓存变量)
74.      * @return 传感器标识名
75.      */
76.     @SuppressWarnings("unchecked")
77.     public static Map<String, String> initConfig() {return null;}
78.     /**
79.      * 修改某个类变量值(只修改类变量值，没同时修改参数配置文件)
80.      * @param apiTag
81.      * @param value
82.      */
83.     public static boolean setApiTag(String apiTag, String value) { return
          false;}
84.
85.     //为类变量 nlecloud 添加 get 和 set 方法
86.     //为其他类变量添加 get 方法
87. }
```

步骤 2：完成缓存并保存物联网云服务平台地址和端口号。

```
1.   /**
2.    * 保存物联网云服务平台地址和端口号
3.    * @param address 物联网云服务平台地址
4.    * @param port 物联网云服务平台端口号
5.    * @return
6.    */
7.   public static boolean saveECloudConnParam(String address, String port) {
8.       DataCache.address = address;
9.       DataCache.port = port;
10.
11.      Map<String, String> values = new HashMap<>();
12.      values.put("address", address);
13.      values.put("port", port);
14.
15.      boolean isOk = ConfigFileUtil.saveOrUpdateData(FILE_PATH, values);
16.      return isOk;
17.  }
```

步骤 3：完成缓存并保存传感器标识名。

```
1.   /**
2.    * 缓存并保存传感器标识名
3.    * @param gatewayDeviceIdLoRa LoRa 网关 ID
4.    * @param well_ultrasonic 井盖超声波传感器标识名
5.    * @param methane 井内甲烷传感器标识名
6.    * @param triaxial 井盖三轴传感器标识名
7.    * @param cover_fault 井盖状态传感器标识名
8.    * @param m_ph pH 值传感器标识名
9.    * @param m_turbi 浑度传感器标识名
10.   * @param m_conductivity 电导率传感器标识名
11.   * @param trashcan_ultrasonic 垃圾桶超声波传感器标识名
12.   * @param m_pm2_5 PM2.5 传感器标识名
13.   * @param carbonMonoxideDeviceIdNB 一氧化碳设备 ID(NB-Iot 设备)
14.   * @param carbonMonoxide 一氧化碳传感器标识名(NB-Iot 设备)
15.   * @param flammableGasDeviceIdNB 可燃气设备 ID(NB-Iot 设备)
16.   * @param flammableGas 可燃气传感器标识名(NB-Iot 设备)
17.   * @return 是否保存成功
18.   */
19.  public static boolean saveDeviceIDAndSensorName(String
             gatewayDeviceIdLoRa,
20.          String well_ultrasonic, String methane,
21.          String triaxial, String cover_fault, String m_ph, String m_turbi,
22.          String m_conductivity, String trashcan_ultrasonic, String m_pm2_5,
23.          String carbonMonoxideDeviceIdNB, String carbonMonoxide,
24.          String flammableGasDeviceIdNB, String flammableGas) {
25.
26.      DataCache.gatewayDeviceIdLoRa = gatewayDeviceIdLoRa;
27.
28.      DataCache.well_ultrasonic = well_ultrasonic;
29.      DataCache.methane = methane;
30.      DataCache.triaxial = triaxial;
31.      DataCache.cover_fault = cover_fault;
```

```
32.     DataCache.m_ph = m_ph;
33.     DataCache.m_turbi = m_turbi;
34.     DataCache.m_conductivity = m_conductivity;
35.     DataCache.trashcan_ultrasonic = trashcan_ultrasonic;
36.     DataCache.m_pm2_5 = m_pm2_5;
37.
38.     DataCache.carbonMonoxideDeviceIdNB = carbonMonoxideDeviceIdNB;
39.     DataCache.carbonMonoxide = carbonMonoxide;
40.     DataCache.flammableGasDeviceIdNB = flammableGasDeviceIdNB;
41.     DataCache.flammableGas = flammableGas;
42.
43.
44.     Map<String, String> values = new HashMap<>();
45.     values.put("gatewayDeviceIdLoRa", gatewayDeviceIdLoRa);
46.     values.put("well_ultrasonic", well_ultrasonic);
47.     values.put("methane", methane);
48.     values.put("triaxial", triaxial);
49.     values.put("cover_fault", cover_fault);
50.     values.put("m_ph", m_ph);
51.     values.put("m_turbi", m_turbi);
52.     values.put("m_conductivity", m_conductivity);
53.     values.put("trashcan_ultrasonic", trashcan_ultrasonic);
54.     values.put("m_pm2_5", m_pm2_5);
55.
56.     values.put("carbonMonoxideDeviceIdNB", carbonMonoxideDeviceIdNB);
57.     values.put("carbonMonoxide", carbonMonoxide);
58.     values.put("flammableGasDeviceIdNB", flammableGasDeviceIdNB);
59.     values.put("flammableGas", flammableGas);
60.
61.     boolean isOk = ConfigFileUtil.saveOrUpdateData(FILE_PATH, values);
62.     return isOk;
63. }
```

步骤 4： 完成读取配置文件参数，并初始化类变量(缓存变量)。

```
1.  /**
2.   * 读取配置文件参数，并初始化类变量(缓存变量)
3.   * @return 传感器标识名
4.   */
5.  @SuppressWarnings("unchecked")
6.  public static Map<String, String> initConfig() {
7.      Map<String, String> values = ConfigFileUtil.loadData(FILE_PATH);
8.      if(values == null) return null;
9.
10.     address = values.get("address");
11.     port = values.get("port");
12.
13.     gatewayDeviceIdLoRa = values.get("gatewayDeviceIdLoRa");
14.
15.     well_ultrasonic = values.get("well_ultrasonic");
16.     methane = values.get("methane");
17.     triaxial = values.get("triaxial");
18.     cover_fault = values.get("cover_fault");
19.     m_ph = values.get("m_ph");
```

```
20.     m_turbi = values.get("m_turbi");
21.     m_conductivity = values.get("m_conductivity");
22.     trashcan_ultrasonic = values.get("trashcan_ultrasonic");
23.     m_pm2_5 = values.get("m_pm2_5");
24.
25.     carbonMonoxideDeviceIdNB = values.get("carbonMonoxideDeviceIdNB");
26.     carbonMonoxide = values.get("carbonMonoxide");
27.     flammableGasDeviceIdNB = values.get("flammableGasDeviceIdNB");
28.     flammableGas = values.get("flammableGas");
29.
30.     return values;
31. }
```

步骤 5： 完成修改某个类变量值(只修改类变量值，没同时修改参数配置文件)。

```
1.  /**
2.   * 修改某个类变量值(只修改类变量值，没同时修改参数配置文件)
3.   * @param apiTag
4.   * @param value
5.   */
6.  public static boolean setApiTag(String apiTag, String value) {
7.      try {
8.          Field field = DataCache.class.getDeclaredField(apiTag);
9.          field.set(null, value);
10.         return true;
11.     } catch (NoSuchFieldException e) {
12.         e.printStackTrace();
13.     } catch (SecurityException e) {
14.         e.printStackTrace();
15.     } catch (IllegalArgumentException e) {
16.         e.printStackTrace();
17.     } catch (IllegalAccessException e) {
18.         e.printStackTrace();
19.     }
20.     return false;
21. }
```

步骤 6： 为类变量 nlecloud 添加 get 和 set 方法。

```
1.  public static void setNlecloud(NewLandECloud nlecloud) {
2.      DataCache.nlecloud = nlecloud;
3.  }
4.  public static NewLandECloud getNlecloud() {
5.      return nlecloud ;
6.  }
```

步骤 7： 为其他类变量添加 get 方法。

```
1.  public static String getTitle() {return TITLE;}
2.  public static String getAddress() {return address;}
3.  public static String getPort() {return port;}
4.  public static String getGatewayDeviceIdLoRa() {return gatewayDeviceIdLoRa;}
5.  public static String getWell_ultrasonic() {return well_ultrasonic;}
6.  public static String getMethane() {return methane;}
7.  public static String getTriaxial() {return triaxial;}
8.  public static String getCover_fault() {return cover_fault;}
```

```
9.   public static String getM_ph() {return m_ph;}
10.  public static String getM_turbi() {return m_turbi;}
11.  public static String getM_conductivity() {return m_conductivity;}
12.  public static String getTrashcan_ultrasonic() {return trashcan_ultrasonic;}
13.  public static String getM_pm2_5() {return m_pm2_5;}
14.  public static String getCarbonMonoxideDeviceIdNB(){return
        carbonMonoxideDeviceIdNB;}
15.  public static String getCarbonMonoxide() {return carbonMonoxide;}
16.  public static String getFlammableGasDeviceIdNB() {return
        flammableGasDeviceIdNB;}
17.  public static String getFlammableGas() {return flammableGas;}
```

(5) AccessToken 拦截器。

创建 TokenInterceptor.java，通过拦截方式添加 AccessToken。登录成功后云平台会返回 AccessToken，以后每次向云平台发送 HTTP 请求都需要带上 AccessToken，所以要使用拦截器拦截 HTTP 请求数据并附加上 AccessToken 后再发送给云平台。

```
1.      public class TokenInterceptor implements Interceptor {
2.      private static final String TOKEN_KEY = "AccessToken";
3.      public static String accessToken;
4.      @Override
5.      public Response intercept(Chain chain) throws IOException {
6.         Request request = chain.request();
7.
8.         if (accessToken != null){
9.            request = request.newBuilder()
10.                   .addHeader(TOKEN_KEY, accessToken)
11.                   .build();
12.        }
13.        Response response = chain.proceed(request);
14.        return response;
15.     }
16. }
```

(6) HTTP 请求工具类。

步骤 1：创建 HttpHelper.java，实现 get 和 post 请求。

```
1.  public class HttpHelper {
2.    private static final MediaType JSON = MediaType.get("application/json;
       charset=utf-8");
3.    private static OkHttpClient client = new OkHttpClient();
4.    static{
5.         client = new OkHttpClient().newBuilder()
6.                .addNetworkInterceptor(new TokenInterceptor())
7.                .build();
8.    }
9.    public static String get(String url) throws IOException {return null;}
10.   public static String post(String url, String json) throws IOException
       {return null;}
11.   public static String post(String url, String json, Interceptor
    interceptor) throws IOException {return null;}
12. }
```

步骤 2：完成 get 请求功能。

```
1.  public static String get(String url) throws IOException {
2.    Request request = new Request.Builder().url(url).build();
3.
```

```
4.      try (Response response = client.newCall(request).execute()) {
5.         return response.body().string();
6.      }
7.   }
```

步骤 3：完成 post 请求功能。

```
1.   public static String post(String url, String json) throws IOException {
2.      RequestBody body = RequestBody.create(JSON, json);
3.
4.      Request request = new Request.Builder().url(url).post(body).build();
5.
6.      try (Response response = client.newCall(request).execute()) {
7.         return response.body().string();
8.      }
9.   }
```

(7) 物联网云服务平台请求操作类。

步骤 1：创建 NewLandECloud.java 实现云平台的相关操作。

```
1.      public class NewLandECloud {
2.      private static Logger logger = Logger.getLogger("NewLandECloud");
3.      private final String baseURL;
4.      private static final String PROTOCOL = "http://";
5.      private static final String LOGIN = "/Users/Login";
6.      private static final String DEVICES = "/Devices";
7.      private static final String DATAS = "/Datas";
8.      public NewLandECloud(){
9.         this(null, null);
10.     }
11.     public NewLandECloud(String ip, String port){
12.        if(ip != null && port != null) {
13.           baseURL = PROTOCOL + ip + ":" + port;
14.        }else {
15.           baseURL = "http://api.nlecloud.com";
16.        }
17.     }
18.     /** 登录物联网云服务平台 */
19.     public boolean login(String loginname, String password) throws
          RuntimeException, IOException {return false;}
20.     /** 批量查询设备最新数据 */
21.     public Map<String, String> getDevicesData(String ... deviceId) throws
          IOException {return null;}
22. }
```

步骤 2：完成登录的 post 请求功能。

```
1.       /**
2.    * 登录物联网云服务平台
3.    * @param loginname
4.    * @param password
5.    * @return
6.    * @throws RuntimeException
7.    * @throws IOException
8.    */
9.  public boolean login(String loginname, String password) throws
       RuntimeException, IOException {
```

```
10.     String url = baseURL + LOGIN;
11.     String json = "{'Account':'" + loginname + "', 'Password':'" + password
          + "', 'IsRememberMe':'true'}";
12.
13.     String response = HttpHelper.post(url, json);
14.     logger.info(response);
15.
16.     JSONObject jsonObj = JSONObject.parseObject(response);
17.     Integer status = jsonObj.getInteger("Status");
18.     if(status.equals(0)) {
19.        TokenInterceptor.accessToken = jsonObj.getJSONObject
             ("ResultObj").getString("AccessToken");
20.        logger.info("accessToken = " + TokenInterceptor.accessToken);
21.        return true;
22.     }else {
23.        throw new RuntimeException("您的账号密码输入有误，请查证后再次尝试登录！");
24.     }
25. }
```

步骤 3： 完成查询数据的 post 请求功能。

```
1.       /**
2.     * 批量查询设备最新数据
3.     * @param deviceId 多个设备编号
4.     * @return 多个传感器值
5.     * @throws IOException
6.     */
7.   public Map<String, String> getDevicesData(String ... deviceId) throws
        IOException {
8.      if(deviceId == null || deviceId.length == 0) {
9.         return null;
10.     }
11.
12.     StringBuilder url = new StringBuilder(baseURL).append(DEVICES).
          append(DATAS);
13.     url.append("?devIds=").append(deviceId[0]);
14.     for (int i = 1; i < deviceId.length; i++) {
15.        url.append(", ").append(deviceId[i]);
16.     }
17.
18.     String response = HttpHelper.get(url.toString());
19.     logger.info(response);
20.
21.     JSONObject jsonObj = JSONObject.parseObject(response);
22.     Integer status = jsonObj.getInteger("Status");
23.     if(!status.equals(0)) {
24.        throw new RuntimeException("您的请求操作失败，请检查您传入的参数");
25.     }
26.     //保存采集到的数据
27.     Map<String, String> data = new HashMap<>();
28.     //解析云平台返回的 JSON 数据
29.     JSONArray deviceAry = jsonObj.getJSONArray("ResultObj");
30.     int deviceSize = deviceAry.size();
31.     //遍历多个设备：LoRa 网关、NB-Iot(一氧化碳)、NB-Iot(可燃气)
32.     for (int i = 0; i < deviceSize; i++) {
```

```
33.         JSONObject deviceObj = deviceAry.getJSONObject(i);
34.         JSONArray sensorAry = deviceObj.getJSONArray("Datas");
35.         int sensorSize = sensorAry.size();
36.         //遍历获取每个设备中多个传感器的值，并保存在 Map 类型的对象中
37.         for (int j = 0; j < sensorSize; j++) {
38.            JSONObject sensorObj = sensorAry.getJSONObject(j);
39.
40.            String apiTag = sensorObj.getString("ApiTag");
41.            String value = sensorObj.getString("Value");
42.            data.put(apiTag, value);
43.         }
44.      }
45.
46.      return data;
47.  }
```

(8) 登录云平台并保存 AccessToken。

步骤 1：创建 LoginCtrl.java 实现登录功能，包含 Login.fxml 界面中的控件、初始化方法、登录方法、显示或隐藏设置面板方法。

```
1.       /** 登录界面控制类，为界面提供各种功能实现。*/
2.       public class LoginCtrl {
3.       /** 登录界面路径 */
4.       public static final String VIEW_URL = "/application/view/Login.fxml";
5.       @FXML // 云平台登录名输入框
6.       private TextField loginnameTxf;
7.       @FXML // 云平台登录密码输入框
8.       private PasswordField passwordField;
9.       @FXML //云平台地址输入框
10.      private TextField addressTxf;
11.      @FXML //云平台端口输入框
12.      private TextField portTxf;
13.      @FXML //云平台参数设置面板
14.      private AnchorPane settingPane;
15.      @FXML // 显示/隐藏云平台参数设置面板
16.      private Button settingBtn;
17.      /** 初始化界面 */
18.      @FXML
19.      public void initialize() {}
20.      /** 实现云平台的登录功能 */
21.      @FXML
22.      public void login(ActionEvent ae) {}
23.      /** 显示或隐藏云平台参数设置界面 */
24.      @FXML
25.      public void setting(ActionEvent ae) {}
26.  }
```

步骤 2：完成界面初始化。

```
1.   /**
2.    * 初始化界面
3.    */
4.   @FXML
5.   public void initialize() {
6.       // 为云平台参数面板的输入框设置默认值
```

```
7.      addressTxf.setText(DataCache.getAddress());
8.      portTxf.setText(DataCache.getPort());
9.
10.     // 界面初始化：设置停止按钮默认无效
11.     settingPane.setVisible(false);
12. }
```

步骤 3：实现显示或隐藏设置面板。先为 Login.fxml 界面中的按钮注册执行事件，再实现此方法，如图 13-30 所示。

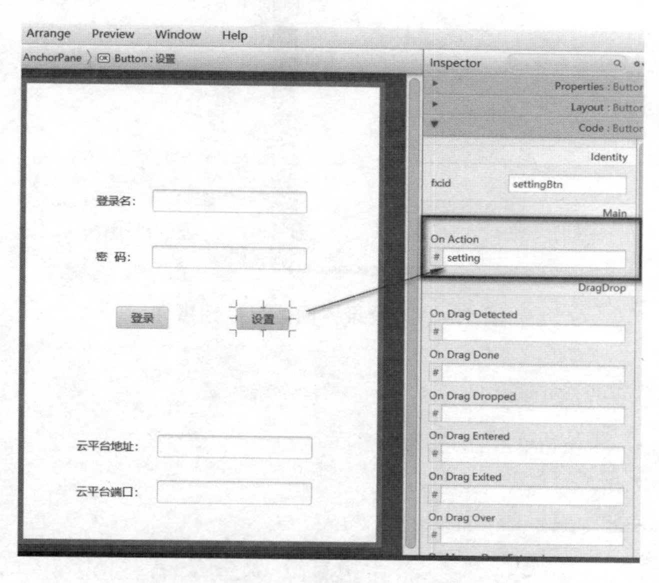

图 13-30　登录界面设置按钮事件

```
1.   /**
2.    * 显示或隐藏云平台参数设置界面
3.    *
4.    * @param ae 按钮单击的动作事件
5.    */
6.   @FXML
7.   public void setting(ActionEvent ae) {
8.       Stage loginStage = (Stage) loginnameTxf.getScene().getWindow();
9.       if(loginStage.getHeight() == 616) {
10.          //隐藏云平台参数设置界面
11.          settingPane.setVisible(false);
12.          loginStage.setHeight(400);
13.          settingBtn.setText("设置");
14.      }else {
15.          //显示云平台参数设置界面
16.          settingPane.setVisible(true);
17.          loginStage.setHeight(616);
18.          settingBtn.setText("隐藏");
19.      }
20.  }
```

步骤 4：完成界面登录功能。先为 Login.fxml 界面中的按钮注册执行事件，再实现此方法，如图 13-31 所示。

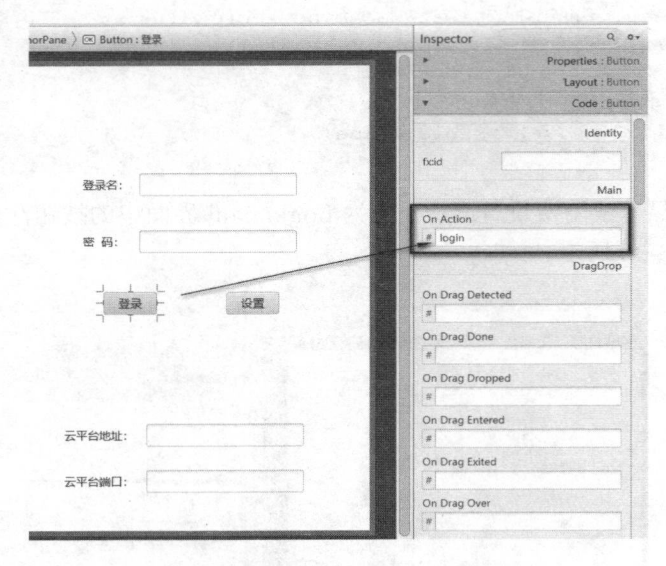

图 13-31　登录界面登录按钮事件

```
1.   /**
2.    * 实现云平台的登录功能
3.    *
4.    * @param ae 按钮单击的动作事件
5.    */
6.   @FXML
7.   public void login(ActionEvent ae) {
8.       String loginname = loginnameTxf.getText().trim();
9.       String password = passwordField.getText().trim();
10.      //获取界面上输入框中值：云平台地址和端口号
11.      String address = addressTxf.getText().trim();
12.      String port = portTxf.getText().trim();
13.
14.      String errorMsg = null;
15.      if(loginname == null || password == null) {
16.          errorMsg = "请输入用户名和密码";
17.      }else if(address == null || port == null) {
18.          errorMsg = "请在设置面板中设置云平台地址和端口号";
19.      }
20.
21.      if(errorMsg != null) {
22.          Alert alert = new Alert(AlertType.WARNING);
23.          alert.setTitle("登录提示");
24.          alert.setContentText(errorMsg);
25.          alert.show();
26.          return;
27.      }
28.
29.      // 以下完成云平台登录
30.      NewLandECloud nlecloud = new NewLandECloud(address, port);
31.      try {
32.          // 登录云平台
33.          boolean isOk = nlecloud.login(loginname, password);
34.          if (isOk) {
35.              //保存云平台操作对象
```

```
36.             DataCache.setNlecloud(nlecloud);
37.             //保存云平台地址和端口号
38.             DataCache.saveECloudConnParam(address, port);
39.
40.             //切换至主界面，显示传感器数据，并实现实时刷新
41.             String viewURL = "/application/view/MainOperation.fxml";
42.             FXMLLoader loader = new FXMLLoader();
43.             loader.setLocation(Class.class.getResource(viewURL));
44.             Parent root = loader.load();
45.             Scene scene = new Scene(root);
46.
47.             Stage stage = new Stage();
48.             stage.setTitle(DataCache.TITLE);
49.             stage.setScene(scene);
50.             stage.show();
51.
52.             //关闭当前登录界面
53.             Stage loginStage = (Stage) loginnameTxf.getScene().getWindow();
54.             loginStage.close();
55.
56.             return;
57.         }
58.     } catch (RuntimeException e) {
59.         errorMsg = e.getMessage();
60.
61.     } catch (UnknownHostException e) {
62.         errorMsg = "无法连接此云平台: " + e.getMessage();
63.
64.     } catch (IOException e) {
65.         e.printStackTrace();
66.         errorMsg = "系统操作异常，异常信息为:" + e.getMessage();
67.     }
68.     // 提示错误信息
69.     Alert alert = new Alert(AlertType.ERROR);
70.     alert.setTitle("登录提示");
71.     alert.setContentText(errorMsg);
72.     alert.show();
73. }
```

(9) 传感器标识名参数配置与保存。

步骤 1：创建 Setting.java 实现参数配置功能，包含 Setting.fxml 界面中的输入框控件、初始化方法、保存参数方法、重置方法、同时修改缓存值方法。

```
1.  /** 参数设置界面控制类，为界面提供各种功能实现 */
2.  public class SettingCtrl {
3.      /** 参数设置界面路径 */
4.      public static final String VIEW_URL = "/application/view/Setting.fxml";
5.
6.      @FXML //LoRa 网关 ID
7.      private TextField gatewayDeviceIdLoRaTxf;
8.      @FXML //井盖超声波传感器标识名
9.      private TextField well_ultrasonicTxf;
10.     @FXML //井内甲烷传感器标识名
11.     private TextField methaneTxf;
12.     @FXML //井盖状态传感器标识名
13.     private TextField cover_faultTxf;
14.     @FXML //井盖三轴传感器标识名
```

```
15.     private TextField triaxialTxf;
16.     @FXML //pH 值传感器标识名
17.     private TextField m_phTxf;
18.     @FXML //浑度传感器标识名
19.     private TextField m_turbiTxf;
20.     @FXML //电导率传感器标识名
21.     private TextField m_conductivityTxf;
22.     @FXML //PM2.5 传感器标识名
23.     private TextField m_pm2_5Txf;
24.     @FXML //垃圾桶超声波传感器标识名
25.     private TextField trashcan_ultrasonicTxf;
26.
27.     @FXML //一氧化碳设备 ID(NB-Iot 设备)
28.     private TextField carbonMonoxideDeviceIdNBTxf;
29.     @FXML //可燃气设备 ID(NB-Iot 设备)
30.     private TextField flammableGasDeviceIdNBTxf;
31.     @FXML //一氧化碳传感器标识名(NB-Iot 设备)
32.     private TextField carbonMonoxideTxf;
33.     @FXML //可燃气传感器标识名(NB-Iot 设备)
34.     private TextField flammableGasTxf;
35.
36.     /** 初始化界面 */
37.     @FXML
38.     public void initialize() {}
39.     /** 将界面参数保存到缓存类和文件中 */
40.     @FXML
41.     public void save() {}
42.     /** 重置界面输入框 */
43.     @FXML
44.     public void reset() {}
45.     /** 修改界面输入框数据同时修改缓存 */
46.     @FXML
47.     public void changeValue(KeyEvent ke) {}
48. }
```

步骤 2：完成重置功能。先为 Setting.fxml 界面中的按钮注册执行事件，再实现此方法，如图 13-32 所示。

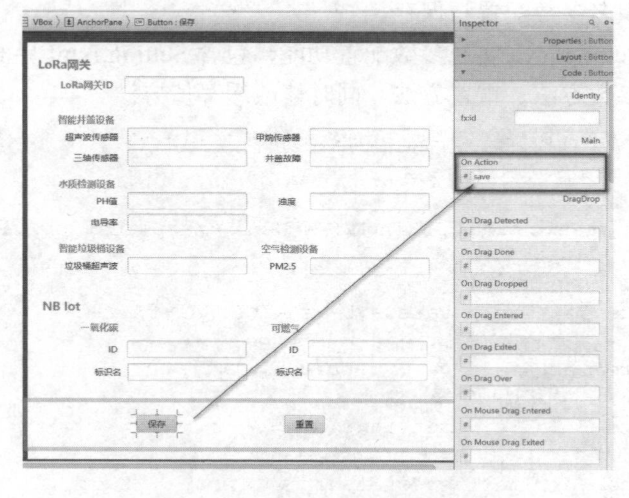

图 13-32　重置界面按钮事件

```
1.   /**
2.    * 重置界面输入框
3.    */
4.   @FXML
5.   public void reset() {
6.       Map<String, String> params = DataCache.initConfig();
7.       if(params == null) return;
8.   
9.       gatewayDeviceIdLoRaTxf.setText(params.get("gatewayDeviceIdLoRa"));
10.      well_ultrasonicTxf.setText(params.get("well_ultrasonic"));
11.      methaneTxf.setText(params.get("methane"));
12.      cover_faultTxf.setText(params.get("cover_fault"));
13.      triaxialTxf.setText(params.get("triaxial"));
14.      m_phTxf.setText(params.get("m_ph"));
15.      m_turbiTxf.setText(params.get("m_turbi"));
16.      m_conductivityTxf.setText(params.get("m_conductivity"));
17.      trashcan_ultrasonicTxf.setText(params.get("trashcan_ultrasonic"));
18.      m_pm2_5Txf.setText(params.get("m_pm2_5"));
19.  
20.      carbonMonoxideDeviceIdNBTxf.setText(params.get("carbonMonoxide
          DeviceIdNB"));
21.      carbonMonoxideTxf.setText(params.get("carbonMonoxide"));
22.      flammableGasDeviceIdNBTxf.setText(params.get("flammableGas
          DeviceIdNB"));
23.      flammableGasTxf.setText(params.get("flammableGas"));
24.  }
```

步骤 3：完成界面初始化。

```
1.   /**
2.    * 初始化界面
3.    */
4.   @FXML
5.   public void initialize() {
6.       reset();
7.   }
```

步骤 4：完成参数保存功能。先为 Setting.fxml 界面中的按钮注册执行事件，再实现此方法，如图 13-33 所示。

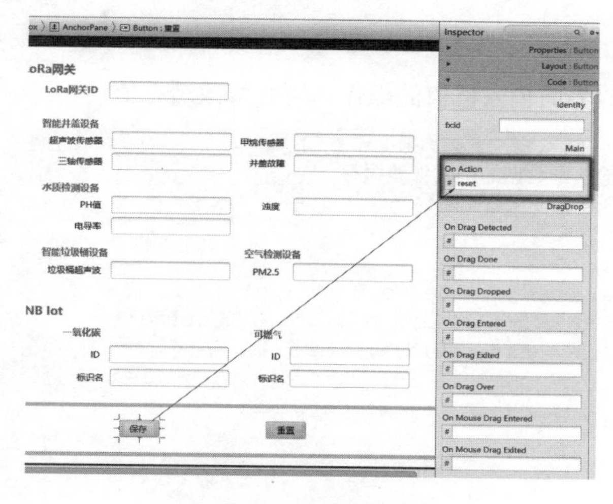

图 13-33　保存按钮事件

```
1.   /**
2.    * 将界面参数保存到缓存类和文件中
3.    */
4.   @FXML
5.   public void save() {
6.       String gatewayDeviceIdLoRa = gatewayDeviceIdLoRaTxf.getText();
7.       String well_ultrasonic = well_ultrasonicTxf.getText();
8.       String methane = methaneTxf.getText();
9.       String cover_fault = cover_faultTxf.getText();
10.      String triaxial = triaxialTxf.getText();
11.      String m_ph = m_phTxf.getText();
12.      String m_turbi = m_turbiTxf.getText();
13.      String m_conductivity = m_conductivityTxf.getText();
14.      String trashcan_ultrasonic = trashcan_ultrasonicTxf.getText();
15.      String m_pm2_5 = m_pm2_5Txf.getText();
16.
17.      String carbonMonoxideDeviceIdNB = carbonMonoxideDeviceIdNBTxf.getText();
18.      String carbonMonoxide = carbonMonoxideTxf.getText();
19.      String flammableGasDeviceIdNB = flammableGasDeviceIdNBTxf.getText();
20.      String flammableGas = flammableGasTxf.getText();
21.
22.      boolean isOk = DataCache.saveDeviceIDAndSensorName(gatewayDeviceIdLoRa,
             well_ultrasonic,
23.          methane, triaxial, cover_fault, m_ph, m_turbi,
24.          m_conductivity, trashcan_ultrasonic, m_pm2_5,
25.          carbonMonoxideDeviceIdNB, carbonMonoxide,
26.          flammableGasDeviceIdNB, flammableGas);
27.
28.      Alert alert;
29.      if(isOk) {
30.          alert = new Alert(AlertType.INFORMATION);
31.          alert.setTitle("保存提示");
32.          alert.setContentText("参数保存成功");
33.      }else {
34.          alert = new Alert(AlertType.ERROR);
35.          alert.setTitle("保存提示");
36.          alert.setContentText("参数保存失败");
37.      }
38.      alert.show();
39.  }
```

步骤 5：实现修改界面输入框值的操作，同时修改缓存值。

```
1.   /**
2.    * 修改界面输入框数据的同时修改缓存
3.    * @param ke
4.    */
5.   @FXML
6.   public void changeValue(KeyEvent ke) {
7.       TextField txf = (TextField) ke.getSource();
8.       String id = txf.getId();
9.       String apiTagName = id.substring(0, id.length() - 3);
10.      String apiTagValue = txf.getText();
11.
12.      DataCache.setApiTag(apiTagName, apiTagValue);
13.  }
```

(10) 环境的实时监测。

步骤 1： 创建 LoRaNBIotCtrl.java 实现参数配置功能，包含 LoRaNBIot.fxml 界面中的标签控件、开始采集方法、停止采集方法。

```
1.  /** 环境参数监测主界面控制类，为界面提供各种功能实现 */
2.  public class LoRaNBIotCtrl {
3.      /** 环境参数监测主界面路径 */
4.      @FXML //井盖超声波
5.      private Label well_ultrasonicLab;
6.      @FXML //井内甲烷
7.      private Label methaneLab;
8.      @FXML //井盖三轴中 z 轴
9.      private Label triaxialLab_z;
10.     @FXML //井盖三轴中 x 轴
11.     private Label triaxialLab_x;
12.     @FXML //井盖三轴中 y 轴
13.     private Label triaxialLab_y;
14.     @FXML //井盖状态
15.     private Label cover_faultLab;
16.     @FXML //pH 值
17.     private Label m_phLab;
18.     @FXML //浑度
19.     private Label m_turbiLab;
20.     @FXML //电导率
21.     private Label m_conductivityLab;
22.     @FXML //PM2.5
23.     private Label m_pm2_5Lab;
24.     @FXML //垃圾桶超声波
25.     private Label trashcan_ultrasonicLab;
26.     @FXML //一氧化碳(NB-Iot 设备)
27.     private Label carbonMonoxideLab;
28.     @FXML //可燃气(NB-Iot 设备)
29.     private Label flammableGasLab;
30.     @FXML // 开启自动采集云平台数据按钮
31.     private Button startBtn;
32.     @FXML // 停止自动采集云平台数据按钮
33.     private Button stopBtn;
34.     // 定时器，实现定时采集云平台数据
35.     private Timer timer;
36.     /**初始化界面 */
37.     @FXML
38.     public void initialize() {}
39.
40.     /**
41.      * 开启自动采集云平台数据
42.      *
43.      * @param ae 按钮单击的动作事件
44.      */
45.     @FXML
46.     public void start(ActionEvent ae) {}
47.
48.     /** 停止自动采集云平台数据 */
49.     @FXML
50.     public void stop(ActionEvent ae) {}
51.
```

```
52.  }
```

步骤 2：完成界面初始化功能。

```
1.    /**
2.     * 初始化界面
3.     */
4.    @FXML
5.    public void initialize() {
6.        // 界面初始化：设置停止按钮默认无效
7.        stopBtn.setDisable(true);
8.    }
```

步骤 3：完成开始采集功能(从物联网云服务平台获取传感器数据)。先为 LoRaNBIot.fxml 界面中的按钮注册执行事件，再实现此方法，如图 13-34 所示。

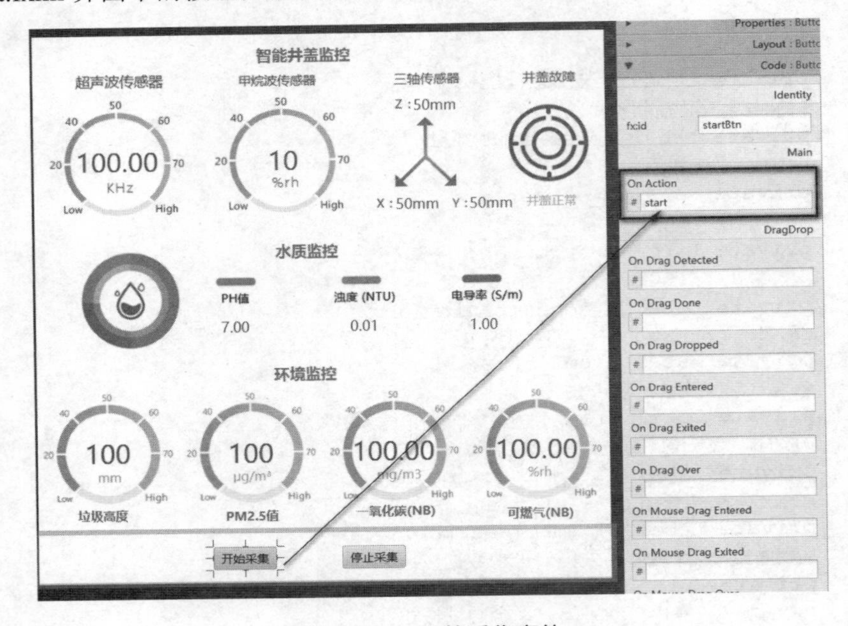

图 13-34　开始采集事件

```
1.    /**
2.     * 开启自动采集云平台数据
3.     *
4.     * @param ae 按钮单击的动作事件
5.     */
6.    @FXML
7.    public void start(ActionEvent ae) {
8.        //如果定时器还在执行，不用重新创建定时器
9.        if (timer != null)  return;
10.
11.       timer = new Timer(true);//创建定时器对象
12.       timer.schedule(new TimerTask() {//设置定时任务
13.           private Map<String, String> map;//传感器的数据值
14.           @Override
15.           public void run() {
16.               try {
17.                   NewLandECloud ecloud = DataCache.getNlecloud();
```

```
18.             String LoRaGateId = DataCache.getGatewayDeviceIdLoRa();
19.             String carbonMonoxideIdNB = DataCache.
                    getCarbonMonoxideDeviceIdNB();
20.             String flammableGasIdNB = DataCache.getFlammableGasDeviceIdNB();
21.             //获取多个设备中多个传感器的数据值
22.             map = ecloud.getDevicesData(LoRaGateId, carbonMonoxideIdNB,
                    flammableGasIdNB);
23.             //更新UI
24.             Platform.runLater(() ->{
25.                 //更新井盖水位和甲烷值
26.                 well_ultrasonicLab.setText(map.get(DataCache.getWell_
                      ultrasonic()));
27.                 methaneLab.setText(map.get(DataCache.getMethane()));
28.
29.                 //更新井盖三轴值
30.                 String triaxial = map.get(DataCache.getTriaxial());
31.                 if(triaxial != null && !triaxial.isEmpty()) {
32.                     String[] triaxial_xyz = triaxial.split("\\|");
33.                     if(triaxial_xyz.length == 3) {
34.                         triaxialLab_x.setText(triaxial_xyz[0] + "mm");
35.                         triaxialLab_y.setText(triaxial_xyz[1] + "mm");
36.                         triaxialLab_z.setText(triaxial_xyz[2] + "mm");
37.                     }
38.                 }
39.
40.                 //更新井盖转态
41.                 boolean coverIsOk = Boolean.parseBoolean(map.get
                      (DataCache.getCover_fault()));
42.                 if(coverIsOk) {
43.                     cover_faultLab.getStyleClass().remove("cover_fault");
44.                     cover_faultLab.setText("井盖正常");
45.                 }else {
46.                     cover_faultLab.getStyleClass().add("cover_fault");
47.                     cover_faultLab.setText("井盖故障");
48.                 }
49.
50.                 //更新空气环境、水质环境监测数据
51.                 m_phLab.setText(map.get(DataCache.getM_ph()));
52.                 m_turbiLab.setText(map.get(DataCache.getM_turbi()));
53.                 m_conductivityLab.setText(map.get(DataCache.getM_
                      conductivity()));
54.                 m_pm2_5Lab.setText(map.get(DataCache.getM_pm2_5()));
55.                 //更新垃圾桶超声波监测数据
56.                 trashcan_ultrasonicLab.setText(map.get(DataCache.
                      getTrashcan_ultrasonic()));
57.                 //更新NB-Iot设备监测数据(一氧化碳、可燃气)
58.                 carbonMonoxideLab.setText(map.get(DataCache.
                      getCarbonMonoxide()));
59.                 flammableGasLab.setText(map.get(DataCache.
                      getFlammableGas()));
60.             });
61.         } catch (IOException e) {
62.             e.printStackTrace();
63.         }
64.     }
65. }, 0, 1000);
66. // 切换按钮状态
67. stopBtn.setDisable(false);
```

```
68.       startBtn.setDisable(true);
69.  }
```

步骤 4： 完成停止采集功能。先为 LoRaNBIot.fxml 界面中的按钮注册执行事件，再实现此方法，如图 13-35 所示。

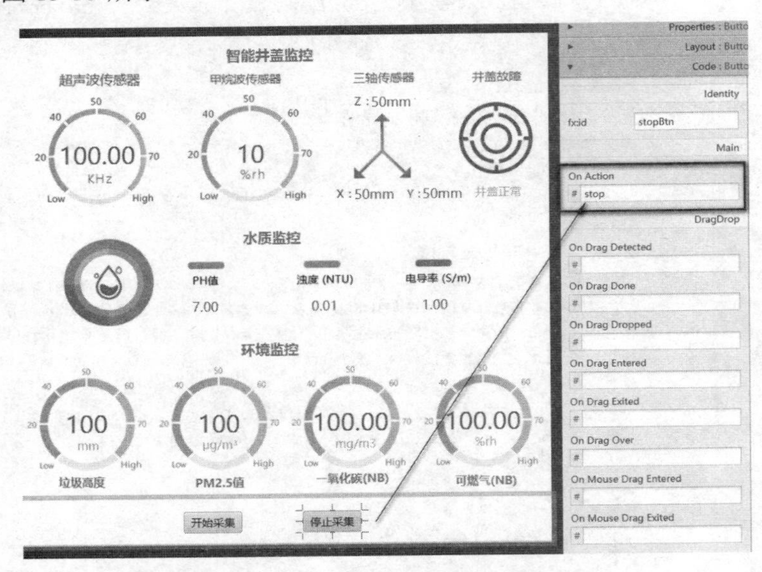

图 13-35　停止采集事件

```
1.    /**
2.     * 停止自动采集云平台数据
3.     *
4.     * @param ae 按钮单击的动作事件
5.     */
6.    @FXML
7.    public void stop(ActionEvent ae) {
8.       if (timer != null) {
9.          timer.cancel();
10.         timer = null;
11.      }
12.      startBtn.setDisable(false);
13.      stopBtn.setDisable(true);
14.  }
```

5. 运行结果

检查设备连接无误后，运行 Main.java 程序，先输入云平台的地址和端口号，再输入云平台的账号和密码(见图 13-36)，单击“登录”按钮后，从控制台输出可以看到登录成功后云平台返回了 AccessToken 数据，如图 13-37 所示。

登录成功后会跳转到下一页面，切换到“参数设置”选项卡，确保与云平台设置一样后保存，如图 13-38 所示。

切换到“实时监测”选项卡，单击“开始采集”按钮，结果如图 13-39 所示。

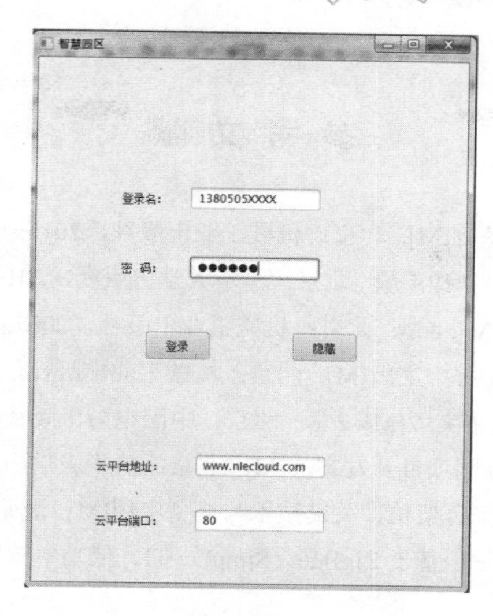

图 13-36　登录

图 13-37　云平台返回了 AccessToken 数据

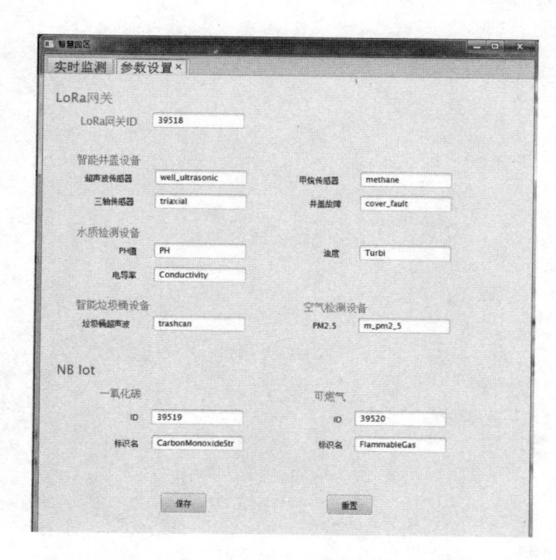

图 13-38　设备标识参数设置

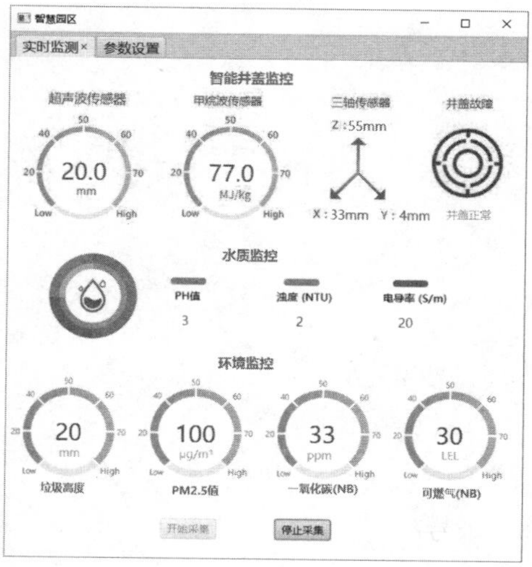

图 13-39　采集数据结果

至此，园区环境实时监测功能已实现完毕，完整的项目源码请参阅随书资料中的 project13_task 源码。

参 考 文 献

[1] 周雯. Java 物联网程序设计基础[M]. 北京：机械工业出版社，2016.

[2] 明日科技. Java 从入门到精通[M]. 5 版. 北京：清华大学出版社，2016.

[3] Bruce Eckel . Java 编程思想[M]. 4 版. 北京：机械工业出版社，2007.

[4] 约书亚・布洛克. Effective Java 中文版[M]. 北京：机械工业出版社，2019.

[5] KathySierra . Head First Java 中文版[M]. 2 版. 北京：中国电力出版社，2017.

[6] Agus，Kurniawan. 智能物联网项目开发实战[M]. 北京：清华大学出版社，2018.

[7] 丁飞. 物联网开放平台——平台架构、关键技术与典型应用[M]. 北京：电子工业出版社，2018.

[8] 廖建尚. 物联网开发与应用——基于 ZigBee、Simplici TI、低功率蓝牙、WiFi 技术[M]. 北京：电子工业出版社，2017.